T0275448

THE SEMANTIC CONCEPTION OF LOGIC

This collection of new essays presents cutting-edge research on the semantic conception of logic, the invariance criteria of logicality, grammaticality, and logical truth. Contributors explore the history of the semantic tradition, starting with Tarski, and its historical applications, while central criticisms of the tradition, and especially the use of invariance criteria to explain logicality, are revisited by the original participants in that debate. Other essays discuss more recent criticism of the approach, and researchers from mathematics and linguistics weigh in on the role of the semantic tradition in their disciplines. This book will be invaluable to philosophers and logicians alike.

GIL SAGI is Lecturer in Philosophy at the University of Haifa, Israel. She has published several articles on logical consequence, the invariance criterion of logicality, and meaning in model-theoretic semantics.

JACK WOODS is University Academic Fellow in Mathematical Philosophy at the University of Leeds. He is the author of numerous articles and book chapters on logical consequence, the invariance criterion of logicality, and logical inferentialism.

THE SEMANTIC CONCEPTION OF LOGIC

Essays on Consequence, Invariance, and Meaning

EDITED BY

GIL SAGI
University of Haifa, Israel

JACK WOODS
University of Leeds

CAMBRIDGE
UNIVERSITY PRESS

University Printing House, Cambridge CB2 8BS, United Kingdom

One Liberty Plaza, 20th Floor, New York, NY 10006, USA

477 Williamstown Road, Port Melbourne, VIC 3207, Australia

314–321, 3rd Floor, Plot 3, Splendor Forum, Jasola District Centre, New Delhi – 110025, India

103 Penang Road, #05–06/07, Visioncrest Commercial, Singapore 238467

Cambridge University Press is part of the University of Cambridge.

It furthers the University's mission by disseminating knowledge in the pursuit of education, learning, and research at the highest international levels of excellence.

www.cambridge.org
Information on this title: www.cambridge.org/9781108422543
DOI: 10.1017/9781108524919

First published 2021

A catalogue record for this publication is available from the British Library.

ISBN 978-1-108-42254-3 Hardback

Contents

List of Figures — *page* vii
List of Tables — viii
List of Contributors — ix
Acknowledgements — x

The Semantic Conception of Logic: Problems and Prospects — 1
Gil Sagi and Jack Woods

PART I INVARIANCE CRITERIA FOR LOGICALITY — 11

1 Invariance and Logicality in Perspective — 13
Gila Sher

2 The Problem of Logical Constants and the Semantic Tradition: From Invariantist Views to a Pragmatic Account — 35
Mario Gómez-Torrente

3 The Ways of Logicality: Invariance and Categoricity — 55
Denis Bonnay and Sebastian G.W. Speitel

4 Invariance without Extensionality — 80
Beau Madison Mount

5 There Might Be a Paradox of Logical Validity after All — 97
Roy T Cook

PART II CRITIQUES AND APPLICATIONS OF THE SEMANTIC APPROACH — 115

6 Semantic Perspectives in Logic — 117
Johan van Benthem

7 Overgeneration in the Higher Infinite 142
Salvatore Florio and Luca Incurvati

8 Propositional Logics of Truth by Logical Form 160
A.C. Paseau and Owen Griffiths

9 Reinterpreting Logic 186
Alexandra Zinke

PART III LOGIC AND NATURAL LANGUAGE 207

10 Models, Model Theory, and Modeling 209
Michael Glanzberg

11 On Being Trivial: Grammar vs. Logic 227
Gennaro Chierchia

12 Grammaticality and Meaning Shift 249
Márta Abrusán, Nicholas Asher and Tim Van de Cruys

Bibliography 277
Index 292

Figures

6.1 Diagram for the forward clause of bisimulation *page* 127
8.1 Wide versus narrow readings 168

Tables

5.1 Consistency of variants of V_1 *page* 102

Contributors

Márta Abrusán, Institut Jean Nicod, CNRS, ENS, EHESS, PSL Research University, Paris, France

Nicholas Asher, Institut de Recherche en Informatique de Toulouse, CNRS, Toulouse, France

Denis Bonnay, Université Paris Ouest Nanterre La Défense

Gennaro Chierchia, Harvard University

Roy T Cook, University of Minnesota

Salvatore Florio, University of Birmingham

Michael Glanzberg, Rutgers University

Mario Gómez-Torrente, Instituto de Investigaciones Filosóficas, Universidad Nacional Autónoma de México (UNAM)

Owen Griffiths, University of Cambridge

Luca Incurvati, ILLC, University of Amsterdam

Beau Madison Mount, University of Oxford

A.C. Paseau, University of Oxford

Gila Sher, University of California, San Diego

Sebastian G.W. Speitel, University of California, San Diego

Johan van Benthem, Stanford University, Tsinghua University and ILLC, University of Amsterdam

Tim Van de Cruys, Institut de Recherche en Informatique de Toulouse, CNRS, Toulouse, France

Alexandra Zinke, University of Tübingen

Acknowledgements

We'd like to thank the Munich Centre for Mathematical Philosophy and the Institut d'Etudes Cognitives at the École normale supérieure for providing generous funding for a conference where we workshopped most of these papers. We'd also like to thank the participants in that workshop and especially Hannes Leitgeb, Paolo Mancosu, and Lavinia Picollo for their helpful comments which substantially improved this volume. We'd like to thank Hilary Gaskin for her patience with us getting this volume into shape, and Bana Saadi for technical support.

Jack would like to also thank Bill Hanson and John Burgess for encouraging his interest in the subject of this volume and Gil for her patience while we brought it to fruition.

Gil would like to thank Ran Lanzet, David Kashtan, and Shaull Almagor for insightful discussions on the contents of this volume and Jack for making this happen.

The Semantic Conception of Logic: Problems and Prospects

Gil Sagi and Jack Woods

The semantic tradition makes languages and their interpretations the objects of formal study. It has flourished through the development of model theory, initially used by Tarski for the formal explication of the notions of truth and logical consequence. Since the midst of the twentieth century, model theory has had tremendous impact in mathematics, computer science, linguistics, and philosophy. In mathematics, model theory started as a foundational discipline and was typically concerned with the study of consistency, compactness, and completeness. Since then, it has turned its focus to the systematic organization of mathematical theories through applications in various fields from number theory to algebraic geometry.[1]

In computer science, the study of the syntax–semantics interface is paramount in both theoretical and applied topics. The study of formal languages in computer science and of computational models more generally is rooted in the semantic tradition. Database theory and specifically finite model theory are examples of fields using high-powered model theory in computer science.[2] In linguistics, model theory permeates modern-day natural language semantics exemplified in the Montagovian tradition. Philosophy interacts with all the aforementioned disciplines, but also has its own connections to model theory in metaphysics and philosophy of science, in particular in studying reference and its connection to realism.[3]

The semantic tradition in the philosophy of logic, which is our primary concern here, develops and scrutinizes the Tarskian definitions of truth and consequence using philosophical notions as well as mathematical tools, including model theory itself. Since model theory prevails in

[1] For the shifting role of model theory in mathematics, see Manders (1987); Baldwin (2018).

[2] For a survey of some of the interaction between model theory and computer science, see Makowsky (1995).

[3] For an extensive presentation of the uses of model theory in philosophy, see Button and Walsh (2018).

so many other fields, it's important to reflect on its presumptions, how it works, and whether it depends in any significant way on facts from other disciplines, such as science or mathematics. These questions permeate this collection as the reader will shortly see.

The wider historical context of the present collection emerges at the turn of the twentieth century, which marks the advent of formal tools in logic. The first stabs at contemporary rigor in logic involved formalizing syntactic notions, such as formation and transformation rules for formulas: the former concern defining well-formed formulas, and the latter concern laws of deduction. Such rules are completely indifferent to the intended semantic content of their objects. The priority of syntactic approaches over semantic approaches seems to have arisen from the suspicion that semantic notions weren't well understood, were rather metaphysical (when that was a term of disapprobation), and smelt of paradox.

Tarski's "The concept of truth in formalized languages" (1933) changed all of this. It provided a rigorous account of truth for sentences which can then be used to define other important semantic notions like consistency and logical consequence. This went a long way towards demystifying semantic approaches to logical properties. It also paved the way for Tarski's "On the concept of logical consequence" (1936) where he analyzed logical consequence in terms that roughly corresponds to the model-theoretic approach we use today.[4]

This definition works as follows: divide expressions into 'logical' expressions and 'non-logical'. Then a sentence φ follows from a collection of sentences Γ if, no matter how we (uniformly) interpret the meaning of the non-logical expressions contained within φ and Γ, φ is true in a model $\mathfrak{M}$ whenever all of Γ are.[5] So 'there are cats' follows from 'there are cats and dogs' since no matter what 'cats' and 'dogs' mean, the first sentence is true whenever the second is. We don't vary the meaning of the quantifier 'there are' since it's intuitively a logical expression.

[4] There are some interpretational issues surrounding the relationship between the Tarskian conception of consequence and our current conception. These tend to center around whether Tarski already was thinking in terms of "varying domains" or whether his account of consequence drew upon a fixed background of objects. See Ray (1996); Gómez-Torrente (1996, 2009); Mancosu (2006, 2014). One may also note that in 1936, Tarski's definition uses an essentially stronger metalanguage, whereas later Tarski and Vaught (1957) definitions are couched in set theory, and do not resort to an essentially stronger metalanguage.

[5] There's been a lively discussion about exactly what this definition requires. In particular, there's a question about the breadth of the necessity captured by 'no matter how'. See Sher (1991), Hanson (1997), and Gómez-Torrente (1999) for discussion.

This definition presumes that we can divide the logical expressions from the non-logical, but it does not itself give a way to do that. The topic of logicality (of expressions) is thus central in the philosophical literature on the semantic conception of logic. In Tarski's initial work, he expressed skepticism that there was a fixed way of doing so, noting that expressions like the set-theoretic relation of inclusion, were sometimes, but only sometimes, treated as logical. His idea seemed to have been that we have, at best, a definition of logical consequence relative to a chosen set of privileged expressions.

Thirty years later, in "What are logical notions?" (a lecture delivered in 1966 and published in 1986), Tarski addresses the issue of logicality from another perspective. He proposes a criterion of *logical notions*, set-theoretic entities which constitute the distinguished subject matter of logic among other mathematical disciplines, in terms of their stability under transformations of the domain. These notions can model the meanings of expressions in a language, and they are susceptible to precise mathematical characterization. The intuitive model-theoretic meaning of 'there are' – the set of all non-empty subsets of the domain – doesn't change when we permute the domain. Although Tarski does not make the connection himself, it is reasonable to identify logical terms as those denoting logical notions, at least at first pass.[6]

The idea is simple: if the meaning of some predicate or quantifier can be disturbed by a permutation, then it somehow depends on the nature of the objects it's about. Any expression that is so worldly doesn't have the requisite 'formal' character that's supposed to characterize logical expressions (McGee 1996). So this formal property of invariance under permutations is used to test for the intuitive property of formality which characterizes logical expressions. Even if one doesn't think this exhausts the analysis of logicality, it's a very plausible necessary condition on an expression being logical.

This kind of definition of logicality can be made precise by assigning each expression a "meaning" drawn from a type-hierarchy built over a domain of objects. We assign the predicate 'cat' the function from the domain to {T,F} which returns true if an object is a cat and false otherwise. Simplifying, we treat the meaning of 'cat' as the set of cats in the domain. We assign the two place predicate '=' the set of ordered pairs $\langle d, d \rangle$ for d

[6] Corcoran, in his editorial introduction to Tarski (1966/86), presents Tarski as filling the lacuna left in his account of logical consequence of 1936. However, for reservations of this interpretation of Tarski, see Sagi (2017).

in the domain. And so on. Then it's easy to see that '=' is a logical expression since the set of ordered pairs $\langle \pi(d), \pi(d) \rangle$ for d in the domain and π a permutation is just the set of ordered pairs $\langle d, d \rangle$ from the domain. Whereas there's no guarantee that $\pi(d)$ for d a cat is a cat.

The *Tarski-Sher thesis* identifies the logical expressions as those which denote logical notions in the sense just defined – so invariance under certain transformations is both a necessary and sufficient condition for logical notions, and denoting logical notions is a necessary and sufficient condition for something to be a logical expression, if it is adequately incorporated in a syntactic-semantic system of logical consequence.[7] Several writers, among them Vann McGee, William Hanson, Mario Gómez-Torrente, and Timothy McCarthy, have criticized this approach for its "extensionality": invariance criteria look only at denotations, and are blind to the level of sense, or of what connects an expression to its denotation.[8]

There are further concerns about this approach. As Feferman pointed out (1999), the formal approaches to identifying logical expressions rely on the existence (or not) of certain permutations and isomorphisms. Yet 'isomorphism' is not an "absolute" property, so changing the background set-theoretic assumptions can change whether or not a notion is logical. Such problems generalize, showing that the Tarskian analysis of logical consequence is dependent on mathematical facts. Since facts about logical consequence are supposed to be prior to mathematical facts, these "entanglements" have bothered many theorists.[9]

On the positive side, logicians have identified interesting connections between a notion being logical and that notion being definable (van Benthem 1982; McGee 1996), between invariant notions and interpolation theorems (Barwise and van Benthem 1999), and between invariance properties and compositionality (Keenan and Westerstahl 1997).[10] Linguists have also identified interesting connections between an expression being logical, in the sense of being invariant under permutation, and grammatical properties.[11] So regardless of the status of the worries mentioned above, the semantic conception of logic has proved a fruitful way of investigating

7 See Sher's chapter in this volume for her latest take on the issue. Griffiths and Paseau (2022) is a forthcoming book on logical monism and the Tarski-Sher thesis.

8 See Gómez-Torrente's contribution to this volume, as well as his (2002), along with McGee (1996); McCarthy (1981); Hanson (1999); Warmbrōd (1999); Feferman (1999).

9 See Florio and Incurvati's contribution to this volume for an investigation of similar "entanglement" phenomena.

10 See van Benthem's contribution to this volume for a survey of other interesting connections.

11 See Chierchia's contribution to this volume, in particular, for an overview of these connections.

a number of issues. Moreover, the semantic conception itself might be separable from the approach just discussed to logicality (see Gómez-Torrente's contribution to this volume.)

In the present volume, we see the semantic account applied to modal notions (van Benthem), fragments of pure mathematics (Florio and Incurvati), grammaticality (Abrusán et al.; Chierchia), validity in natural language (Glanzberg), and purely formal logical validity (Paseau and Griffiths; Cook). We see the broader implications of the semantic conception explored in the contributions of Bonnay and Speitel, Gómez-Torrente, Sher, Madison Mount, and Zinke. Together these provide a balanced and nuanced picture of both the applications and limitations of the semantic approach.

Contributions

Our volume opens on the use of invariance criteria to characterize logicality. Gila Sher's essay "Invariance and Logicality in Perspective" introduces her seminal approach to using invariance properties to characterize formality and thereby logicality (1991). Sher's view is that logical constants are expressions that pick out properties which are isomorphism invariant as well as satisfying a few other conditions concerning how they are picked out. This approach, as mentioned above, has been very influential over the last 30 years. Sher then surveys how developments of the last 35 years have affected her project. This involves distinguishing her approach from pragmatic alternatives (represented by Hanson (1997) and Gómez-Torrente (2002, this volume)) and argues that her approach maintains desirable connections between logicality, formality, and necessity.

Sher's essay is followed immediately by Gómez-Torrente's defense of his pragmatic alternative. Gómez-Torrente argues that the semantic conception of logic, as described above, works perfectly well even with an account of logicality shot through with local pragmatic choices. In developing his pragmatic alternative, he usefully revisits some of his earlier objections to Sher's non-pragmatic account of logicality (2002), such as worries about properties which are invariant, but which are necessarily denoted by expressions which do not seem logical. Consider 'is a regular heptahedron' or 'is a male widow'. He then suggests that there are a number of other marks of logicality, such as broad applicability in our systematic theorizing. His pragmatic suggestion leaves the notion of logicality intentionally a bit vague, rendering as clearly logical those expressions which score sufficiently well on all the marks, and clearly non-logical those which score sufficiently badly on them. This allows a more degreed notion of the logicality of expressions.

We then turn to the contribution of Bonnay and Speitel which attempts to solve some of the problems with the definition of logicality mooted above by Sher and criticized by Gómez-Torrente. While Bonnay and Speitel agree with treating invariance under bijections as a necessary condition on logicality, they hold that it needs to be augmented with certain syntactic constraints. Drawing on an idea dating back to Carnap (1943), they want logical expressions to be *categorical* in the sense of their meaning being uniquely given by their inferential role.[12] Their overall suggestion is that logical expressions are those which denote a unique bijection-invariant notion by means of their inferential role.[13] This novel suggestion seems a worthy contender for a mixed semantic-syntactic account of logicality, improving (as they note) on another suggestion for a mixed account by Feferman (2015).

Most of the discussion of logicality concerns *extensional* contexts, like those found in mathematics and other informal applications of logic. Yet if we truly want an account of logicality that applies to (rigorized) natural language, we should be willing to consider intensional contexts as well (as argued in Woods 2016). Madison Mount's contribution takes up this challenge, developing an account of logicality in the context of the intensional type theory of Muskens (2007). The account is a useful first step towards developing an intensional criterion of logicality, especially as Madison Mount gives a careful treatment of the challenges and technology necessary to adequately extend invariance-style criteria from extensional to intensional contexts.

Part I closes with Cook's contribution. Cook is concerned with the so-called "validity paradox". This problem, which has gained a lot of contemporary attention, concerns what happens when you add an expression '$\Rightarrow$' expressing logical consequence to our language. Under a number of natural rules governing this expression, and the assumption that these rules are "logically valid", the resulting logic is inconsistent (as Cook (2014) and Ketland (2012) demonstrated in prior work). Here, Cook additionally shows that the meaning of '$\Rightarrow$' is permutation invariant only if it's trivial (in the sense that if anything $\Rightarrow$ anything, then everything $\Rightarrow$ everything). These two observations motivate Cook to develop a more permissive notion of logicality which maintains much of the interest of

[12] This meaning of 'categorical', though related to the more usual notion, is a bit narrower.

[13] Note that this suggestion allows that other notions would satisfy the inferential role of these expressions, but that these notions must not be bijection-invariant.

the permutation invariance account while allowing '$\Rightarrow$' to be a consistent and non-trivial logical expression. He closes his contribution by mulling on whether the resulting notion of logicality is too permissive – and what this might mean for those pushing the validity paradox as a significant problem for ordinary approaches to logical consequence.

The volume then turns to further basic aspects of the semantic approach. Van Benthem's chapter extends the philosophical discussion of invariance to a broader look at the functioning of semantics in logic. In particular, he discusses connections between the various notions of invariance used in logic and definability in different formal languages, and the fundamental further issue of coherence: how the resulting different views of structure can still be connected. A second main theme is the ubiquitous entanglement of invariance with inference, where, for instance, model-theoretic preservation and interpolation theorems describe a generalized notion of valid inference, going from truths in one model to truths in other suitably related models. Finally, the chapter connects to other fields by drawing attention to the pervasive role of computation and games beneath the surface of semantics.

Incurvati and Florio investigate a problem mentioned above, the entanglement of logic with mathematics. It's a well-known result (see Shapiro 1991) that there's a sentence of full second-order logic (with a standard model-theoretic semantics[14]) that's valid just in case Zermelo-Fraenkel set theory with the axiom of choice (ZFC) proves the continuum hypothesis (CH). And there is another sentence of full second-order logic that's valid just in case ZFC proves the negation of CH. Yet both CH and its negation are provably consistent with ZFC. So the validities of second-order logic seem to determine what ZFC proves, violating what Incurvati and Florio call "neutrality principles". These principles, which claim logic should be dialectically and informationally neutral, are very intuitive, so these *overgeneration* arguments are rather troubling. Incurvati and Florio defend full second-order logic by arguing that the proponent of the semantic conception of logic can adopt a *higher-order* model theory instead of a first-order one. Once they've done that, the troubling overgeneration arguments disappear. They close by exploring whether related arguments can be made against the higher-order approach, tentatively concluding that it doesn't appear to be so.

[14] This means that bound function and relation variables range over all the functions and relations on the domain, not just some collection of such. The latter sort of semantics is typically called a "Henkin semantics".

Paseau and Griffiths's contribution explores an issue we've neglected so far. It's natural to take logical truth to be truth in virtue of logical form. The Tarskian approach to logical truth – truth in virtue of logical form – analyzes logical form in terms of holding fixed the logical expressions of a language. It uses the logical expressions occurring in a language to explicate logical form, as explained above. But the distinction between logical and non-logical expressions already presumes another division, a principled division of expressions of the language into grammatical categories. Paseau and Griffiths investigate this presumption in the context of specifying the logic of logical truth, understood as truth in virtue of logical form. That is, in the context of investigating which axioms hold of the concept of logical truth. They show that this depends on whether we take a very fine-grained or coarse-grained approach to treating two expressions as being of the same rough grammatical category. For instance, if G and F are atomic (or basic) predicates, then when we have a claim like 'it's logically true that "$Fa \vee \neg Fa$"', we need to consider whether we mean this to entail that 'it's logically true that "$(Fa \wedge Ga) \vee \neg(Fa \wedge Ga)$"' or only substitutions of atomic predicates like G for F. As they show, the logics of logical truth that result from each way of graining are quite different, one being the normal modal logic $S5$, the other perhaps $S4$. Furthermore, if 'it's logically true that' is not a logical constant, the resulting logic is non-normal. Their contribution helps to show exactly what's being presumed in the background of even something as seemingly straightforward as the semantic approach.

Zinke rounds off this Part 2 of the volume with her related discussion of neglected presuppositions of the semantic approach. As she reminds us, the Tarskian approach mooted above makes substantial assumptions about what sort of reinterpretations of the non-logical terms are admissible. These assumptions give rise to presumptions about the division between analytic and logical truths as well as presumptions about the division between metaphysical and logical truths. Semantic entanglement of this kind is problematic without a rationale since the Tarskian account is supposed to give something like the most basic *formal* truths that make no presumptions about meaning or the world. Zinke explores a number of potential rationales for these presumptions and finds them all wanting. This leaves the proponent of the semantic approach with a challenge: find a rationale for what is presumed or accept that the division between logical and non-logical expressions, and thereby the division between logical and non-logical truths and implications, is somewhat arbitrary. Albeit in a way that usually goes unnoticed.

The third and final Part of our volume addresses connections between the semantic approach to logic and natural language. Glanzberg opens Part 3 by discussing the relationship between logic and the semantics of natural language. His particular focus is on formal model-theoretic semantics for natural language. His earlier work (Glanzberg 2015) attacks the idea that there's a neat connection between logic and the semantics of natural language by arguing that natural language "entailments" either aren't formal or they aren't necessary. His essay here is more constructive, describing ways in which model-theoretic semantics *can*, in fact, be seen as modeling aspects of natural language. The most novel of these is the idea that model-theoretic semantics can function like analogical models in the sciences. That is, as formally precise structures having certain structural properties that are analogous to some kind of structure in the modeled phenomenon. The match need not be perfect, but in order to be successful something *like* the structural property must be present in the modeling target. This is a development of the attractive "logic as modeling" approach first mooted by Shapiro (1998) and developed in Cook (2002).

Chierchia introduces and defends the idea that many cases of ungrammatical sentences can be explained in terms of their logical falsity (an idea originally suggested and developed by Gajewski (2002)). Drawing on recent work by Del Pinal (2019), Chierchia shows how we can identify a class of sentence – the *G-trivial* sentences – which are logically inconsistent under any contextually specified interpretation of their non-logical vocabulary. These sentences, surprisingly, turn out to largely match sentences we intuitively judge to be ungrammatical. He then argues for a four-property account of the distinction between logical and non-logical vocabulary. Functional vocabulary, and in particular logical vocabulary, is typically of a high semantic type, based in inferential patterns, works in ways identifiable across a broad class – sometimes even a universal class – of languages, and is permutation invariant. So this exciting – in Chierchia's words, game-changing – account of natural language ungrammaticality draws heavily on the semantic tradition, starting with Tarski's account of logical consequence and, in particular, the model-theoretic distinction between logical and non-logical expressions.

Our concluding paper, by Abrusán, Asher, and Van de Cruys, raises issues with Gajewski's view. Using methods of distributional semantics, they find problem cases which suggest that the logical/non-logical distinction does not work cleanly, at least for natural language. They thus reject the "logical falsity" explanation of ungrammaticality that Gajewski offers. Instead, they hold that facts about semantic composition in context explain ungrammatical unacceptability. They illustrate this by drawing on the examples of weak island ungrammaticality such as 'How tall isn't John?' and exceptives 'Some boys but John smoke'. Their essay neatly interfaces with both Glanzberg and Chierchia's discussion, adding important examples and exceptions that should be carefully considered in looking at the relationship between logic and natural language. Especially when doing so through the lens of the semantic conception.

These papers jointly show how lively the semantic tradition remains. In interpreting natural language, analyzing the difference between logic and mathematics, and getting to the bottom of what follows from what, the semantic tradition provides an important starting place for many philosophers, logicians, and mathematicians. Moreover, the connections to issues of definability, paradox, and grammaticality continue to draw interest to the fundamental issue that Tarski analyzed in 1936. We hope that these papers help to fuel a continued interest in this topic by displaying explicitly exactly how fruitful the semantic tradition can be.

PART I

Invariance Criteria for Logicality

CHAPTER 1

Invariance and Logicality in Perspective

Gila Sher

Although the invariance criterion of logicality first emerged as a criterion of a largely mathematical interest (Mostowski 1957, Lindström 1966, Tarski 1966/86), it has developed into a criterion of considerable philosophical significance. As a philosophical criterion, invariance has been studied and developed from several perspectives. Two of these are the natural-language perspective and the theoretical-foundational perspective, centered on logic's role in knowledge. My own work (Sher 1991 to 2016) has focused on the second perspective. I have argued that the invariance criterion of logicality makes important contributions to the development of a theoretical foundation for logic focused on its contribution to knowledge – a dual, normative-descriptive foundation centered on (i) the veridicality of logic and (ii) its strong modal force. Those who focus on the natural-language perspective concentrate on the descriptive adequacy of this criterion for the study of natural language. Here we have on the one hand philosophers and linguists who study the criterion's contributions to linguistic semantics (see Peters and Westerståhl 2006 and references there). On the other hand, there are critics of the criterion who base their criticisms on its purported linguistic and intuitive inadequacy (see, e.g., Hanson 1997, Gómez-Torrente 2002, MacFarlane 2005, and Woods 2016). Thus, Woods opens his nuanced criticism by saying:

> I argue that in order to apply the most common type of criteria for logicality, invariance criteria, to *natural language*, we need to [require] both invariance of content . . . and invariance of character . . . . If we do not require this, then old objections . . . suitably modified, demonstrate that content invariant expressions can display *intuitive* marks of *non-logicality*. [2016: 778, my emphases]

These critics commonly focus on natural-language inferences whose logical validity is allegedly sanctioned by the invariance criterion but challenged by speakers' intuitions (either raw or theory-laden intuitions). Some criticisms are directed at the prevalent version of the invariance criterion, while others are directed at the very idea of an invariance criterion. Still others are directed at the more general idea of a precise, systematic criterion of, or necessary-and-sufficient condition for, logicality, regardless of whether it involves invariance. Among the latter, some opt for a purely pragmatist approach to logicality.

Naturally, there is room for misunderstandings between philosophers who evaluate the invariance criterion of logicality on different grounds and from different perspectives. In particular, there is room for misunderstandings between (i) those who evaluate this criterion on theoretical grounds and those who evaluate it on intuitive grounds, and (ii) those who evaluate it from the point of view of its contribution to a philosophical foundation of logic focused on logic's veridicality and role in knowledge and those who evaluate it from the point of view of its descriptive adequacy with regard to natural language. In this paper I will try to remove a few misunderstandings concerning the theoretical-foundational perspective on the invariance criterion of logicality. To avoid repetition, I will focus on certain aspects of invariance that I have not expanded on in the past as well as on certain points concerning the theoretical approach to invariance and logicality that have led to misunderstandings. I hope that the clarification of these points will help alleviate the tensions between the theoretical-foundational approach to logicality and the natural-linguistic approach.

1.1 The General Idea of Invariance

Invariance in general is a relation of the form "X is invariant under all variations Y" (where "variations" can be understood as "changes", "transformations", "replacements", and similar expressions, and "Y" can be read as "in Y", "of Y", "of type Y", "of type Y in Z", etc.). Invariance, in this general sense, is a very fruitful notion. Three examples (on different levels) of claims involving an invariance relation, taken from logic, mathematics, and physics, are:

1. A sentence is logically true iff (if and only if) its truth is invariant (preserved) under all replacements of one model by another.
2. The different geometries can be characterized in terms of the transformations of space under which their concepts are invariant.
3. The laws of physics are invariant under all changes of inertial frames of reference.

The first example is a reformulation of the standard semantic (model-theoretic) definition of logical truth. Spelled out in more detail, it says that a sentence is logically true iff it is true (in the/a model representing the actual world or even just true in some model) and its truth is preserved under all variations in models (replacements of any model by another). The second example is based on Klein's 1872 *Erlangen* program of classifying geometries and explaining the relations between them in terms of the transformations of space under which their characteristic notions are invariant (Klein 1987). Thus, the notions of "rigid-body" geometry are invariant under all transformations of space that preserve *distance* between points, while the notions of Euclidean geometry are invariant under all transformations of space that preserve *ratios of distance* between points. Since the latter condition involves invariance under more transformations than the former, Euclidean geometry is more general than rigid-body geometry. One of the most general geometries is topology, whose notions are invariant under all transformations that preserve closeness (open sets). And in principle, geometry G_1 is *more general* than geometry G_2 iff the notions of G_1 are *invariant* under *more transformations* of space than the notions of G_2. The third example is taken from special relativity, whose laws are invariant under all variations in inertial frames of reference.

What does invariance mean? What is its significance? What does it amount to? We may say that when X is invariant under all variations Y, X "does not notice", "does not pay attention to", "is blind to" changes in Y, "is immune" to changes of type Y, or "is not affected" by changes in Y and "cannot be undermined" by discoveries concerning features that vary from one Y to another. Thus, if we regard models as portraying all possible ways the world could have been (in some relevant sense of "possible"), then we may say that logical truths "do not pay attention" to whether the world is as portrayed by one model or by any other. In a similar way, the property of being a Euclidean triangle "is blind" to transformations of space that change distances between points so long that they preserve ratios of distances. (The image of any Euclidean triangle under such transformations is also a Euclidean triangle.) The laws of physics "are immune to changes" in inertial frames, or "are not affected" ("cannot be undermined") by discoveries concerning the distinctive features of given inertial frameworks, those that vary from one inertial framework to another. And so on.

Accordingly, one of the ways in which invariance is highly significant is that the *stronger* the invariance conditions a given notion satisfies (or the characteristic notions of a given field satisfy), the *stronger* or more *stable* the notion (field of knowledge) is, in relevant respects. "Stronger", in the cases we consider here, can be characterized as follows: Invariance condition I_1 is stronger than invariance condition I_2 if the class of transformations associated with I_1 properly includes the class of transformations associated with I_2. But if the stronger the invariance conditions satisfied by X, the stronger (in relevant respects) X is, then it is to be expected that *if X satisfies especially strong invariance conditions, X is especially strong (in relevant respects)*. It would thus not be surprising if we could explain the fact that, and the way in which, logical truths and consequences are *stronger* than other truths and consequences based on their strong invariance. And as we shall see below, it is indeed possible to explain the exceptional modal force of logical truths and consequences based on the fact that they, and/or some of their constituents, satisfy certain *especially strong invariance conditions*.

1.2 The Theoretical Challenges of Logicality and Veridicality

I. The Logicality Challenge

The logicality challenge is the challenge of establishing the theoretical viability of a system of genuine *logical consequences* and explaining how it might be structured. Philosophers may have less and more demanding conceptions of *genuine* logical consequence. Here I am interested in a relatively *demanding* conception, associated with logic's role in knowledge. This role, as I understand it, is to devise a powerful, universal method or system for extending knowledge in any field by moving us from truths – robust, correspondence-like truths – that we may already know to truths (of the same kind) that we may not yet know. In this spirit, I require that a genuine logical consequence satisfy the following strong conditions:

(T) A logical consequence transmits truth from premises to conclusion (where truth is a demanding notion: correspondence in a broad yet robust sense, rather than mere coherence, pragmatic justification, disquotation, etc.).[1]

(M) The transmission of truth is guaranteed with an especially strong modal force

[1] (i) I understand "disquotational truth" in this paper as exemplifying the view that truth in general takes into account only facts (such as disquotation) concerning language. I understand "robust" as involving demanding requirements concerning the world (generally, the extra-linguistic world).
(ii) In this broad sense, correspondence is free from its traditional association with the naive and simplistic idea of copy, mirror-image, or direct isomorphism. For further explanation of this broad (yet robust) conception of correspondence see, e.g., Sher (2016).

II. The Veridicality Challenge

The veridicality challenge is the challenge of truth and justification of the logical theory (system) itself. To be adequate, a logical theory has to say *true* things about logical truths and consequences. It should *not* say that a sentence S follows logically from a set of sentences Γ *unless* S *in fact* follows logically from Γ, i.e., unless the sentences of Γ in fact transmit correspondence-truth – truth *in the world* – to S and do so with an especially strong modal force. It is not sufficient that our intuitions tell us, or give us the impression, that this is the case; this has to *be* the case, and we need to theoretically *justify* the claim that it is the case.

Now, ideally, there would be no need to treat the *veridicality* challenge as a separate challenge. It would go without saying that an adequate system of logical consequence satisfying the logicality challenge produces consequences that *truly* or *in fact* transmit truth from premises to conclusion with an especially strong modal force. But in contemporary philosophy, as we have noted above, philosophers sometimes focus on intuitive rather than theoretical justification.[2] So it is important to indicate that this is not sufficient. An adequate account of logicality must *show* that the requisite conditions on logical consequence are in fact satisfied, and this "showing" must be *theoretical* rather than *merely intuitive* in the everyday sense of the word.

The critical question concerning the invariance criterion of logicality, as a theoretical-foundational criterion, is, then, whether it enables us to establish, theoretically, the viability of a system of consequences that affirms all and only patterns of consequence that in fact transmit truth from premises to conclusions with an especially strong type of necessity.

1.3 Preliminaries

I. Methodology

The challenges of logicality and veridicality are foundational challenges, challenges that have to do with fundamental philosophical questions concerning logic. But the attempt to deal systematically with such foundational questions raises methodological problems that have to be treated

[2] For example, Hanson rejects "modal and formal accounts" of logicality on the ground that they "fail to satisfy our *intuitions* about logical consequence" (1997: 386, my emphasis). He denies the logicality of a term alleged to be logical by the invariance account on the ground that "it seems *bizarre*", i.e., *counter-intuitive*, "to treat" it as logical (1997: 392, my emphasis). And so on.

with care. Traditionally, philosophers assumed that the only methodology for dealing with foundational questions is the *foundationalist* methodology. But the foundationalist methodology makes a theoretical foundation of logic impossible. I have discussed some of the problems it raises and proposed an alternative methodology elsewhere (Sher 2013, 2016), so here I will be very brief. One problem with the foundationalist methodology is its requirement that in giving a foundation for a field of knowledge K we limit our epistemic resources to those produced by *more fundamental* fields than K (fields lying lower than K in the foundationalist hierarchy). But no basic field of knowledge can be given a theoretical foundation under these conditions. Since logic is classified by foundationalists as a basic field, this problem applies to logic. In the literature, many philosophers focus on a particular aspect of this problem: due to the basicness of logic, we cannot provide a theoretical foundation for it without circularity or *infinite regress*. Since all forms of circularity and infinite regress violate the foundationalist strictures, we cannot provide a theoretical foundation for logic at all.

To investigate logicality theoretically, therefore, we need a different methodology. The methodology I will use here is a *holistic* methodology of a *special kind*, called "foundational holism" (see Sher 2016). This methodology is holistic rather than foundationalist, but it differs from various other types of holism in being *geared toward foundational* studies. Thus, this holistic methodology is *world oriented* rather than coherentist, it emphasizes the *inner complexity of structures* rather than totalities or wholes, and so on. Its holistic nature is reflected in its attentiveness to large and open-ended networks of connections between diverse elements. It recognizes that there are many ways to reach the world cognitively, both on the level of discovery and on the level of justification. In particular, both discovery and justification may exhibit multiple patterns, some hierarchical, others not. Accordingly, not all forms of circularity are forbidden: some occurrences of circularity are innocuous, and some are even constructive. The paradigmatic metaphor of foundational holism is Neurath's Boat. In trying to meet the logicality challenge we go back and forth between various kinds of considerations on various levels, using whatever resources are available to us at the moment.

II. Philosophical Theory and Mathematical Background-Theory

In studying logicality theoretically from a philosophical point of view we are faced with a special problem. On the one hand, we aim at a *philosophical* rather than a mathematical account, and in particular, we wish to avoid commitment to any particular mathematical background-theory. On the other hand, using the resources of some mathematical background-theory may have considerable benefits: expressing philosophical ideas using precise terms-of-art, bringing clear examples and counter-examples, answering questions that are difficult to answer without mathematical resources, and so on. The usefulness of a mathematical background-theory is especially significant in the philosophies of logic and mathematics, due to the formality of the disciplines they study. But using a specific mathematical theory as a background-theory might introduce complications. Whereas our philosophical ideas are devoid of problematic mathematical commitments, using the resources of a specific mathematical theory to express them can easily create the false impression that they do carry such commitments. To avoid such false impressions, I prefer to divide my discussion of logicality into two parts. In Sher (2016) I started by formulating and explaining my ideas philosophically, without using mathematical terms-of-art. Once this account was completed, I presented a *precisified* version of the account, helping myself to the resources of a specific mathematical theory, ZFC. Throughout the discussion I stressed that in principle one could use a different mathematical background-theory, with different mathematical commitments, so ZFC's commitments are *not inherent* in the account.

Due to limitations of space I will not be able to be as thorough in separating the two accounts here. But to avoid misunderstandings, it is important to be aware of this point. In particular, it is important to realize that the explanation of invariance and logicality given in the present paper is *philosophical* rather than set-theoretical. It is not committed to ZFC; nor does it carry its commitments.

1.4 Two Invariance Principles of Logicality

In the philosophical literature on logicality, talk of *invariance* is usually directed at one use of invariance – demarcation of *logical constants* – and accordingly, at one type of invariance. But in fact, there is another use, and another type, of invariance in logical semantics as well. This invariance principle appears as my first example of general invariance above. It concerns the use of *models* for demarcating *logical truths and consequences*. I will call it the first invariance principle of logicality, or the model-theoretic invariance principle (I-M).

I. The First Invariance Principle of Logicality: Invariance under (Changes in) Models (I-M)

The first invariance principle of logicality underlies the standard semantic definition of logical consequence, whose roots go back to Tarski (1983b). Consider a collection Γ of sentences of a given language L and a sentence S of L. The standard semantic definition of logical consequence can be formulated as:

(LC) S is a logical consequence of Γ (in L) iff in every model (for L) in which all the sentences of Γ are true S is also true,

without commitment to a specific mathematical construal of models. To capture the requirement that the truth in question is of a robust kind, i.e., a broadly correspondence-truth, we can reformulate LC as:

(LC') S is a logical consequence of Γ (in L) iff in every model (for L) in which all the sentences of Γ are correspondence-true S is also correspondence-true.

Now, although people rarely think of LC as a definition of logical consequence in terms of *invariance*, the idea of invariance (the same idea as in Section 1.1) is implicit in it. We can make this idea explicit by reformulating LC as *Invariance-under-Models*, I-M:

(I-M) S is a logical consequence of Γ iff the transmission of (correspondence-) truth from Γ to S is *invariant* under all variations in (replacements of) models,

Three questions concerning LC, or its reformulation, I-M, concern language, models, and logical constants:

(a) *Language*. What kind of language is assumed by LC/I-M? Since we are interested in a *theoretical* account of logicality, we need to think of this language, which we may identify with L above, as a *theoretical language*, rather than as a natural language. As a theoretical language, L abstracts from those features of language in general that are deemed irrelevant for understanding logicality.[3]

(b) *Models*. How shall we understand models, philosophically? To capture the conception of logical consequence as transmitting correspondence-truth from sentences to sentences with an especially strong modal force, models have to satisfy certain conditions: (i) models should

[3] In principle, L can be either an extensional or an intensional language. But for reasons explained in Sher (1991), the logical constants of L are extensional.

represent all and only ways the actual world could have been, given a relevant understanding of possibility,[4] (ii) there should be a model that represents the way the world actually is in relevant respects,[5] and (iii) the totality of models should be especially large, i.e., the conception of possibility involved should be especially broad, broader than that of physical and even metaphysical possibility. By focusing on the world – the way it is and the ways it could have been – (i) and (ii) ensure that logical consequence transmits the right kind of truth, namely correspondence-truth (truth-in-the-world), and that the transmission of truth occurs in all relevant situations, actual and counterfactual. (iii) ensures that logical consequences have an especially strong modal force, i.e., the modal force of logic is greater than that of physics and even metaphysics. What the relevant conception of possibility is will become clear shortly.

(c) *Logical Constants.* To achieve transmission of truth and exceptional modal force, logical consequence is dependent on a special feature of sentences. This feature is commonly called "logical form", but in fact it could also be called "logical content". Logical consequence takes into account only the logical form or content of the sentences involved, not their non-logical form/content. The logical form/content of sentences has to do with the identity and distribution of constants of a certain kind: constants that, due to their special character, support especially strong and universal consequences. Logical consequence holds fixed the content or denotation of these constants while treating the content or denotation of the non-logical constants as variable (in effect, treating these constants as schematic letters or variables). When it comes to models, logical constants have a fixed denotation (content, satisfaction conditions) in all models, while the denotation (content, satisfaction conditions) of the non-logical constants varies (vary) from model to model.[6]

[4] As we will see below, however, my conception of models as *representational* is subject to constraints that distinguish it from the "representational" conception discussed, and rightly rejected, by Etchemendy (1990). For a more detailed explanation, see Sher (1996).

[5] Some philosophers (e.g., Field 2009) argue that models based on standard set theory as a background-theory are incapable of adequately representing the actual world. Whether they are right or wrong, the fact that our conception of models is not tied up to this particular background-theory (or, indeed, to any other) exempts it from this argument.

[6] (i) See Sher (1991). This amounts to another important constraint on models.
(ii) As explained in Sher (1991), the fixity of logical constants does not mean that they have the same extension in all models (the extension of the universal quantifier in a model with 8 individuals is a set of a set with 8 individuals, while its extension in a model with 9 individuals is a set of a set with 9 individuals). What it means is that their extension is determined for all models in advance, by a fixed principle. (In the case of the universal quantifier, this principle says that its extension in any model is the whole domain of that model.)

Given the conditions T (transmission of truth) and M (especially strong modal force) on *logical* consequence, it is crucial that we set specific requirements on admissible logical constants. This was already noted by Tarski (1983b). If, for example, we treat the material conditional as a non-logical constant, changing its denotation from model to model, *modus ponens* will come out logically invalid. And if we treat "Tarski", "Frege", and "is a logician" as logical constants, "Tarski is a logician; therefore, Frege is a logician" will come out logically valid. Tarski himself did not arrive at any principled criterion for (characterization of, requirements on) logical constants in his 1936 paper, considering it "quite possible that investigations will bring no positive results in this direction" (1936: 420). From the present perspective, the challenge is to find a criterion for logical constants that satisfies, and perhaps even maximizes the satisfaction of, the two conditions on logical consequence, T and M.

These considerations leave the theoretical philosopher of logic with three major tasks:

1. Construct a theoretical criterion for logical constants.
2. Specify a type of possibility suitable for logical consequence (and underlying the totality of models).
3. Explain how (1) and (2) satisfy T and M.

In other words, the theoretical philosopher's task is find, or develop, a theoretical criterion for (or characterization of) logical constants and identify a type of possibility that, together, render LC/I-M an adequate criterion of logical consequence. This brings us to the second invariance principle of logicality and the discussion of formal possibility.

II. The Second Invariance Principle of Logicality: Invariance under Isomorphisms (I-I)

The second invariance principle of logicality is a criterion for logical constants. This criterion is often referred to as the "invariance under isomorphisms" criterion (I-I).[7]

(iii) For an interesting discussion of the fixity of logical constants in the context of current model theory (the current mathematical theory of models), see Sagi (2018).

[7] It is also often referred to as the "invariance under bijections" criterion and the "Tarski-Sher thesis". A related criterion is the invariance-under-automorphisms/permutations criterion (Mostowski 1957, Tarski 1966/86), but depending on how one understands it, this criterion is significantly different from, and inferior to, the invariance-under-isomorphisms/bijections criterion. See McGee (1996) and Sher (1991, 2016).

The Invariance-under-Isomorphisms criterion for logical constants (I-I) that I will discuss here is the criterion developed in Sher (1991) based on earlier mathematical criteria due to Mostowski (1957) and Lindström (1966).[8] I-I has two parts, an objectual part and a linguistic part. The latter concerns the constants of the language L, the former – their objectual denotations, and more generally, objects (in particular, extra-linguistic objects).

A. *Objectual Part of I-I.* The objectual part of I-I applies to objects of a certain kind. One can think of these objects in various ways. Given the present goal, I prefer to think of the relevant objects as properties, where properties include proper properties, relations, and functions of any level and any arity.[9] I-I divides properties into two types: those that do and those they do not satisfy it. Adherents of I-I regard the former as admissible denotations of logical constants, the latter as inadmissible. The formulations of I-I by Mostowski, Lindström, and Tarski are limited to its objectual part.

B. *Linguistic Part of I-I.*[10] The linguistic part of I-I does two things:

1. It tells us that a logical constant must denote a property that satisfies the objectual part of I-I.
2. It sets additional conditions on logical constants, intended to ensure that logical constants satisfying (1) are adequately integrated into a syntactic-semantic system of logical consequence incorporating LC/I-M.

In this paper I will focus on the objectual part of I-I. (For the linguistic entries of I-I see Sher 1991.) In accordance with my second preliminary comment in Section 1.3, I will offer two versions of I-I: one that is not and one that is couched in a mathematical background-theory. I will call the non-mathematical version of the criterion "Invariance under 1-1 replacements of individuals", and I will use the abbreviation "I-R" for this version. I-R is intended to be understood in a way that does not involve specific mathematical (including set-theoretical) commitments. Depending

[8] The 1991 criterion was developed in the mid-80s, before Tarski's 1966 lecture was published. But it can also be construed as a development of the criterion proposed by Tarski.

[9] I think of objects in general as divided into individuals (objects of level 0) and properties (objects of level > 0). The use of properties in the present discussion does not assume any specific theory of properties, and various theories of properties are compatible with this account. For the purpose of the present discussion we can for the most part disregard current controversies concerning properties.

[10] "Linguistic", here, is a theoretical adjective applicable to languages in the sense of Section 1.4(I)(a) above.

on context, "I-I" will name either the mathematical version of the criterion or the broader conception (I-R).

The non-specifically mathematical version of I-I, *invariance-under-replacements-of-individuals*, or I-R, can be presented as follows:

(I-R) An n-place property, $\wp$, of level m, is invariant under all 1-1 replacements of individuals iff for any domain of individuals, D, and any argument, β, of $\wp$ (in D), β has the property $\wp$ (in D) iff the image of β under any 1-1 replacement $\mathbb{R}$ of the individuals in D has the property $\wp$ (in D', the image of D under $\mathbb{R}$).[11]

Consider the 2-place 1st-level property *x-loves-y*.[12] It is quite clear that this property does not satisfy I-R. But the 2-place 1st-level property $x = y$ does satisfy I-R. Consider the 1-place 2nd-level property P-IS-A-PROPERTY-OF-HUMANS, where P is a 1-place 1st-level property. This property does not satisfy I-R, but the 1-place 2nd-level property P-IS-NON-EMPTY – the existential-quantifier property – does.

The mathematical version of I-R, (I-I below) is thought of as a precisification of I-R:

(I-I) An n-place property, $\wp$, of level m, is invariant under all isomorphisms iff for any domains D, D' and any arguments β, β' of $\wp$ in D,D' respectively: if $<D,\beta>$ is *isomorphic* to $<D',\beta'>$, then β has the property $\wp$ (in D) iff β' has $\wp$ (in D').[13]

As noted above, various variants of (I-I) can be introduced using various background mathematical theories. One version of I-I will use ZFC as its background-theory,[14] another may use Russell's theory of types as background-theory,[15] and still others may have other mathematical background theories.

[11] (i) D is any collection of individuals, actual or counterfactual. Since I-R is not formulated in any specific mathematical background-theory, D does not have to be identified as a set, a proper class, or an entity of any other specific mathematical type. For the sake of simplicity we assume that D is non-empty.
(ii) Given a $\wp$ and a D: If $\wp$ is a 1-place 1st-level property, its arguments in D are individuals in D. If $\wp$ is a 2-place 1st-level property, its arguments in D are pairs of individuals in D. And so on. If $\wp$ is a 1-place 2nd-level property of 1-place properties, its arguments in D are 1-place 1st-level properties whose arguments in D are individuals in D). And so on.

[12] I use italics for 1st-level properties and small capital letters for 2nd-level properties.

[13] $<D, \beta>$ is *isomorphic* to $<D', \beta'>$ iff there is a bijection f from D to D' such that β' is the image of β under f.

[14] In this version, D, D' will be proper sets.

[15] In fact, Russellian type-theory was one of the two background theories used by Tarski for his 1966 version of I-I (Tarski 1966/86).

Although in the historical order of discovery I-I was prior to I-R, in the order of philosophical explanation and justification I-R is prior to I-I. This calls for a methodological clarification: My goal in this paper is to explain how the foundational theorist approaches the question of logicality and how invariance enters into her eventual account. To that end, the explanation I provide has the character of a *rational reconstruction* (in a quasi-Carnapian sense). It does not seek to trace the history of the invariance criterion; instead it explains how it is rational to reconstruct it.

I-I as presented so far is, strictly speaking, a criterion for *properties* and *predicates (including quantifiers)*. What about *sentential operators and connectives*? I-I can be generalized to an invariance criterion of logicality for such operators/connectives in several ways. If we assume bivalence, the sentential version of I-I (given in Sher 1991, 2016) coincides with the usual truth-functionality criterion for logical connectives. For the purpose of the present discussion, however, it is sufficient to focus on I-I as a criterion for properties/predicates.

We are now ready to explain why the *Invariance-under-Isomorphisms* criterion is an appropriate criterion for logical constants and to specify the type of possibility that must be represented by models – the models used in logic, which I will call "logical models". This will enable us to explain how the two invariance conditions, I-M (invariance under models, or LC) and I-I, satisfy the two conditions on an adequate notion (system, method) of logical consequence – T (transmission of correspondence-truth) and M (especially strong modal force).

1.5 Invariance-under-Isomorphisms, Formality, and Modal Force

One of the distinctive characteristics of the invariance-under-isomorphism criterion – a characteristic that distinguishes it from other criteria for, and accounts of, logical constants[16] – is that it captures a certain *especially fruitful philosophical idea*. This idea is *formality*. Not formality in the traditional syntactic sense, or the schematic semantic sense, or the substitutional semantic sense, but formality in an *objectual* semantic sense. Objects – specifically properties – satisfying I-I are *formal* in this sense; objects that do not satisfy I-I are *not formal* (in this sense). Any constant can be formal in the syntactic, schematic, or substitutional sense, i.e., be treated as a fixed, distinguished element, partaking in the "form" of sentences (see, e.g.,

[16] From Feferman's (1999, 2010) invariance-under-homomorphisms criterion to pragmatist, non-invariance accounts (see below).

Etchemendy 1990). But only constants that denote properties satisfying I-I are *formal* in a sense that is relevant to the two conditions on logical consequence noted above, T (transmission of correspondence-truth) and M (especially strong modal force).

I-I is connected to formality both extensionally and intensionally. Extensionally, properties satisfying I-I are mathematical and all mathematical objects – individuals, properties, and structures – either satisfy I-I or are systematically correlated with properties that satisfy I-I. Among the mathematical properties that satisfy I-I are identity, the 2nd-level Boolean properties corresponding to the standard logical connectives, the existential- and universal-quantifier properties (NON-EMPTINESS, UNIVERSALITY), ONE, TWO, . . . , FINITELY MANY, INFINITELY MANY, IS-REFLEXIVE/SYMMETRIC/ TRANSITIVE, IS-WELL-ORDERED, etc. Mathematical structures, such as the structure of the natural numbers, are systematically correlated with quantifier-properties satisfying I-I. Mathematical individuals such as the number 1 are correlated with 2nd-level cardinality properties – ONE, . . . – which satisfy I-I. The 1st-level 1-place property *x-is-even* satisfies I-I when construed as a 3rd-level property of 2nd-level cardinality properties, and so on. In contrast, all paradigmatic non-mathematical objects and properties (such as *Archimedes, is-red* and IS-A-PROPERTY-OF-HUMANS) do not satisfy I-I.[17]

Intensionally, I-I captures the idea of formality as *strong structurality*. Take any property $\wp$ of any level, any domain D, and any argument β of $\wp$ in D. Now take the pair <D,β> and take any pair <D',β'> that has exactly the *same structure* as <D,β>. Such a structure can be obtained from <D,β> by a 1-1 replacement of the individuals of D, and if it does, then β satisfies $\wp$ in D iff β' satisfies $\wp$ in D'. I.e., $\wp$ satisfies I-I iff it pays attention *only to highly structural features* of its arguments, iff it is blind to all features of its arguments but (some of) their *highly structural* features. Speaking in terms of invariance, we may say that most properties abstract from some features of their arguments, and as such they satisfy some invariance condition and have some degree of invariance. In this sense, they are at least weakly structural. But I-I is an especially strong invariance condition. Paradigmatically, biological, physical, and other properties do not satisfy this condition; only highly structural properties do. Such highly structural properties are *formal*.

[17] To apply I-I to Archimedes, we identify Archimedes with a property, such as *is-Archimedes*. Clearly, this property does not satisfy I-I.

One ramification of I-I is that the transmission of truth from premises to conclusion by a logical consequence is due to formal relations between the contents of its premises and the content of its conclusion. Semantically, the transmission of truth is due to formal relations between the truth conditions of its premises and its conclusion. Objectually, the transmission of truth is due to formal relations between the situations corresponding to its premises and conclusion, or more precisely, between the formal structures of these situations. For example, the logical consequence

(4) $(\exists x)(Ax \vee Bx), \sim(\exists x)Ax;$ therefore : $(\exists x)Bx$

is based on a relation between two formal structures: a structure of a non-empty union of two properties, P_1 and P_2, the first of which, P_1, is empty, and a structure in which the second property, P_2, is non-empty. This relation is itself formal, so (4) is based on a formal relation between two formal structures, or on a formal relation between formal features of the situations that make (or would make) the premises and conclusion of (4) true. It is due to this relation that (4) transmits (correspondence-) truth from its premises to its conclusion.

Another ramification of I-I is that the transmission of truth from premises to conclusion by a logical consequence has an especially strong modal force. This ramification arises from the fact that the invariance associated with the properties denoted by logical constants – invariance under isomorphisms – is connected to an especially strong type of necessity. The connection between invariance under isomorphisms and strong necessity is based on the fact that properties invariant under all isomorphisms cannot distinguish between individuals of any kinds, actual or counterfactual, and therefore the laws governing such properties cannot distinguish between actual-counterfactual individuals of any kind either. Since the space of such actual-counterfactual individuals is especially large (larger than the space of individuals that physical and even metaphysical properties do not distinguish), the actual-counterfactual scope of the laws governing them is especially large. In other words, these laws have an especially strong modal force. Since logical consequences are grounded in such laws, they have an especially strong modal force. This result has two parts: 1. Logical consequences are grounded in laws governing the properties denoted by their logical constants, namely formal properties. 2. Since these properties have an especially strong degree of invariance, their laws – formal laws – hold in an especially large space of possibilities, hence have an especially strong modal force.

We can finally understand the notion of possibility represented by logical models: logical models represent the totality of formal possibilities, namely, all the ways the world could have been when only formal structure is taken into account. This is the reason invariance under all replacements of models is an adequate criterion of logical consequence.

We have seen how the two invariance criteria of logicality, invariance-under-isomorphisms (I-I), and invariance under models (I-M) establish, theoretically, the viability of an adequate system of logical consequence. Elsewhere (Sher 2016 and works mentioned there) I showed that the formality of logical consequences (in the sense of I-I) also explains their other properties: their considerable generality, topic neutrality, basicness, certainty, and normativity, as well as their quasi-apriority.[18]

This theoretical account of logic employs the foundational holistic methodology. The account is developed in a stage-by-stage (step-by-step) manner, going back and forth in a Neurath-boat style. While in earlier stages we did not have sufficient resources for explaining the relation between logical and metaphysical possibility, at this point we do. The degree of invariance of metaphysical properties is smaller than that of formal properties. Hence the space of logical/formal possibilities is greater than that of metaphysical possibilities. Consider the metaphysical impossibility of being all-red and yellow at the same time. This impossibility is not formal. The property of being both all-red and yellow is not invariant under all isomorphisms. That is to say, the combination of being all-red and yellow is not ruled out on formal grounds. Formal possibility abstracts from most features of individuals, including color and color relationships. Therefore, an individual that is both all-red and yellow is formally possible and as such belongs in the domain of some logical models. There are models that represent individuals that are both all-red and yellow, individuals that are both dead and alive, individuals that do not satisfy the regularities of biology or the laws of physics or the principles of metaphysics. This is the reason the scope of logical possibility is broader than that of other types of possibility and the modal force of preserving truth in all (logical) models is exceptionally high.[19]

[18] Concerning generality, Bonnay (2008) interprets Tarski as saying that I-I is associated with *utmost generality* rather than with *formality*. But for reasons presented both in Bonnay (2008) and in Sher (2008), I-I does not really capture the idea of utmost generality. It captures the idea of formality which, in turn, is associated with *considerable*, yet not *utmost*, generality. For discussion see Sher (2008).

[19] Regarding the comparison between logical and metaphysical necessity/possibility, however, it is important to note that metaphysics is a highly heterogeneous discipline, dealing, on the one hand, with very basic ontological issues, such as what makes something an object, and on the other hand,

Many of the alleged counter-examples to I-I neglect the difference between formal possibility and other kinds of possibility, which is crucial for understanding the philosophical significance of both the invariance-under-models criterion (I-M or LC) and the invariance-under-isomorphisms criterion (I-I). These alleged counter-examples often assume an intuitive or a metaphysical notion of possibility, which is weaker than the notion relevant for I-M and I-I. Therefore, they are not genuine counter-examples. These examples are also usually presented as natural-language examples.

This brings us to the relation between the theoretical, philosophical-foundational, perspective on logicality and the natural-linguistic perspective.

1.6 The Natural-Linguistic and Foundational Perspectives

So far we have discussed the two invariance criteria of logicality – invariance-under-models (I-M) and invariance-under-isomorphisms (I-I) – as criteria designed to satisfy theoretical conditions on logical consequence: transmission of (correspondence-) truth (T) and especially strong modal force (M). We have seen that, from this perspective, the combination of the two invariance criteria, I-M and I-I, fares well: it ensures the satisfaction of T and M, thereby establishing the viability, in principle, of an adequate system of logical consequence. How does this combination, and in particular I-I, fare from a natural-linguistic perspective? Is I-I a descriptively adequate criterion of logicality from this perspective?

To answer this question we need, first, to understand what it means. What, exactly, does *descriptive adequacy* amount to in this case? How do we establish it in principle? It is hard to find a detailed answer to these questions in the critical literature on I-I.

Two co-authors who do raise this question are Peters and Westerståhl (2006). Peters and Westerståhl first formulate this question in a way that is similar to our theoretical question, namely, by asking whether I-I is adequate for a *genuine* logical consequence. Next they ask whether the method commonly used in empirical linguistics, namely, the method of consulting

with less basic issues, such as causality, free will, observable vs. unobservable objects, physical vs. mathematical objects, abstract vs. concrete objects, and so on, including issues like color incompatibilities. These less basic (but still quite basic) issues occupy a much larger space in contemporary metaphysics than the more basic ones, and my references to metaphysics in this paper concern metaphysical possibilities and impossibilities of the less basic kinds. (I leave the relation between logic and the most basic parts of metaphysics to another paper.)

speakers' linguistic intuitions, is appropriate for answering this question. It is widely agreed that this method is appropriate for determining grammaticality; the question is whether it is also appropriate for determining validity and logicality. Peters and Westerståhl are skeptical about a positive answer to this question. While linguistic intuitions have been shown to be reliable with respect to grammaticality, it is easy to see that they are *unreliable* in determining validity and logicality. As a result, Peters and Westerståhl give up the attempt to solve the problem of logicality from a natural-linguistic perspective. They take I-I to be a *necessary* condition on logical constants in natural language, but they do not try to determine whether it is a *sufficient* condition, i.e., whether it is an adequate *criterion* of logicality for natural language.

What they do investigate, instead, is whether the invariance-under-isomorphisms criterion, I-I, enables us to better understand linguistic phenomena that are difficult to understand without it. Their answer to this question is positive. They show that and explain how non-standard logical quantifiers sanctioned by I-I enable us to explain phenomena concerning determiners and complex quantifier-structures in natural language. For example, the non-standard monadic logical operator MOST, sanctioned by I-I, explains the behavior of the determiner "most" in sentences such as "Two critics reviewed *most* films"; the polyadic operator MOST ... -AND-MOST ..., sanctioned by I-I, explains *branching-quantifier* structures in natural language such as "Most of the boys in my class and most of the girls in your class have all dated each other"; and so on.

Peters and Westerståhl's approach is reasonable. On the one hand, studying the ways the invariance-under-isomorphisms criterion, I-I, provides new resources for understanding linguistic structures, both in natural language and in artificial languages, makes good sense. But relying on speakers' intuitions to determine validity and logicality does not. Validity and logicality are significantly different from grammaticality, and employing the same method for both requires careful justification.

Most philosophers, however, do not heed Peters and Westerståhl's warning about the use of linguistic intuitions to determine logicality. Such intuitions are widely used by philosophers as grounds for rejecting I-I without any attempt to justify the use of intuition as an arbiter in this case. In addition, some opponents of I-I appeal to views whose relevance to logicality is questionable. Let me explain these points by reference to two alleged counter-examples to I-I due to Gómez-Torrente (2002, 2003): "unicorn" and "male widow".

Gómez-Torrente claims that the properties *is-a-unicorn* and *is-a-male-widow* are empty "in all possible universes" (2003; 2002, 18). As such, he says, they satisfy the invariance-under-isomorphisms criterion, I-I. Accordingly, the linguistic expressions "is a unicorn" and "is a male widow" come out logical. This, in turn, implies that "There are no unicorns" and "There are no male widows" are logically true. But these sentences "are intuitively not logically true" (2002; 2003, 204). Hence, according to Gómez-Torrente, I-I is not an adequate criterion of logicality.

I explained why this criticism is incorrect in Sher (2003). But there I focused on the fact that the linguistic expressions "x is a unicorn" and "x is a male widow" do not satisfy the extended, linguistic, version of I-I, spelled out in Sher (1991). Here I would like to focus on the properties *is-a-unicorn* and *is-a-male-widow*. I would like to point out certain assumptions underlying Gómez-Torrente's use of these properties to criticize I-I and explain how these assumptions lead us to think that these properties satisfy I-I when in fact they do not. Let me begin with *male widow*.

Gómez-Torrente claims that *male widow* is empty in "all possible universes". What is the basis for this claim? My understanding is that this claim is based on our ordinary intuitions. But this approach to the issue *neglects* the fact that the notion of possibility involved in both the invariance-under-isomorphisms criterion (I-I) and the invariance-under-models criterion (I-M) is a specific and especially broad notion of possibility, namely, the notion of *formal possibility*, whereas the notion of possibility employed in the claim that *male widow* is empty in all possible universes is a non-specific notion of possibility, one that is usually understood in a way that makes it weaker than formal possibility. This explains why this example cannot be used to undermine I-I. The incompatibility between being male and being a widow is *not a formal incompatibility*. Therefore, it does not rule out the *formal possibility* of situations in which the property *male-widow*, like the property *both-all-red-and-yellow*, is not empty. *Male-widow*, then, does *not* satisfy I-I, and "There are no male widows" is *not* true in all logical models, hence does *not* come out logically true on the invariance account of logicality incorporating I-I and I-M.

What about the property *unicorn*? Why would anyone think that this property is empty in all possible universes? The claim that *unicorn* is empty in all possible universes is, if I understand Gómez-Torrente correctly, based not on natural-linguistic intuitions but on a particular philosophical theory that is naturally viewed as belonging to the philosophy of language or to metaphysics, due to Kripke (1980). But this theory does not provide an adequate ground for rejecting I-I. First, this theory is not a theory of *formal* possibility/necessity, but a theory of *metaphysical* possibility/necessity, and as such it is irrelevant to I-I. Second, this theory does not really say that *unicorn* is an empty property in all possible universes. It says that *unicorn*, being a mythological-species "property", is, like all other mythological species "properties", *not* a genuine *property*. I will not go into Kripke's reasons for this claim here. But if one accepts his claim, one cannot bring *is-a-unicorn* as a counter-example to I-I, since I-I does not deal with non-properties.

There are other linguistic/intuitive grounds on which some philosophers have tried to deny I-I. For critical discussions of these grounds see, e.g., Paseau (2014), Sagi (2015), and Sher (1991, 2003, 2016).

1.7 A Pragmatist Approach to Logicality

A number of philosophers – e.g., Hanson (1997, 2002) and Gómez-Torrente (2002, 2003) – prefer a pragmatist approach to logicality over a theoretical approach. Two main weaknesses of the pragmatist approach to logicality are: (i) its neglect of *veridicality*, and (ii) its neglect of *theoretical explanation*. These, I believe, are pragmatism's main weaknesses in all theoretical branches of knowledge. If, and so long as, we view the search for knowledge as a search for *truth* (in a robust, correspondence, sense), if we require *veridical justification* and *evidence* for theoretical claims, and if we aim at *genuinely explanatory* theories, then we cannot be content with a pragmatist approach to knowledge. In the philosophy of logic, or those parts of the philosophy of logic that are discussed in this paper, the question of truth arises in multiple places and on multiple levels: What should logical consequence transmit from premises to conclusion given its role in knowledge? Is a given claim of logical consequence true? (Does it in fact transmit correspondence truth from premises to conclusion with an especially strong modal force?) What is (are) the source(s) of truth of logical-consequence claims? Is it true that a system of logic based on I-I and I-M satisfies the requirements of transmission of truth and modal force on logical consequence? Does the formality of logic, articulated in terms of I-I, provide a theoretical explanation of the special features of logic – necessity, generality, topic-neutrality, etc.? And so on. All these are theoretical questions of truth and explanation that, in principle, require veridical theoretical answers rather than pragmatic answers.

This is not to say, however, that pragmatic considerations cannot play any role in theoretical knowledge. Where can pragmatic considerations enter into the invariantist account of logicality? They can play a partial role in choosing the overall best background theories for the account. (Such a choice is needed when, e.g., we have no decisive veridical basis for choosing between two candidates for a background-theory.) They can play a partial role in deciding which logical system licensed by the invariantist account of logicality to choose in a particular context or given a particular goal. (For example, it is pointless to choose a system that includes high-infinite-cardinality quantifiers if our interest is limited to everyday inferences or even to inferences in physics.) They may be used in deciding on certain details of our system of logical consequence. (For example, the decision whether to limit models to structures with *non-empty* universes.) And so on. But pragmatic considerations should be used alongside, and be balanced by, considerations of veridicality and theoretical explanation, not in lieu of these.

1.8 Conclusion

Invariance plays a central role in many fields of knowledge. In logic, it plays a central role in a theoretical foundational account of logicality, both on the level of logical constants and on the level of logical consequence. Often, however, the invariance criteria of logicality, and in particular I-I, have been evaluated from other perspectives, and this has led to disagreements based on a misunderstanding of their designated role. In this paper I have tried to put some of these disagreements in perspective. In particular, I have explained the foundational-theoretical perspective on logicality as distinguished from the natural-linguistic intuitive perspective.

The foundational-theoretical perspective starts with a conception of logic's role in the advancement of human knowledge, and proceeds to the requirement that logical consequences transmit (correspondence) truth from sentences to sentences with an especially strong modal force. It shows how the two invariance criteria of logicality, invariance-under-models and invariance-under-isomorphisms, give rise to a logical system that grounds logical consequences in a particular facet of the world, *formal laws*, which have the requisite modal force. A central aspect of this account is the connection between invariance, formality, and modal force. Logical constants represent formal properties, properties that have an especially high degree of invariance and as such do not distinguish between most individuals (including metaphysically possible and impossible individuals). Logical consequences are based on laws governing the relations between such properties, laws that hold in all formally possible situations, which are represented by the totality of models. As such, their modal force is greater than the modal force of laws and consequences of other disciplines, whose actual-counterfactual scope is narrower.[20]

[20] I would like to thank the participants in the conference "Model Theory: Philosophy, Mathematics, and Language" (Munich Center for Mathematical Philosophy, LMU, 2017), and in particular Gil Sagi and Jack Woods for very helpful comments.

CHAPTER 2

The Problem of Logical Constants and the Semantic Tradition: From Invariantist Views to a Pragmatic Account

Mario Gómez-Torrente

2.1 Introduction

While the problem of logical constants cuts across lines separating the semantic or model-theoretic tradition in logic from other traditions, in the model-theoretic tradition the problem has been felt with special intensity. This is due mainly to the fact that the notion of a logical constant appears explicitly in the general formulation of Tarski's 1983b method for the construction of definitions of logical consequence, even if a characterization of the notion is not needed for the construction of particular definitions, which can simply rely on particular lists of logical constants. Although there are several characterizations of logical constancy within the model-theoretic tradition, all share the idea that the logicality of an expression is to be understood as consisting in some kind of invariant behavior across some kind of mathematical transformations between some kind of domains. Among the corresponding notions we find the notions of invariance under permutations (Tarski and Givant (1987, ch. 3); Keenan and Stavi (1986) – prefigured in Tarski (1966/86), though not strictly speaking as a notion of logical constancy), of invariance under bijections (van Benthem (1986a, ch. 2)), of invariance under surjective functions of a certain kind (Feferman (1999, 2010)), of invariance under arbitrary surjective functions (Casanovas (2007)), and of invariance under potential isomorphisms in a certain sense (Bonnay (2008)). All these are notions definable in terms of the mathematics of classical model theory, as supplemented with appropriate definitions of the concepts of denotation of an expression in a domain assumed by the different invariance notions. There are also other, less purely mathematical invariance notions that have been proposed as reconstructions of the notion of logicality, such as the notions of rigid invariance under various modalities (metaphysical, epistemic, etc.)

in McCarthy (1981, 1987), the notion of invariance under bijections of domains containing both actual and fictional individuals in Sher (1991, 2003, 2008), and the notion of analytical invariance under bijections in McGee (1996).

Among all these notions of invariance, Sher's notion has some right to be considered the most popular invariance notion of logicality, and has even been called the "received" view in the semantic tradition (Bonnay (2014, 54)). I would hesitate to call Sher's view the received one, even in the semantic tradition,[1] but the notion of invariance under bijections (at least in abstraction from the differences between the related more purely mathematical notions and Sher's notion) is probably the most well-known reconstruction of logicality in the tradition. And it is in any case the invariance theory of logicality that has been more ambitiously and vigorously defended from a philosophical point of view within the tradition, especially by Sher, but also by other recent defenders (see, e.g., Sagi (2015), Griffiths and Paseau (2016), and Woods (2016)). In a critical appraisal of invariantist ideas, it is thus natural to focus on the invariance under bijections idea as presented by Sher and on her defense of it, and this is something I will do in the present paper.

Sher often presents her proposal as having a strongly prescriptive component. To some extent, proposing a definition as prescriptive shields it from philosophical criticism. However, Sher is also much concerned with defending the descriptive adequacy of her proposal and with offering arguments that the proposal is explanatory of a number of philosophically important properties of logical truth. In this paper, I will begin by criticizing Sher's arguments for explanatory richness (Section 2), relying on my earlier objections to her claims of descriptive adequacy of her proposal (in Gómez-Torrente (2002, 2003)). This critique leaves one wondering what is to be made of the problem of logical constants from the point of view of a sympathizer of the semantic conception of logic. Section 3 will suggest that the conception can do perfectly well without a model-theoretic notion of logicality, and that the descriptive and explanatory theoretical roles ascribed to invariance under bijections by Sher can be played by a non-model-theoretic account of logicality, and specifically by one in which some pragmatic properties of expressions play an important role. I will also develop the pragmatic account somewhat beyond the mere hints offered in its defense in my earlier work.

[1] I guess that the vast majority of logicians and philosophers in the tradition simply have no definite view on the matter of logicality, as opposed to the situation with regard to, say, what is the appropriate model-theoretic notion of logical consequence for classical quantificational logic.

2.2 Invariance and Explanatory Power

Let me recall first of all that in the literature there are two main sorts of considerations against the descriptive adequacy of the invariance under bijections idea. Both accuse invariance under bijections of allowing too many constants to be logical. The first sort of consideration is that a lot of constants that appear to have a rich set-theoretical content are invariant under bijections, such as cardinality quantifiers like 'There are infinitely many things such that' and 'There are uncountably many things such that', but they strike many logicians and philosophers as not clearly logical in some sense. The second main sort of consideration is that a lot of non-set-theoretical constants which in some intuitive sense seem clearly non-logical are invariant under bijections. To take three examples from my earlier work (Gómez-Torrente (2002, 2003)), 'unicorn', 'heptahedron' (meaning "regular polyhedron of seven faces") and 'male widow' have the empty extension in all domains, or at any rate in all actual domains, in all metaphysically possible domains, and in other sorts of domains as well, and are thus invariant under bijections of these domains. However, they would seem to be pre-theoretical clear cases of non-logical expressions.[2] Assuming the examples in question are indeed pre-theoretically clear cases of non-logical expressions (and as far as I know, no one has denied this in the literature on the topic), invariance under bijections has a serious descriptive problem given the semantic tradition's typical requirement for theoretical reconstructions of a concept, that they should exclude all pre-theoretical clear non-cases of the concept. (See Gómez-Torrente (2003) for rebuttals of attempts by Sher to play down the importance of these examples for the issue of descriptive adequacy.)

The second main sort of counterexamples to the invariance under bijections proposal are important not just because they show that existing invariantist proposals are descriptively inadequate. Perhaps more importantly, they point the way to several skeptical considerations about Sher's ambitious claims that the invariantist idea provides a foundation or explanation for some of the philosophically important properties that logically correct arguments and logically valid sentences appear to possess. That Sher's claims concerning the explanatory power of her proposal are indeed ambitious is made plain by the following quotation:

[2] In the case of 'heptahedron', it may be worth noting explicitly that geometry has been one of the parts of classical mathematics that have typically not been treated as parts of logic by the modern logical tradition.

> Logic is often characterized by its basicness, generality, topic-neutrality, necessity, formality, strong normative force, certainty, a-priority, and/or analyticity. While, as foundational holists, we reject the purported analyticity of logic and qualify its a-priority, we can explain its other characteristics (including quasi-apriority) based on the Invariance-under-Isomorphism criterion, i.e., explain why the laws of logic and its consequences are as basic, general, topic-neutral, formal, strongly normative, and highly certain as they appear to us to be, and to what degree they are a-priori. (Sher (2008, 316))

(Necessity is omitted in the last enumeration, but this is just an oversight.) We will question, to varying degrees, all the purported explanations offered by Sher, paying closest attention to those of necessity, strong normativity (closely linked to "basicness") and (quasi-)apriority (closely linked to "high certainty").

Sher's claims of explanatory richness are developed most fully in Sher (2008), where she treats logicality as a property of operators, by which she means the class-functions defined over domains which she identifies with expressions in other parts of her work. (An operator is invariant under bijections when, for any two domains and any bijection between them, the operator's value on the first domain is the image under the bijection of its value on the second.) Sher attributes the formality of logic to the invariance of logical operators. The idea is that invariant operators do not distinguish between bijectable domains, which in a reasonable sense can be said to have "the same form". I have little to object to this use of the Protean words 'form' and 'formal', and if invariance under bijections is indeed a necessary condition of logical operators (or of logical constants),[3] then there is no objection to the claim that logic is formal in the sense that its constants signify (or are identified with) operators which are invariant under bijections. However, this is barely more than a choice of usage, and has little explanatory value in and of itself. The traditional distinction between the form and the matter of an argument or proposition is essentially just the distinction between its logical constants and its non-logical constants. A theory of logicality is thus automatically a theory of formality, and so there is little to be objected to the trivial claim that a certain theory of logicality is a (good) theory of formality.[4]

[3] This can actually be questioned, at least for some extensional constants that appear to have some degree of logicality, such as 'is a part of' and 'is true'. But it may well be that all extensional *clearly* logical constants are invariant under bijections. We will say a bit about this in Section 3 below.

[4] As Gil Sagi points out to me in personal communication, Sher evidently sees as virtues of her account that it characterizes logical constants in terms of properties of their contents, the logical operators, and that it characterizes logical operators in terms of a property that has some independent right to get the Protean word 'formal' applied to it. However, as I see things, this doesn't give

According to Sher (see Sher (2008, 205ff.)), her identification of logicality or formality with invariance explains the generality and topic-neutrality of logic, because bijections are highly unrestricted transformations compared to other transformations. Thus she recalls Tarski's 1966/86 comparison of permutations of a domain of geometric points with other transformations of such domains in which more relations between items in the domain have to be preserved. But Sher notes that there are other transformations between domains that require even less than bijections, the most unrestricted type of transformation being perhaps that of an arbitrary function; so invariance under bijections does not characterize "utmost generality" of notions. Nevertheless, insofar as bijections are a highly unrestricted type of transformations, invariance under bijections does capture generality in some appropriate sense, and indirectly topic-neutrality, even if it doesn't capture "utmost generality".

However, the level of unrestrictedness of the transformations involved in a notion of invariance has little to do with what is normally understood by the generality and the topic-neutrality of logic. The counterexamples above indicate a way of seeing this. That an expression is general means that it has wide application, in all or nearly all areas of discourse. That an expression is topic-neutral means that its topic is somehow common to all or not peculiar to any of the specific areas of discourse. Expressions such as 'unicorn', 'heptahedron', and 'male widow' are not general or topic-neutral in these senses, despite being invariant under bijections, so Sher's explanation of the generality and topic-neutrality of logic is of dubious value. (I take it for granted here that all these examples are invariant under bijections of arbitrary domains, on any theoretically viable way of understanding this notion, assuming that, as argued in Gómez-Torrente (2003), Sher's rejection of the examples by appeal to domains with fictional elements and "conventionally designated" elements is based on catastrophic theoretical choices.)

She might protest that she is offering an explanation of the generality and neutrality of logical operators, not an explanation of the generality and topic-neutrality of expressions in other senses, e.g., in the sense of full-fledged expressions with all the aspects of their content settled. An operator in Sher's sense will be general and topic-neutral if it is invariant under bijections, even if many expressions in the intuitive sense, some of them intuitively non-general and non-topic-neutral, can correspond to one

us a distinctive explanation of anything we wanted explained in advance. As pointed out in the text, in the relevant pre-theoretical sense of 'formal', any theory of logical constancy that is not blatantly inadequate provides some explanation of the formality of logic.

and the same operator. This may be an acceptable reply, but whether it is acceptable or not will depend on whether there is a reasonable intuitive sense in which operators as such can be classified as general or non-general, or topic-neutral or non-topic-neutral, and such that it corresponds approximately to their being invariant or non-invariant. Such a sense surely exists for linguistic expressions as usually understood, expressions endowed with full meaning or content – it's this meaning or content that determines their intuitive topic, for example. But it is unclear to me that the appropriate sense exists for operators as Sher understands them. After all, an operator in Sher's purely extensional sense is supposed to be defined over every domain,[5] regardless of the nature of its objects, whether it turns out to be invariant under bijections or not, and, as noted, one same operator is associated with many intuitively different topics.

Be this as it may, we will now see that Sher's explanations of the necessity, strong normativity, and (quasi-)apriority of logic suffer from related problems, which in these cases certainly cannot be shrugged off by an appeal to the distinction between full-fledged linguistic expressions and operators.

Let's begin with Sher's explanation of the quasi-apriority of logic. She holds a Quinean view on which logic is not *a priori*, because some empirical discoveries might lead one to revise one's logic (she cites the suggestion of Birkhoff and von Neumann, Putnam, and others that quantum-mechanical experiments provide an empirical justification for a change in logic). However, logic is "quasi-a-priori",

> in the sense that the logical laws themselves are unlikely to be refuted by our empirical discoveries. The Invariance-under-Isomorphism criterion explains why logic is immune to refutation in this sense. Since most of our empirical discoveries do not concern the formal regularities in the behavior of objects – i.e., regularities governing features of objects that are invariant under isomorphism – logic is not affected by most of these discoveries, and in this sense it is resistant to refutation and, furthermore, a-priori-like. (Sher (2008, 317))

Even granting for the sake of argument that logic is not *a priori* but quasi-a-priori in some sense, one problem with this alleged explanation is that not all "regularities governing features of objects that are invariant under isomorphism" need be "unlikely to be refuted by our empirical discoveries". Many "features of objects" which are invariant under bijections will appear

[5] It may be worth noting, however, that on Sher's own standards it is unclear that her operators are well defined, not just in view of the vagueness of the terms with the help of which they are defined in the metalanguage, but because Sher (2003) claims that vast numbers of terms from natural language are undefined for many domains; and it is just impossible to get out of natural language if we want to define metalinguistically the operator for 'red', say.

in the statement of "regularities" which need not be particularly resistant to undermining by new empirical information. Take 'unicorn'. Suppose it is indeed the case that 'unicorn' has the empty extension in all (metaphysically possible) domains. It then corresponds to a "feature of objects" which is invariant under bijections, and which appears in the "regularity" '∀x~unicorn(x)'. But this doesn't mean that it must then be unlikely that new empirical discoveries will undermine our belief in this regularity. If skeletons of equine creatures with something that looks like a horn on their heads are discovered around a place where it is believed that (the earliest ancestor of) the word 'unicorn' originated, our belief that there never were any unicorns would be undermined (even if the belief is ultimately correct). I don't know how likely or unlikely it is that such a discovery will be made, but the important point is in any case that the invariance under bijections of 'unicorn' clearly doesn't guarantee that the discovery is unlikely. (Of course, much less could it guarantee the full apriority of '∀x~unicorn(x)', for this, or its attitudinal content, is evidently *a posteriori*.)

Sher might again reply that she is offering an explanation of the (quasi-)apriority of regularities involving logical operators, not an explanation of the (quasi-)apriority of regularities involving full-fledged expressions or their contents. But now the reply is evidently not adequate, for a "regularity", if it is to be something of which it makes sense to say that it is (quasi-)*a priori* or (quasi-)*a posteriori*, has to be something of which it makes sense to say that it is susceptible of being the object of propositional attitudes such as knowledge and (justified) belief. An operator in Sher's sense is pretty clearly not something that can be the object of propositional attitudes, or at least can only be such an object through the mediation of some corresponding attitudinal content, a kind of thing that is normally associated with the idea of a full-fledged linguistic content. A "regularity" involving merely an operator in Sher's sense, whatever such a regularity might be, is thus not something that could be justified or undermined as a part of a belief system, whether by empirical or non-empirical information. Only a regularity involving something essentially analogous to a full-fledged content, the content of a full-fledged linguistic expression, is susceptible of being *a priori* or *a posteriori*, and this renders Sher's attempted explanation of the (quasi-)apriority of logic vulnerable to our objection above.

A related problem affects Sher's purported explanation of the "basicness" and strong normative force of logic. The phenomenon to be explained is that logic appears to hold some kind of overarching authority over other less "basic" disciplines.

> Chemistry, biology, and geography have to attend to the strictures of logic, but logic need not attend to their strictures. Logic has normative authority over these disciplines, but not vice versa. The Invariance-under-Isomorphism criterion explains why this is so: Since chemical properties are not preserved under isomorphisms, logic has a stronger invariance property than chemistry. As a result, logic does not distinguish chemical differences between objects and is not subject to the laws governing chemical properties. But chemistry does distinguish formal differences between objects; for example, it distinguishes between one atom and two atoms. So chemistry is subject to the laws of formal structure. (Sher (2008, 317))

Now, here again "laws of formal structure", if they are to be something that thinkers must abide by, have to be things susceptible of being accepted or rejected by them. "Laws" involving merely operators in Sher's sense are not things that can be obeyed or disobeyed as such; only laws involving full-fledged contents can be obeyed or disobeyed. And then the problem is that many laws involving full-fledged contents that determine operators invariant under bijections are nevertheless intuitively not "basic" or authoritative over non-logical disciplines in general. There is no intuitive sense in which '∀x~heptahedron(x)' is authoritative over arithmetic or geometry, and there is similarly no sense in which '∀x~male widow(x)' is authoritative over physics or biology, even though 'heptahedron' and 'male widow' correspond to operators which are invariant under bijections.

The situation is somewhat different in the case of Sher's explanation of the necessity of logic. This is condensed in the following passage:

> The totality of models represents the totality of formal possibilities; logical consequences preserve truth across all models; they do so due to the logical structure of the sentences involved; this logical structure reflects the formal skeleton of the situations described by those sentences; therefore the preservation of truth is due to connections that hold between the formal skeletons of the situations involved in all formal possibilities; and formal connections persisting through the totality of formal possibilities are laws of formal structure. It follows that consequences satisfying Tarski's definition are formal and necessary, as required by the intuitive constraints (however strong the necessity constraint is taken to be). (Sher (2008, 316))

While Sher's presentation is perhaps less than fully clear, the relevant train of thought seems to be reconstructable as follows. (Compare the related argument in McGee (1992) and its generalization in Gómez-Torrente (2000, ch. 8).) (i) That a sentence is a Tarskian logical truth means that it is true in all its (actual set-theoretical) models (interpretations of its non-logical constants); similarly, that an argument is a Tarskian logically correct argument means that it is truth-preserving in all its (actual set-theoretical) models (interpretations of its non-logical constants) (ii) That a sentence is necessary means that it is true in all possible worlds; similarly, that an argument is necessary means that it is truth-preserving in all possible worlds (iii) Suppose that a sentence S is not necessary and an argument $P_1, \ldots, P_k/C$ is not necessary. Then S is false in some possible world *w* that provides an interpretation for its non-logical constants, and $P_1, \ldots, P_k/C$ is non-truth-preserving in some possible world *w'* that provides an interpretation for its non-logical constants (iv) But then, since items in logical form do not distinguish between bijectable domains, S will be false in every isomorphic interpretation in *w* (or in any other world), and $P_1, \ldots, P_k/C$ will be non-truth-preserving in every isomorphic interpretation in *w'* (or in any other world) (v) Models consisting just of pure sets include models of all the forms that can be drawn from all the different possible cardinalities, so S will be false in a model consisting just of pure sets in *w*, and $P_1, \ldots, P_k/C$ will be non-truth-preserving in a model consisting just of pure sets in *w'* (vi) But these pure sets models are also actual set-theoretical models, so S will be false in some actual set-theoretical model and $P_1, \ldots, P_k/C$ will be non-truth-preserving in some actual set-theoretical model (vii) Therefore, logical truths and logically correct arguments are necessary.

Now I agree that, with some probably surmountable caveats,[6] this is a valid derivation of the necessity of sentences and arguments from the assumptions that they are Tarski-valid and that their logical constants are invariant under bijections in *arbitrary possible* worlds. I stress 'arbitrary possible' both because invariance under bijections in arbitrary possible worlds and not merely invariance under bijections of actual domains is needed in the derivation and because the fact that invariance under bijections in arbitrary possible worlds is what is needed deprives the derivation of much value as an explanation of necessity in terms of invariance. Let me explain these two points.

That an expression is invariant under bijections in arbitrary possible worlds, or necessarily invariant under bijections, means simply that given an arbitrary world, any two domains of things existing in that world, and

[6] For example, (i) is plausible for higher-order sentences only if we assume that all non-set-theoretical interpretations of them are somehow adequately represented by models, which are normally assumed (and assumed by Sher) to be set-theoretical. For another example, (vi) arguably presupposes that the universe of pure sets is strongly uniform from world to world, i.e., that it's not richer in some worlds than in others (a version of the traditional view that mathematical objects are necessary existents).

a bijection between these domains, the extension of the expression in the second domain is the image under the bijection of its extension in the first domain. Intuitively, all standard logical constants of classical quantificational languages are necessarily invariant under bijections. But that logical constants are necessarily invariant under bijections is evidently required in step (iv) of the derivation above. If a sentence S is made false by a model existing in some other possible world, and we can only assume that it is invariant under bijections of actual domains, or even under bijections of domains restricted to some particular possible world, then we will not be able to conclude that S is made false also by a model existing in the actual world. This is the first point.

The second, more important point, arises from the first. The point, to put it in a nutshell, is that the purported explanation of necessity in terms of invariance is not really such, insofar as the derivation, however valid, deduces the desired modal property from a closely related modal property of the relevant expressions, not from invariance or formality by itself. To take a simple case, consider the predicate '$\varnothing$', defined to be synonymous with 'non-identical to itself'. The sentence '$\forall x{\sim}\varnothing(x)$' is intuitively a Tarskian logical truth, as '$\varnothing$' is intuitively a logical constant and, under the assumption that it (and '$\forall$' and '$\sim$') is a logical constant, '$\forall x{\sim}\varnothing(x)$' is true in all actual set-theoretical models. And '$\forall x{\sim}\varnothing(x)$' is also intuitively necessary, e.g., in the sense that intuitively no object from any domain in any possible world is non-identical to itself. However, that '$\forall x{\sim}\varnothing(x)$' is necessary doesn't follow from the facts that '$\varnothing$' is invariant under bijections of actual domains and that '$\forall x{\sim}\varnothing(x)$' is true in all actual set-theoretical models; it only follows from the assumption that '$\varnothing$' is invariant under bijections of domains in arbitrary possible worlds. (The reasoning is again this: if '$\forall x{\sim}\varnothing(x)$' were false in some possible world, it would be false in a domain from that world, hence false in a domain of pure sets – this is the step that uses invariance under bijections in arbitrary possible worlds – hence false in some actual model.) But the assumption that '$\varnothing$' is invariant under bijections of domains in arbitrary possible worlds is just too short a distance apart from the claim that '$\varnothing$' has the empty extension in all domains irrespective of the possible world they come from. And this is just too short a distance apart from the very claim that '$\forall x{\sim}\varnothing(x)$' is necessary. Sher's derivation of the necessity of Tarskian logical truths surreptitiously uses intuitive modal properties of expressions in order to derive closely related modal properties of the sentences they appear in; but this is no real explanation of those modal properties, let alone one in terms of invariance or formality by itself. A real explanation must provide a substantial account of what it is that determines that '∅' is necessarily empty, and thus invariant under bijections of domains in arbitrary possible worlds.[7]

2.3 The Semantic Conception of Logic and the Proper Outlook on Logical Constancy

Invariantist theories are not alone among theories of logicality in facing serious objections. In my earlier work, I have argued that there appear to be difficulties of principle for all theories of logical constancy that are formulated purely in terms of mathematical or semantic notions. I would like to stress that this is a weaker view than the view of some detractors of the semantic conception of logic, who have argued against the possibility of a non-arbitrary divide between logical and non-logical constants as a part of their case against Tarski's method for defining logical consequence (see e.g., Etchemendy (1990, ch. 9); Read (1994)). It's also important to note that, if I am right and an adequate semantic or mathematical theory of logical constancy were in fact impossible, this would not really be a serious problem for the semantic conception of logic. Tarski would have undoubtedly welcomed a substantive mathematical explication of the notion of logical constancy that could have permitted its elimination from the general statement of his method for constructing definitions of logical consequence for particular languages. However, he himself noted that his method could get by without such a general explication, defining 'logical constant' by enumeration for particular languages (see Tarski (1987)). Furthermore, when he presented his idea of defining 'logical notion' in terms of invariance under permutations in Tarski (1966/86), he made it clear that in his view the idea did not settle once and for all questions as to the scope of logicality, and that what concepts counted as logical in a language still depended on arbitrary decisions of some kinds. This was the latent thought in Tarski all along, especially in the seminal paper on logical consequence, where he says that "no objective grounds are known to me which permit us to draw a sharp boundary between the two groups of terms [logical and extra-logical]" (Tarski (1983b, 418–19)), and that

[7] For another example reinforcing this point, consider again the predicate 'unicorn'. The necessity of '∀x∼`unicorn(x)`' is surely a consequence of the facts that 'unicorn' (and '∀' and '∼') is invariant under bijections of arbitrary possible domains and that '∀x∼`unicorn(x)`' is true in all its models when one takes invariance under bijections as one's criterion of logicality. But this doesn't mean that these facts provide a real explanation of the necessity of '∀x∼`unicorn(x)`'; a real explanation must involve mention of aspects of the meaning and/or the mechanism of reference fixing of 'unicorn' that determine what its extension is in a possible world. (Thanks to Gil Sagi for pushing me to clarify my critique of Sher's explanation of the necessity of logical truth.)

> I also consider it to be quite possible that investigations will bring no positive results in this direction, so that we shall be compelled to regard such concepts as 'logical consequence', 'analytical statement', and 'tautology' as relative concepts which must, on each occasion, be related to a definite, although in greater or less degree arbitrary, division of terms into logical and extra-logical. (Tarski (1983b, 420))

Clearly Tarski did not think that the value or the correctness of his method for defining logical consequence would have been threatened by this possibility.

My positive view on the question of the characterizability of the notion of logical constant is, in fact, more optimistic than Tarski's latent view or his detractors' explicit view. I think that a descriptively adequate characterization of logical constancy may well be possible if it is given in terms of properties which, unlike those involved in the usual characterizations, involve reference to certain interests of human beings as argumentative and truth-seeking creatures. While I cannot propose a characterization or list of necessary and sufficient conditions for logical constancy with which I am completely satisfied, I think I can give some informative indications on the form that an appropriate characterization would take. These indications make it clear already that, if a characterization of this form is correct, then the distinction between logical and non-logical constants is not arbitrary. Besides, the distinction that would emerge from such a characterization would be descriptively adequate in all the ways in which invariantist and other characterizations are not. Furthermore, the indications I will give will also suggest how a characterization of this form might be coupled with a certain kind of account of how it is that logical truths and logically correct arguments have the philosophically important properties they are traditionally thought to have. And finally, the indications also suggest an explanation for the recurring difficulties of usual characterizations.

Much of the *raison d'être* for logic is the need to isolate and study relatively uncontroversial, theory- and topic-neutral bases for argumentation and reasoning, or at least for argumentation and reasoning that is in some sense important or relevant theoretically or practically. Parties in a discussion or in the evaluation of claims of any sort need a common ground that is not under dispute or under examination. Such a ground need not always be logical in nature, but logic arises as the study of a special need arising from this general need; its aim is to identify important sentential forms and forms of argument whose instances are not just indisputable in particular contexts, but in all contexts or as many contexts as possible, and specifically independently as much as possible of the nature of the claims or of the topic which is being discussed or evaluated.[8]

These motivations are vague and some are pragmatic. It is thus natural to conjecture that the notion of a logical constant is correspondingly vague along various dimensions, some of them pragmatic, which parallel these motivations. To begin with, logical constants must be words which are generally applicable regardless of the theory or the topic under discussion. But 'generally applicable' is vague, there being no clear-cut line separating the generally applicable linguistic expressions from the non-generally applicable. The right view would seem to be that there is a dimension of degrees of general applicability on which expressions are ranked, with the ranking influencing the degree of logicality that an expression is perceived to have.

Expressions such as 'is a part of', 'is true', and others are generally applicable to a great degree, but they are logical in some degree that appears intuitively inferior to the degree in which expressions such as 'not', 'and', and 'there are things such that' are logical. At least one philosopher sympathizing with a pragmatic outlook (Warmbrōd (1999)) has suggested that logical constants must satisfy a necessary condition stricter than general applicability: they must appear necessarily in the systematization of deductive scientific reasoning. Perhaps expressions such as 'is a part of' and 'is true' fail to meet this requirement, but the requirement seems too strict anyway. First of all, it doesn't motivate appropriately the restriction to scientific reasoning; surely other kinds of reasoning might be kinds of reasoning that logic (even logic as traditionally conceived) would aspire to systematize. Second, it leaves unexplained the fact that expressions such as 'is a part of' and 'is true' appear to have at least some degree of logicality. Third, read literally, the requirement might leave out all expressions altogether, for perhaps no expression in particular is really necessary for the systematization of deductive scientific reasoning; perhaps every particular expression can be appropriately replaced without this affecting the relevant qualities of the systematization.

Warmbrōd (1999) has also proposed that his condition of appearing necessarily in the systematization of deductive scientific reasoning is merely

[8] This is true at least of logic as traditionally conceived and understood. Surely the concept of logic has undergone changes through time, and especially over the last fifty years or so, when the feasibility of logic as traditionally conceived has been called into question and the word 'logic' has become applicable to sets of sentential forms and forms of argument whose validity is restricted to highly particular contexts or topics. I assume, with most discussions of logical constancy (including Sher's), that the traditional logical ideal is not unrealizable, and that it has been realized, however partially, with the identification and description of things such as propositional logic and first-order logic.

a necessary condition on logical constancy, and has rejected the project of characterizing the concept in terms of necessary and sufficient conditions. While I think it may be wise in general to be skeptical about the possibility of characterizations of pre-theoretical concepts in terms of jointly necessary and sufficient conditions, in the special case of logical constancy I don't see compelling reasons to give up the idea of a characterization. In particular, I don't think Warmbrōd provides any discernible reason why we should think that if his postulated requirement were correct, it should not be, besides necessary, also sufficient for logical constancy.

The most natural hypothesis, absent some strong specific motivation to the contrary, is that the other intuitive dimensions corresponding to the vague motivations for the singling out and study of logically valid sentences and arguments play the required role here. First, in rounding out the basic intuition of general applicability into a characterization of logicality in terms of (vague) necessary and sufficient conditions, and second, in explaining the intuitively lower degree of logicality of expressions such as 'is a part of' and 'is true'.

As noted above, the intuitions that drive the singling out of expressions for logical study implicitly require that logical constants be important or relevant theoretically or practically (thus in scientific reasoning in particular, but not exclusively), where this importance or relevance may be measured in several ways, e.g., by the degree of usefulness of paying attention to the expression in the resolution of significant problems or disputes in reasoning in general. Again it would seem that there is a dimension of degrees of importance or relevance on which expressions are ranked, with the ranking influencing the degree of logicality that an expression is perceived to have. This reasonably introduces one kind of vagueness that might assign expressions such as 'is a part of' and 'is true' a lesser degree of logicality than that of paradigmatic logical constants, even if their degree of general applicability is comparable.

A third dimension of variation corresponding to one of the intuitive motivations for logic as a distinguished field of study is the dimension of degrees of controversiality of the principles apparently governing a linguistic expression. The higher the degree of uncontroversiality of the principles apparently governing an expression the higher its perceived degree of logicality could be expected to be. Thus, for example, the law of universal instantiation apparently governing the universal quantifier is presumably less controversial than the standard axiom for the first-order quantifier 'There are uncountably many things such that' to the effect that if a union of sets is uncountable then either the set of the united sets is uncountable or at least one of the united sets is uncountable.

In my view, it is not implausible to think that these dimensions, along with perhaps other related ones, play a constitutive role in the pre-theoretical concept of a logical constant and determine its (vague) extension in terms of (vague) necessary and sufficient conditions. A logical constant is, I conjecture, an expression that ranks high in all the relevant dimensions. A clearly logical constant must then rank very high in all dimensions, and a clearly non-logical constant will rank very low in at least some of the dimensions.

The vagueness of the ensuing concept of logical constancy is then in direct correspondence with the vagueness of concepts such as those of general applicability, importance and relevance to general reasoning, and uncontroversiality. Some of these concepts, furthermore, are pragmatic, in the sense that they involve notions that make implicit reference to relations between expressions and the practical interests of human beings, such as importance, relevance, and uncontroversiality. But, as advanced above, the vagueness and the pragmatic nature of the postulated concept of a logical constant provide several things the theorist of logical constancy wants: a refutation of the attempts to dissolve the problem of logical constants; an account that is arguably descriptively correct *vis-à-vis* the examples that have been considered in the literature; a certain account of how it is that logical truths and logically correct arguments come to have the philosophically important properties they are traditionally thought to have; and an explanation of the failure of the usual attempts to solve the problem, and in particular of the failure of invariantist proposals.

In the first place, the vagueness of the characterization refutes the mentioned attempts to dissolve the problem, which postulate that the distinction between logical and non-logical constants has no real difference behind. Why is this? What vagueness does is to leave a large borderline area of expressions that don't clearly rank high on (some of) the relevant dimensions but also don't clearly fail to rank high. Are these constants in the borderline area of (some of) these vague dimensions logical constants? The arbitrariness theorist is right that in these cases there is no clear answer. The vagueness of the concept is compatible with many mutually incompatible ideas about what expressions can be considered logical. It's therefore true that this vagueness implies that we cannot suppose without further ado that every argument has its logical form. If an argument contains expressions which are neither clearly logical nor clearly non-logical, then we cannot speak of a single logical form for that argument. This has motivated in part the attempts to dissolve the problem of logical constants by claiming that the question of which expressions are logical is arbitrary. But the fact that the question of which expressions are logical is vague doesn't imply that it is arbitrary, because the dimensions involved in the vague concept of logical constancy are not compatible with just any idea about which constants can be considered logical. The vagueness of a concept does not imply that its application is arbitrary, and in fact it excludes this, as it is based on the existence of clear cases of application and clear cases of non-application. What is arbitrary to some degree is which constants of the borderline area to consider as logical or non-logical. But if an argument contains only words which are either clearly logical or clearly non-logical, then the logical form of that argument is fully determined by the intuitive conditions on logical constants. So our characterization makes it clear that the distinction between logical and non-logical constants is not arbitrary.

In the second place, our characterization is also descriptively adequate, at least insofar as the examples we have considered are concerned. Expressions such as 'not', 'and', '=', 'is non-self-identical', and 'there are things such that' rank high in all the identified dimensions and are thus presumably logical on any appropriate completion of the proposal. Expressions such as 'is a part of' and 'is true', and also many expressions with a peculiarly set-theoretical content, such as 'belongs' (in the set-theoretical sense), 'set', 'function', 'there are infinitely many things such that', and 'there are uncountably many things such that' don't seem to rank as high, at least in some of the dimensions. For example, undoubtedly all these constants have a somewhat less general applicability than the clearly logical constants. They appear also to be less relevant in reasoning in general, and their study appears less useful for the solution of problems about the validity of arguments in general than the study of the paradigmatic logical constants. And finally, the principles governing them appear to be more controversial than the principles governing these constants. (Our characterization provides an explanation for the fact that (contra Sher (2008)), the question about where to locate the border between logic and mathematics does seem to be vague. This is explained by the vagueness of the dimensions relevant to logical constancy: plausibly, the more intuitive mathematical content a constant has, the farther away it is from the clear cases of logical constancy along (some of) these dimensions.[9]) On the other hand, 'unicorn', 'heptahedron',

[9] The same sort of observation may go a long way toward providing also an explanation for the feeling, emphasized by Bonnay (2014, 62f.) and Feferman (2010, 17), that the choice between the several invariantist notions in the literature on logical constancy may be arbitrary to a good extent.

and 'male widow' presumably rank pretty low in all or most dimensions (perhaps some of them don't rank low in the dimension of uncontroversiality), and are thus appropriately declared (clearly) non-logical by our characterization.

In the third place, our characterization provides accounts of how it is that logical truths and logically correct arguments come to have at least some of the philosophically important properties they are traditionally thought to have. The generality and topic-neutrality of logic is automatically guaranteed by the characterization, insofar as this requires that logical constants, the constants responsible for the logical validity of truths and arguments, rank high in the dimension of general applicability. And it's worth noting that generality and topic-neutrality presumably explain that (many or all) clear logical constants are invariant under bijections (not the other way round, as in Sher). That a constant is invariant under bijections follows from the requirement that a clearly logical constant must be generally applicable to a very high degree, under the assumption that a constant generally applicable to a very high degree must not distinguish between bijectable extensions (between individuals, in the simplest case).[10]

The basicness and strong normativity of logic is presumably a consequence of its generality and topic-neutrality coupled with the requirement of high uncontroversiality. For high uncontroversiality guarantees the epistemic strength that is required of a principle if it is to hold the needed authority in discussions and evaluations of material from non-logical disciplines. But generality and topic-neutrality are also needed, for otherwise the authority in question would not be overarching. On the other hand, generality, topic-neutrality, and uncontroversiality seem to provide intuitively jointly sufficient conditions for strong normativity in the sense that concerns us.

[10] It may be worth noting that I see interesting connections between John Burgess's (2015) recent suggestions concerning structuralism and the present proposal. According to Burgess, the source of the apparent prevalence of a structuralist outlook on the part of mathematicians is a byproduct of their search for rigor: as soon as one sets out to derive one's theorems from basic principles about certain notions, these principles and the theorems derived from them will also hold for any structure isomorphic to one in which the principles hold, in virtue of the fact that the usual logical constants are invariant under isomorphisms. (And this will be so independently of whether some particular structure was the one that mathematicians originally intended to describe.) While I suspect Burgess's suggestion is right, it leaves unexplained the very fact that the usual logical constants are invariant under isomorphisms. On the present view, this is again a byproduct of a more fundamental fact, one that, like the search for rigor, answers to pragmatic features of a certain human endeavor: the interest in isolating expressions that can serve as the basis of logical reasoning because they are, among other things, generally applicable and topic-neutral to a very high degree.

Let's now focus on apriority and necessity. Unlike Sher's invariantist proposal, the account I have in mind does not seek to derive the apriority and necessity of logical truths and logically correct arguments from the features of the proposed notion of logical constancy. In my view, it is doubtful that the explanation of the apriority or necessity of logical truths and logically correct arguments can be a peculiar explanation that relies on the peculiarities of a group of expressions. Apriority and necessity presumably have homogeneous sources across different kinds of sentences and arguments. The explanation I have in mind takes it for granted that such an explanation must exist, whatever it is, both for logical truths and logically correct arguments and for other kinds of *a priori* and necessary truths and arguments, and instead suggests an account of why some of those truths and correct arguments are selected as special concerns of logic.

The account is based on the idea that logic seeks to serve the need for a special study of the (largely) uncontroversial bases for argumentation and reasoning in all contexts. At the level of the expressions it studies, this has the effect that logic seeks to study expressions governed by principles which are as safe as possible; this is reflected in the condition on logical constants that the principles governing them should be uncontroversial to a large degree. But it also has another effect, at the level of the sentences and arguments: it has the effect that the logical truths and the logically correct arguments must be uncontroversial to a large degree, since they are true or correct in virtue of the principles governing the non-schematic expressions in their logical forms. Indirectly, this ultimately determines that logical truths and logically correct arguments are *a priori* and necessary, as uncontroversial truths and forms of arguments valid for all contexts or as many contexts as possible will have those properties. Note, for example, that if logic is to provide a largely uncontroversial ground for reasoning in all contexts, it must do so for arithmetical reasoning; and in this context the uncontroversial ground for reasoning cannot be constituted by contingent propositions or principles or by *a posteriori* propositions or principles. A reasoning basis which included contingent or *a posteriori* propositions or principles would not be as strong a basis for arithmetic as reasoners would desire one such basis to be.

In the fourth place, the pragmatic nature of at least some of the intuitive dimensions involved in the concept of logical constancy explains to a good extent the descriptive problems of the traditional attempts to characterize them. The difficulties are due to the fact that logicians and philosophers of logic have almost always tried to offer their characterizations exclusively in terms of mathematical properties, invariantist proposals being a prime example of this phenomenon. It is very implausible that the pragmatic intuitions underlying the intuitive concept of a logical constant can be captured via a search for mathematical peculiarities. I can't show, of course, that it is impossible to offer a good characterization of logical constants (that is, one that contains in its extension the clear positive cases and leaves out the clear negative cases) exclusively in terms of mathematical properties. But I think it's obvious that it must be extremely difficult to characterize a notion that makes reference to certain practical interests of human beings in terms of properties of this kind, and I think that the counterexamples of Section 2 bear witness to the difficulties.[11] (It may be noted that other concepts that the recent logical tradition has characterized relatively successfully (such as the concepts of truth, validity (with respect to the standard quantificational constants), computable function, etc.) seem to be, already at an intuitive level, fully objective concepts lacking *a priori* connections with concepts involving the practical interests of human beings.)

It is of course to be expected that some conditions definable exclusively in terms of mathematical properties be necessary conditions of the intuitive logical constants, or of significant groups of them, and this is actually the case. We have argued, in fact, that invariance under bijections is presumably a property of clearly logical constants that follows from their high degree of general applicability and topic-neutrality. (On the other hand, 'is a part of' and 'is true', while widely applicable, and possessing an intuitively high degree of logicality, are not invariant under bijections.) But it is not to be expected that this mathematical property should be a sufficient condition for logical constancy in the intuitive sense, because it is not to be expected that all the constants with that mathematical property can be generally applicable (or that they should all satisfy the other intuitive pragmatic conditions on logical constants). This idea is fully confirmed by our critical appraisal of invariantist theories.

Thus, a pragmatic characterization of the kind we are envisioning promises a founded response to the attempts to dissolve the problem of logical constants, an adequate approach to descriptive problems, a certain account of how it is that logical truths and logically correct arguments have some of the philosophically important properties they come to have, and an explanation of the vulnerability of other characterizations to objections such as the ones mentioned in Section 2. It remains to be seen whether a pragmatic characterization of this kind is liable to weighty

[11] I must insist, however, that 'extremely difficult' does not mean impossible.

objections of other kinds. It's not easy to imagine clear counterexamples to its descriptive adequacy, due precisely to its vagueness and its pragmatic character. Perhaps some critics will want to claim that its vagueness, or its pragmatic nature, or its non-mathematical nature, make it unacceptable to the semantic tradition in logic. But this would be confused. The reconstructions of concepts in the semantic tradition use vague and non-mathematical notions: all the Tarskian semantic definitions use the general notion of an expression and notions of particular expressions; and Tarskian definitions of satisfaction and truth for object languages containing vague and non-mathematical expressions use those same expressions in the appropriate metalanguages. Furthermore, a vague and pragmatic notion of logical constancy has, as I hope to have shown, intrinsic explanatory virtues that can be appropriated by the semantic conception. (Here it's good to recall how Tarski himself, when considering the implications of the possibility that the distinction between logical and non-logical constants is arbitrary, suggested that this might have the explanatory virtue of accounting for what he perceived as "the fluctuation in the common usage of the concept of consequence" (Tarski (1983b, 420).) In any case, and finally: if, as noted at the beginning of this section, the value and the correctness of the semantic conception of logical truth and logical consequence are not threatened by the possibility that the distinction between logical and non-logical constants is arbitrary, much less could they be threatened by the possibility that the distinction is vague and pragmatic. I conclude, therefore, that a pragmatic characterization such as the one sketched here has the potential of providing a descriptively adequate and explanatorily rich account of logical constancy within the Tarskian semantic conception of logic.[12]

[12] For very helpful comments on earlier versions of this material, I am grateful to the participants in my seminar on the nature of logic at the UNAM in 2016, to the participants in a class I gave at Gillian Russell's seminar on the philosophy of logic at UNC Chapel Hill in 2017, to the audience of the 6th International Symposium of Research in Logic and Argumentation, and to the participants in the workshop The Semantic Conception of Logic at the MCMP in Munich in 2018; special thanks are due to Ariel Campirán, Daniel Garibay, Carla Merino-Rajme, Gil Sagi, Johan van Benthem, Melisa Vivanco, and Jack Woods. Research was supported by the Mexican CONACyT (CCB 2011 166502), by the PAPIIT-UNAM project IA 401015, and by the Spanish MINECO (research project FFI2015-70707-P).

CHAPTER 3

The Ways of Logicality: Invariance and Categoricity

Denis Bonnay and Sebastian G.W. Speitel[1]

3.1 Introduction

A. Tarski's celebrated model-theoretic definition of *logical consequence* according to which

> [t]he sentence X follows logically from the sentences of the class K if and only if every model of the class K is also a model of the sentence X (Tarski 1983b, 417)

relies on a division of the expressions of the language under consideration into two importantly different categories:

> Underlying our whole construction is the division of all terms of the language discussed into *logical* and *extra-logical*. (Tarski 1983b, 418, our emphasis)

The attempt to delineate the logical expressions of a language has given rise to the perennial issue of the demarcation of the class of logical constants, the search for mathematically precise and philosophically informative criteria by means of which to decide whether to count an expression part of the logical lexicon.

In the tradition of devising such criteria one can distinguish between rule-based and denotation-based approaches. The former take inferential behavior as decisive for the logical status of an expression and attempt to delineate the class of logical constants by providing conditions that characterize the kinds of inferential roles logical constants can occupy. The latter, denotation-based approach, by contrast, locates logicality at the level of reference, in terms of properties of the semantic values of the expressions

[1] The authors would like to thank audiences at UC San Diego and UC Davis, the MCMP in Munich, the ILLC in Amsterdam, the HSE in Moscow, and the Stockholm Philosophy Department for helpful feedback and discussion of the contents of this paper. We would like to express special thanks to Gil Sagi and Jack Woods for commenting on various drafts, and to Dag Westerståhl for making things right regarding a conjecture we had made in an earlier version of this paper.

of a language. Expressions are then, derivatively, designated as logical in virtue of denoting an object that possesses these properties.

Both approaches have some intuitive appeal, and both approaches face some serious problems. In particular, the semantic approach is often accused of overgenerating and of mixing the logical with the mathematical, whereas the inferential approach often stops short of classical first-order logic. But then, maybe, there is hope in combining these two approaches, so as to produce a philosophically well-motivated criterion capable of resisting the criticisms brought forward against semantic and inferential criteria in isolation. This is the hope we shall pursue in this paper. Our focus will, first and foremost, be semantic, and consist in trying to understand what is special about the denotations of logical operations. Inferential, or proof-theoretic, constraints enter the picture as aids for 'getting to' or 'pinning down' those denotations in a particular way. On the account of combined criteria provided here it is thus natural to see the model-theory as providing the *meaning* of the logical expressions, and the proof-theory as *constraining that meaning* in some way, or, better, as constraining the way a certain kind of designation is characterized.[2]

The outline of the paper is as follows: in Section 3.2 we will briefly review invariance-based criteria for logicality to isolate what we, following the semantic tradition, take to be a core feature of the logicality of extensions, and which will also feature prominently in our account, to then rehearse some of the criticisms directed against the semantic approach. In Section 3.3 we will outline the inferentially motivated notion of *uniqueness*, the ability of inference rules to uniquely determine an expression, and sketch its relevance to a combined criterion of logicality. In Section 3.4 we discuss and assess Feferman's *Semantical-Inferential Necessary Criterion for Logicality*, one of the few extant and worked-out combined criteria that emphasizes the primarily semantic nature of logicality, to then, in Section 3.5, formulate and introduce our own criterion (3.5.1), discuss some of its consequences (3.5.2), and compare it with Feferman's (3.5.3).

3.2 Invariance and Its Limits

Analyzing logicality at the level of denotations seems rather natural. There is a prima facie obvious difference between words which refer to things in

[2] We use the term "semantic" in the context of this paper interchangeably with "model-theoretic", in deference to the tradition that assigns priority to model-theoretic methods and criteria. This is somewhat unfortunate terminology, given *proof-theoretic semantics*, but very much in line with the model-theoretic perspective and approach taken here, which sees meanings as being provided by the model-theory. We describe requirements provided in terms of proof-theoretic constraints as *inferential* or *syntactic*.

the world and logical particles which are merely instrumental in helping us articulate contents. The noun "birds" refers to birds, the adjective "yellow" is about those things whose color is similar to the color of the sun at sunset or to the color of a canary's feathers. By contrast, the conjunction "and", the determiner "all" do not have something to do with particular things in the world; "and" is about a specific way of combining statements, "all" is about how we are going to speak of the things referred to by the noun in the noun phrase. The difference between logical and non-logical words seems to readily appear when we ask what those words denote.

However, it cannot simply be said that "birds" and "yellow" refer whereas "and" and "all" do not. Modern model-theoretical semantics has built its success on attributing denotations to every expression occurring in a sentence in a way which makes it possible to compute the truth-conditions of that sentence. Just like, in an extensional setting, "birds" and "yellow" may be taken to respectively denote functions from objects to truth-values – indicating whether an object is a bird or whether an object has color yellow – "and" will denote a truth-function mapping pairs of true sentences to **true** and other pairs to **false**, and "all" will denote a somewhat more complicated higher-order function, which, when treated as a type ⟨1⟩ quantifier, is a purely set-theoretical function checking whether a characteristic function is the characteristic function of the whole domain. So both candidate logical words and non-logical words have denotations which model-theoretic semantics model as functions in a type hierarchy over one or more basic domains. But one may still get the idea that the kind of semantic functions involved in each case are very different. In keeping with the first intuition we have hinted at, the functions that are indicative of sets of birds or yellow things essentially make differences between objects, whereas truth- or set-theoretical functions are formal functions whose action is not sensitive to the identity of objects.

How may we get a firmer grip on the distinction between functions which are sensitive to objects' identity and functions which are not, the latter being candidate logical denotations, the former providing denotations for more mundane expressions? This is where invariance under permutation comes into play. As bijective functions from the domain onto itself, permutations arbitrarily replace objects with other objects. A function is invariant under permutation if it gives the same result no matter whether objects have been switched under a permutation. The characteristic function for the set of birds will not be invariant under a permutation which switches Tweety Bird and Sylvester the cat, and the same goes for the function which yields **true** if and only if it is applied to a yellow object. By contrast, truth-functions remain of course unaltered, as does the characteristic function for the whole domain, as well as the function testing whether a function is that function, which is our interpretation for "all" as a type ⟨1⟩ quantifier. However, permutations operate within a fixed domain. In order to allow comparisons between different domains, one should replace permutations with bijections. Invariance under bijection may then be taken to provide an explication for what it means for a function to be formal, in the sense of being insensitive to the identity of objects.

The question is then whether such an explication of formality as a property of semantic values readily provides a demarcation for logic itself.[3] If logic is thought of as being concerned with everything which is formal, there is no obstacle to a positive answer. But the assumption that logicality and formality coincide is put in doubt by the consequences of adopting such a stance. Testing whether a set is non-empty, or has at least two members come out as formal functions, but so does testing whether a set is the size of the continuum, or whether a relation is well-founded. This has sparked a debate over the so-called overgeneration problem. On one side, Sher (1991) and Griffiths and Paseau (2016) wish to stand their ground and acknowledge whatever comes out as invariant under bijection as logical. On the other side, critics such as Feferman (1999), MacFarlane (2000), and Bonnay (2006) consider that invariance under bijection lets too many functions, loaded with too much set theory, pass the logicality test, suggesting that something is wrong with the criterion and that it should be amended.

Our take on this issue is that overgeneration is a genuine problem, even though it appears difficult to solve within the invariance framework. Part of the appeal of invariance under bijection lies in its simplicity, by contrast with various proposals to complicate invariance in order to alleviate the overgeneration problem. Invariance can be phrased so as to allow for more stringent criteria, along the lines of Feferman (1999) and Bonnay (2006). The trick is, first, to see invariance under bijection as simply one way to decide which structures are similar (namely, two structures are similar if one can be obtained from the other via a bijection), and then to consider other possible similarity criteria (for example, two structures can be considered as similar if they are potentially isomorphic, or if they are elementary equivalent, and so on and so forth). However,

[3] There is, however, more than meets the eye when it comes to whether invariance under bijection is an adequate explication of formality and how exactly it should be spelled out. See, for example, Woods (2014) for a treatment of logical indefinites. Choice operators also raise issues for the categoricity constraint to be advocated in Section 3.5: is there a sense in which essentially free choice functions could still be said to be fully determined?

the point has rightly been made by Feferman (2010) and Button and Walsh (2018) that such complications make it hard to find a natural stopping point. Invariance criteria may be devised that would make logical notions consist only in trivial notions (operations always yielding constant truth-values), or only in first-order definable notions, or only in notions definable in a given infinitary generalization of first-order logic, but such choices seem to lack explanatory power in that they appear to be essentially grounded in the desire to have these or those notions come out as logical.

However, failure to refine invariance in a non-circular manner does not mean that we should be happy with recognizing any expression denoting a bijection-invariant operation as logical. Cardinality quantifiers expressing properties of higher infinites which cannot be made sense of independently of set-theory still seem to go beyond the natural realm of logic, even though they admittedly are as formal, and as bijection-invariant, as any simpler cardinality quantifier. Griffiths and Paseau (2016) have argued against this very diagnosis of overgeneration. Considering that raw intuitions about the logicality of operations or expressions are scarce, they direct their main point towards worries about logical truths, and more particularly worries about making problematic set-theoretic truths into logical truths. According to them, being able to find a purely logical sentence which is true if and only if the continuum hypothesis is true is not significantly different from being able to find a purely logical sentence which is true if and only if $7 + 5 = 12$. Despite this parallel, one may have the lingering feeling that admitting an expression synonymous with "there are at least $\aleph_{17}$ objects such that ..." as a logical quantifier is at odds with the beautiful simplicity of logic.

A possible way to reconcile those who believe in overgeneration and those who do not, or adversaries and proponents of bijection-invariance, would be to grant that there may be no ontological difference between mathematical operations and logical operations, but insist that logic may still be special because of its relationship to inference. Whereas mathematical expressions are confined to the specialized language of mathematical theories, logical expressions, or their informal counterparts, show up in every sphere of discourse, as a means to articulate statements and to reason. It then seems plausible that there should be at least a difference in terms of access, if not of nature: logical expressions should not be any kind of expression denoting formal operations, but expressions denoting formal operations that may be fully characterized in terms of inference rules. Invariance under bijection captures the formality constraint and formality is a necessary feature of logical expressions. However, the formality constraint may not be sufficient to single out our reasoning tools within the broader realm of formal non-empirical operations. The suggestion is then to supplement invariance criteria ensuring the formality of logical denotations with syntactic criteria connecting logical denotations with inferential behavior. This is the lead we shall follow in the next sections.

3.3 Inferential Constraints on Logicality

On canonical semantic approaches to logicality, say the one pursued by G. Sher (1991), the logicality of an expression derives from the logicality of the operation it denotes. Determining the logicality of an operation is independent of any inferential concerns. The logical extension of an expression has to eventually vindicate the inferences that the expression licenses, but all this somewhat curiously remains outside of the picture. The match between inferential behavior and semantic denotation is not part of the logicality criterion. By contrast, the syntactic/inferential considerations of a combined approach mean to add to the semantic perspective by putting this match back into the picture. The intuition is that not only should the denotation of a logical expression be such that the rules governing its use are sound, but also that this is the way we pin down what the denotation of a logical expression is.

Adopting syntactic constraints because of their function as meaning- or sense-conferring and -determining has a long history in proof-theoretic investigations into the demarcation of the logical constants. Originating in a remark of G. Gentzen that "[t]he introduction[-rules] constitute so to say the 'definitions' of the respective signs, and the elimination[-rules] are, in the end, only their consequences" (Gentzen 1934, 5.13, our translation), there have been several influential attempts to tie, or at least make receptive, the logicality of an expression to its ability of determining (via its inferential behavior) a semantic value,[4] or a sense.[5]

While it is commonly acknowledged that, by themselves, proof-rules cannot supply definitions of the logical constants in a full-blown sense, but rather provide something like an implicit definition, explication, or characterization, this status has been used to justify the imposition of various formal constraints on the kinds of rules that can legitimately be

[4] See Hacking (1979).
[5] See Peacocke (2004), Hodes (2004).

claimed to determine a logical notion. One of the most prominent constraints that emerged from this line of investigation, in particular as a response to A. Prior's infamous connective *tonk* (Prior 1960), is the requirement of *conservativity*.[6] Loosely put, *conservativity* amounts to the demand that, after the introduction of a new constant into a language $\mathcal{L}$, no new theorems not containing the new notion, i.e., no new theorems in the old language-fragment without the new constant, become provable. This guarantees that a new constant introduced by a set of rules maintains *consistency*, and can thus be regarded as actually having been defined successfully by these rules.

Having ensured that *some* notion has been defined through obeying the conservativity-constraint it remains to be established that it is possible to talk about *the* notion thus defined. A further formal constraint deriving from Gentzen's rules-as-definitions comment – already mentioned in N. Belnap's response (Belnap 1962, 133) to Prior's *tonk* – that emerges as the definitional complement of conservativity is therefore the property of *uniqueness*. Uniqueness has been translated into the demand of interderivability for two connectives obeying the same rules: two identical sets of rules used to define a connective and its twin should be such that whatever holds of the connective holds of its twin.[7]

In a model-theoretic setting where logical constants are individuated not by their inferential roles but in terms of their contribution to the truth-conditions of the formulas in which they occur, uniqueness will concern the uniqueness of truth-conditions. The kind of implicit definability appropriate in this context will therefore relate to the *semantic value* of a constant, rather than (primarily) to its inferential behavior. This shift in perspective is accompanied by a particular case of underdetermination – *Carnap's Problem*: the underdetermination of the semantics of logical constants by their inferential roles.[8] For while the provision of inference rules for, say, the classical conditional suffices to establish its uniqueness in the sense of interderivability, Carnap showed that the classical

[6] As stated by Belnap (1962) and Hacking (1979).

[7] See (Humberstone 2011, 578ff) for a precise definition and general discussion of this idea; Zucker and Tragesser (1978) adopt a similar requirement under the label of *implicit definability* by introduction- and elimination-rules.

[8] First discovered by R. Carnap (Carnap 1943) the problem consists in the fact that the standard natural deduction proof-rules of the first-order constants underdetermine their semantics, i.e., that there are multiple semantic values of the constants consistent with the standard first-order consequence relation. While initially attracting little attention (see, e.g., Church (1944)), the issue later resurfaced in the context of (moderate) inferentialist accounts, e.g.: Garson (2013), Hjortland (2014), and Rumfitt (2000). For a recent semantic approach to Carnap's Problem, see Bonnay and Westerståhl (2016).

single-conclusion natural deduction rules for the conditional are not sufficient to uniquely determine its classical semantics. Both its standard semantics as given by its usual truth-table and the valuation rendering a conditional statement true if, and only if, it is a tautology are fully consistent with its proof-rules.[9] By themselves, the proof-rules therefore do not suffice to determine a unique semantics and to narrow down the space of possible valuations to recover the standard meaning for the classical conditional (and similarly for the other constants). They do not exhaust its meaning and mere interderivability thus fails to capture the sense of uniqueness at issue in the classical truth-based semantics.

Carnap's Problem nicely illustrates two requirements that emerge when changing the perspective from a purely inferential to a syntactic-semantic point of view: on the one hand, it demonstrates that additional (semantic) constraints are needed in order to rule out non-standard interpretations for the logical constants – as soon as proof-rules cease to constitute the entire meaning of a constant, purely proof-theoretic means are (in general) no longer sufficient to be able to speak about *the* connective defined by a set of rules. Relatedly, it shows that when truth-conditions become essential to the identity of a constant, wholly inferential ways of capturing uniqueness can no longer guarantee the identity of the notion involved: mere interderivability is no longer a guarantee that what is talked about is the same connective, and that what is pinned down actually a unique operation.

It is important to note that the uniqueness requirement here is indeed a logicality requirement. In general, there would be no reason to assume that the role a given expression plays in our inference suffices to uniquely fix its denotation. This should be clear from an example. One may provide axioms, say the three laws of movement and the law of gravitation, to capture the way we reason about force, mass, and movement in a Newtonian setting. However, it would be very strange to demand that this axiomatic set-up uniquely fixes the extension of the symbols involved. What the vectors denoting force and acceleration at a given point in a space are, or what the value of the gravitational constant is, is constrained by our understanding of the relationships between mass, force, and acceleration. But at the end of the day, the repartition of mass is an empirical property of physical systems, and the gravitational constant is an empirical physical constant. Denotations for empirical symbols depend on the world, not just on rules of reasoning, even when our conceptual scheme is such that there are rules for reasoning about those symbols. Admittedly,

[9] See Belnap and Massey (1990).

it would seem equally strange to consider that the denotation of conjunction or the universal quantifier is an empirical property of physical systems and depends on the world. As logical symbols, those are uniquely determined by the role they play in inferences, and there should not be further room for empirical determination. The uniqueness requirement thus captures an *analyticity constraint*, that the denotation of logical symbols should be fully determined by inference rules (unless one wishes to admit that the meanings of the connectives are underdetermined). What we have just said follows Carnap's motivation for framing the issue we referred to as Carnap's Problem. Early on, Carnap was worried that the realm of logical symbols could not be demarcated merely by saying that these are the symbols we have rules for. There are many symbols the use of which we have rules for which are clearly empirical and not logical. In the syntactic context of (Carnap 1937), this was implemented as a completeness requirement (the set of logical symbols would be such that any sentence only containing logical symbols would be either derivable or have its negation be derivable).

So far for uniqueness. What about existence and its syntactic counterpart, conservativity?[10] On a semantic approach, the provision of a semantic value already suffices to ensure existence, thereby guaranteeing consistency from the outset. This was indeed an early response proposed by Stevenson (1960) to the problem of *tonk*: not any inference rules may be offered to define a new connective, but instead of trying to find an inferential test to rule out tonk-like connectives, one may be content with requiring that the rules are semantically interpretable, that is, that there exists a semantic interpretation such that the rules come out as sound with respect to that interpretation.

In Section 3.5, we shall spell out our combined criterion, building on the renderings of existence and uniqueness we have outlined. Before that, we shall pause and examine in Section 3.4 an earlier attempt at a combined approach by Feferman.

3.4 Feferman's Combined Criterion

In his recent article, S. Feferman (2015) favors a *combined approach* to logical constants over a purely semantic criterion, using both semantic and inferential constraints in order to address the perceived shortcomings of purely semantic approaches. Contrary to solely inferential criteria that

[10] In the inferential setting of Belnap (1962), the conservativity requirement functions as an *existence condition*, as it ensures that a definition successfully defined something.

privilege inferential behavior over model-theoretic denotation, and in line with traditional semantic views, Feferman maintains that "the meaning of given connectives and quantifiers is to be established semantically in one way or another *prior* to their inferential role" (Feferman 2015, 21), while "what is needed to bring inferential considerations into play is to explain which quantifiers have axioms and rules of inference that completely govern its forms of reasoning" (Feferman 2015, 21). More precisely, in devising his criterion he adopts from the inferentialists the idea that the *uniqueness* of a notion is necessary for it to ultimately qualify as logical.

Following an insight of J.I. Zucker (1978), Feferman accepts that a crucial condition for a notion to be considered logical is that its meaning be "completely contained" in its axioms and inference rules – that its rules and axioms 'fully determine' this meaning – and that more than mere consistency[11] with the rules is required. On these grounds, Feferman proposes and adopts the constraint of *implicit definability*: the demand that a logical constant must be *implicitly defined* by its axioms and inference rules. Since Feferman regards "the meaning of a quantifier to be provided from the outside so to speak, i.e., to be given in model theoretic terms prior to the consideration of any rules of inference that may be in accord with it" (Feferman 2015, 28), his implementation of the demand for uniqueness cannot be met through mere interderivability and has to be sensitive to the model-theoretic meanings of the constants under consideration. A standard *model-theoretic* analysis of the notion of implicit definability is the following:[12]

> For languages $\mathcal{L}$ and $\mathcal{L}^+$, a theory T in $\mathcal{L}^+$, and a symbol Λ of $\mathcal{L}^+$, we say that Λ is *implicitly defined* by T in terms of $\mathcal{L}$, if, whenever we have models $\mathcal{M}, \mathcal{N}$ of T with $\mathcal{M}|_{\mathcal{L}} = \mathcal{N}|_{\mathcal{L}}$: $\Lambda^{\mathcal{M}} = \Lambda^{\mathcal{N}}$.

And it is indeed this kind of characterization that Feferman, following Zucker, adopts for capturing the *uniqueness-constraint.*

In order to apply this analysis and to investigate the question of what are *all* the logical notions obeying the uniqueness-constraint interpreted as implicit definability, Feferman considers a second-order language $\mathcal{L}_2(Q)$ with quantifiers ranging over individuals, propositions, and n-ary relations. $\mathcal{L}_2(Q)$ contains all the usual connectives and quantifiers of standard

[11] If $\vdash$ is a consequence relation and $\mathcal{M}$ a model, we say that $\mathcal{M}$ is *consistent* with $\vdash$ if, whenever $\Gamma \vdash \varphi$, if $\mathcal{M} \models \Gamma$, then $\mathcal{M} \models \varphi$; see Bonnay and Westerståhl (2016). Analogously, we say that a model $\mathcal{M}$ is *consistent* with a set of rules if $\mathcal{M}$ is consistent with the consequence relation presented by the rules.

[12] See, e.g., Hodges (1997, 149).

first-order logic and possesses, in addition to the second-order universal and existential quantifier-symbols, a second-order relation constant Q as its only non-logical symbol.[13] The rationale for considering and adopting such a language is that, taking all the usual operations of first-order logic for granted, it is then possible to capture all potential inference rules and axioms of a first-order generalized quantifier by a sentence of $\mathcal{L}_2(Q)$ in a canonical manner (so long as it possesses a finitary axiomatization), allowing us to talk *about* inference rules and making this talk receptive to the application of the notion of implicit definability as outlined above. While Zucker opts for a *full semantics* for the second-order language he considers, Feferman's meta-theory is more modest in that he uses a Henkin-style semantics[14] for $\mathcal{L}_2(Q)$ in which the second-order relation variables can range over *any* collection of relations of the appropriate adicity over the domain.[15] The meaning of a generalized quantifier $\mathcal{Q}$ in a model is then naturally taken to be the restriction of its (local) meaning to the appropriate domain(s) of relations that are taken to 'exist' in $\mathcal{M}$.

Adapting the demand of *uniqueness*, considered a necessary condition for logicality and implemented as implicit definability, to this framework Feferman states that "the criterion for accepting a quantifier $\mathcal{Q}$ given by such rules is that they implicitly define $[\mathcal{Q}(M)]$ in each model of $A(Q)$ [the $\mathcal{L}_2(Q)$-sentence formalizing the inference-rules for $\mathcal{Q}$] (more precisely, the restriction of $[\mathcal{Q}(M)]$ to the predicates of the model)" (Feferman 2015, 24). That is, Feferman's *Semantical-Inferential Necessary Criterion for Logicality* (Feferman 2015, 24) asserts that:

> A global quantifier $\mathcal{Q}$ of type $\langle k_1, \ldots, k_n \rangle$ is logical only if there is a sentence $\Lambda[Q]$ of $\mathcal{L}_2(Q)$, s.t. for each $\mathcal{L}_2$-model $\mathcal{M}$, $\mathcal{Q}(M)$ is the unique solution of $\Lambda[Q]$ when restricted to the predicates of $\mathcal{M}$.

Feferman's criterion thus amounts to the following: a quantifier $\mathcal{Q}$ is logical only if there exists a sentence $\Lambda[Q]$ of $\mathcal{L}_2(Q)$, s.t., for all $\mathcal{L}_2$-models $\mathcal{M}$,

(F1) $\mathcal{Q}^{\mathcal{M}}$ is a solution to $\Lambda[Q]$ in $\mathcal{M}$, i.e., $\langle \mathcal{M}, \mathcal{Q}^{\mathcal{M}} \rangle \models \Lambda[Q]$; and
(F2) $\mathcal{Q}^{\mathcal{M}}$ is the unique solution,

[13] To simplify, we assume in the following that Q is monadic.

[14] In a *full* semantics the second-order variables range over all possible relations of appropriate adicity over the domain, i.e., the complete power-set of the appropriate Cartesian product of the domain. In a Henkin-style semantics second-order variables may range over subsets of this power-set; see, e.g., van Dalen (2008, 143ff).

[15] Feferman bases the choice of this very permissible semantics on what he calls the LOCALITY PRINCIPLE, according to which whether a quantifier applies to a tuple of relations of appropriate adicity depends only on those relations and not on which relations exist in general, and which he takes to be in keeping with Lindström's original idea concerning generalized quantifiers (Lindström 1966).

where (F2) is captured by the demand that $\Lambda[Q]$ implicitly define Q in the model-theoretic sense given above. Taken together, (F1) and (F2) say that $\Lambda[Q]$ uniquely determines $\mathcal{Q}^{\mathcal{M}}$ as semantic value for Q on each $\mathcal{L}_2(Q)$-model $\mathcal{M}$. When both conditions hold, we say that $\Lambda[Q]$ *implicitly defines* $\mathcal{Q}$.

With this framework in place, the question of "[w]hich quantifiers $\mathcal{Q}$ in general have formal axioms and rules of inference that uniquely characterize it" (Feferman 2015, 22) becomes: for which values of Q is there a sentence of $\mathcal{L}_2(Q)$ that is 'solved' by such a value, and which that sentence implicitly defines? Feferman (2015, 24ff) proceeds to show that whenever a quantifier $\mathcal{Q}$ satisfies his criterion, it is, in fact, equivalent to a quantifier definable in standard first-order logic.

Feferman's criterion has much to recommend itself: not only does it deliver a mathematically precise demarcation of the logical second-order predicates in terms of the purely logical part of a second-order language whose choice rests on its ability to properly talk about inference rules, but it also justifies the pre-theoretical intuition "that the classical first-order predicate logic has a privileged role in our thought" (Feferman 1999, 32). However, while remarkable in its elegance, simplicity, scope, and results, there is also something deeply unsatisfactory about it. This discomfort stems from the fact that, in effect, all the criterion tells us is that nothing is logical beyond the operations already presupposed as logical in the formulation of the criterion (and those notions definable in terms of them), for in re-writing inference rules as higher-order sentences of $\mathcal{L}_2(Q)$ it is necessary to make use of first- and higher-order constants, such as the quantifiers.[16]

Yet, the criterion does not tell us anything about *why* these presupposed notions qualify as logical in the first place, and especially why the universal and existential quantifier occupy the special position they do with regard to logicality among all the other second-order predicates. Feferman's criterion thus collapses into another form of if-then-ism for type-level 2 expressions: *if* we accept the standard first-order quantifiers as logical, *then* nothing will be logical apart from them and notions definable in terms of them. This still is informative, since we might have discovered that if we accept the standard first-order quantifiers as logical, some other non-first-order definable operations are as well, but one might wish for more.

In order to make this issue more vivid, we shall twist Feferman's criterion and ask what happens if the first-order existential and universal quantifiers

16 "[t]aking for granted that the standard operations of FOL are logical" (Feferman 2015, 28).

were not granted special status, but were judged by the same principles that are applied to other putative logical constants.

At first blush, it seems easy to show that the universal quantifier $\mathcal{Q}(M) = \{M\}$ is implicitly defined by the sentence (U) of $\mathcal{L}_2(Q)$ that Feferman proposes[17] and therefore qualifies as logical. More precisely, let (U) be the sentence[18]

$$\forall X^1 \forall X^0 [(\forall x(X^0 \rightarrow X^1 x) \rightarrow (X^0 \rightarrow Q(X^1))) \wedge (\forall x(Q(X^1) \rightarrow X^1 x))]$$

and consider a *(normal)* $\mathcal{L}_2$*-structure* $\mathcal{M}$, where a *normal structure* is a structure in which the interpretation of the first-order universal quantifier is the usual one, that is, $\mathbf{\forall}(M) = \{M\}$ for all domains M. Then,[19]

PROPOSITION 1 Over normal structures, (U) implicitly defines $\mathcal{Q}(M) = \{M\}$.

As a logicality certificate, this is not fully satisfactory, in so far as the result is obtained by making essential use of the fact that the universal quantifier occurring in (U) already possesses its standard interpretation. To break the circularity, it is worth asking whether Feferman's criterion would still deliver adequate results for the universal and existential quantifier when their standard interpretations are not taken for granted and already assumed from the outset. To this end, we move from *normal structures* to *general structures*, i.e., structures for $\mathcal{L}_2(Q)$ in which the interpretation of the universal quantifier is any unary generalized quantifier which makes classical consequence sound. We get the following:[20]

PROPOSITION 2 Over general structures, (U) does not implicitly define $\mathcal{Q}(M) = \{M\}$.

When circularity is alleviated and the standard meaning of the first-order universal quantifier for the meta-language in which inference rules are described is not presupposed, the universal quantifier ceases to be implicitly defined by (U). Furthermore, the following proposition establishes that there is no way in the current framework of $\mathcal{L}_2(Q)$ to strengthen/replace

17 See (Feferman 2015, 22).

18 X^1 is a predicate variable which gets assigned subsets of the domain, X^0 is a sentence variable which gets assigned truth-values; see Feferman (2015) or the Appendix for details. (U) is written down so as to mimick usual introduction and elimination rules for $\forall$, though it could be further simplified.

19 Proof-sketches and full definitions for the results can be found in the Appendix.

20 Underdetermination comes as no surprise. It is well-known that the universal quantifier is not categorical. See Bonnay and Westerståhl (2016) for a full characterization of interpretations of $\forall$ compatible with classical consequence.

(U) in such a way as to recover the logicality of $\mathcal{Q}(M) = \{M\}$ when considering general structures, and thus that the non-logicality of universal quantification under Feferman's criterion with respect to general structures is not just due to a bad choice of characteristic $\mathcal{L}_2(Q)$-sentence:

PROPOSITION 3 There is no sentence φ of $\mathcal{L}_2(Q)$ that implicitly defines $\mathcal{Q}(M) = \{M\}$ over general structures.

The difficulty encountered here might be blamed on Feferman's strict interpretation of the LOCALITY CONSTRAINT, according to which *any* subset of the totality of relations over a domain M forms a permitted second-order domain of a model. One could, as suggested by F. Engström (2014), for example, require that the existence of at least some relations in this domain be guaranteed. A canonical way of doing this consists in requiring the validity of all instances of the *Comprehension Schema*[21]

$$(\text{COMP})\ \exists X^n \forall x_1 \ldots \forall x_n [\varphi(x_1, \ldots, x_n) \leftrightarrow X^n(x_1, \ldots, x_n)]$$

This ensures the existence of at least all parametrically definable relations over a domain M, weakening Feferman's interpretation of the LOCALITY CONSTRAINT. Let $\vdash_2$ be the consequence relation for the resulting logic and call the structures for it, in which the interpretation of the universal quantifier is allowed to vary, *general models*. There might be good reasons for looking at *models*, rather than *structures*,[22] but, unfortunately, this does not mitigate the problem we are concerned with:

PROPOSITION 4 There is no sentence φ of $\mathcal{L}_2(Q)$ that implicitly defines $\mathcal{Q}(M) = \{M\}$ over general models $\mathcal{M}$.

While it might in principle be possible to find restrictions on the class of models that render the standard quantifiers logical according to Feferman's criterion, the above exclusion of some of the most natural constraints calls into doubt the availability of philosophical motivations for potential additional restrictions. An open question concerns what would happen if we

[21] Where X^n is a variable of any adicity n, and φ a formula of $\mathcal{L}_2(Q)$ in which X^n does not occur free (though other variables may).

[22] In general, allowing *any* set of relations over a domain M to form the second-order universe of a structure results in a rather misbehaved and disordered class of structures that lacks systematicity and nice logical properties (such as, e.g., soundness for standard second-order quantifier-rules). Concerning the present case, one might worry that it is these 'unruly' structures that make it impossible to have a sentence that renders the universal quantifier logical, and that these models constitute unfair counter-examples given how logically unconstrained they are.

were to move to *full models*, i.e., models whose second-order domains contain *all* possible relations, and in how far this would alleviate the underdetermination in an acceptable manner.

To be fair to Feferman, however, his criterion is meant as a way to tell apart logical and non-logical *quantifiers*, as clearly stated in the title of his paper, "Which Quantifiers are Logical?", where the definition of a quantifier is provided by Lindström's generalization of the first-order quantifiers (Feferman 2015, 20). Non-standard interpretations consistent with the rules may not be Lindström quantifiers, which encapsulate an invariance requirement. Thus, one could reply that the upshot of the previous propositions should be reconsidered when available interpretations for the universal quantifier are restricted to Lindström quantifiers. Indeed, as we shall see in Section 3.5, invariance under bijection successfully supplements consistency with the rules. The problem, however, is that Feferman's set-up covers up the role played by invariance in fixing the interpretation of quantifiers, getting implicit definability in an all too easy and circular way. In order to get the combined approach straight, implicit definability and invariance under bijection need to be put on an equal footing, and standard meanings must not be presupposed. This is the line we shall pursue next.

3.5 Formality and Categoricity: A Novel Criterion

3.5.1 The Criterion

We are now ready to state the combined criterion we have been aiming for. Our conceptual take on what makes an expression logical consists in the following three principles:

(double life) Logical symbols are identified both in terms of inferential behavior and semantic denotation.
(formality) Logical symbols have purely formal denotations.
(analyticity) The denotations of logical symbols are fully determined by their inferential behavior.

Double life is our initial claim that logicality should not be analyzed in either semantic or inferential terms exclusively. That claim may be vindicated on the basis of logicians' practice. It is a fact to be acknowledged that logical symbols show up as coming both with special semantic denotations and with inference rules. Deciding that one of these two aspects should be discarded or reduced to the other would be at odds with what logicians do. The claim may also be vindicated on the basis of intuitions about logic. As discussed in Sections 3.2 and 3.3, the logical part of our vocabulary distinguishes itself both from a purely denotational perspective – logical symbols having denotations which are very different from those of usual empirical terms – and from a broader cognitive perspective – logical symbols expressing concepts which shape our reasoning.

Formality is what logicality amounts to at the semantic level. Logical symbols do not denote any kind of operation, they denote formal operations which do not say anything about any object in particular. *Analyticity* is what logicality amounts to at the level of sense. Logical symbols are tools in reasoning, there is nothing more to what they are than their behavior in inferences, so that that behavior should suffice to fix their denotation.

Note that *formality* and *analyticity* are to be thought of as complementary rather than redundant. *Analyticity* should not be reduced to *formality*. One may consider expressions endowed with formal denotations which could not be grasped solely by means of inference rules determining their role in reasoning. Typically, we expect an expression such as "being uncountable" to be formal but not analytic in that sense. Neither should *formality* be reduced to *analyticity*. One may, for example, fully specify *by fiat* that a predicate applies to this and only to this object. Such a predicate would qualify as purely analytic but not as formal.[23]

We shall now see how to capture those three principles, building on the works discussed earlier. Let $\mathcal{L}$ be a relational first-order language with predicate variables, $\vdash$ a consequence relation for $\mathcal{L}$ and Q a second-order predicate symbol of $\mathcal{L}$ endowed with an interpretation $\mathcal{Q}$ providing for each domain M a denotation $\mathcal{Q}^M$ of the appropriate semantic type for Q. In the context of a consequence relation $\vdash$, a symbol Q equipped with an interpretation $\mathcal{Q}$ is a logical symbol if and only if it satisfies the following three properties, which shall constitute our *Combined Criterion of Logicality* (CCL):

(consistency) For every domain M, there is a model $\mathcal{M}$ consistent with $\vdash$ such that the interpretation of Q in $\mathcal{M}$ is $\mathcal{Q}^M$.

(invariance) $\mathcal{Q}$ is invariant under bijection.

(Carnap-categoricity) $\mathcal{Q}$ is the unique interpretation for Q satisfying *consistency* with respect to $\vdash$ and *invariance*.

[23] We can think of, for example, an interpreted language equipped with a constant symbol a, interpreted by an object a. One can then fix the interpretation of P inferentially by demanding consistency with $\forall x\ (Px \leftrightarrow x = a)$. The interpretation of P is categorically fixed as $\{\mathsf{a}\}$, yet P is not invariant under permutation.

In keeping with the terminology used earlier, $\mathcal{M}$ is *consistent* with $\vdash$ iff, whenever $\Gamma \vdash \varphi$ and $\mathcal{M} \models \Gamma$, then $\mathcal{M} \models \varphi$. $\mathcal{Q}$ is *invariant under bijection* iff for every bijection $f : M \mapsto N, f[\mathcal{Q}^M] = \mathcal{Q}^N$.

Note that *double life* is captured by the very framing of our criterion, which takes logicality to be a property that symbols with an interpretation possess in the context of a consequence relation, together with the consistency constraint, which requires that semantic interpretation and inferential behavior cohere. *Formality* is captured by *invariance* as discussed in Section 3.2. *Analyticity* is captured by *Carnap-categoricity*, which is categoricity in the sense of Carnap as discussed in Section 3.3. Note the interplay between *invariance* and *Carnap-categoricity*: uniqueness is not required with respect to *all* consistent interpretations, but only with respect to *formal* consistent interpretations. This is in keeping with the idea that *formality* and *analyticity* complement each other. Logical symbols are categorical as logical. It may still be the case that rules are compatible with non-invariant interpretations because analyticity is not supposed to force formality.

One last remark before moving on to consequences and applications of (CCL) in the next section, and to comparing our criterion with Feferman's in the last. Asking for consistency with respect to a consequence relation is in general weaker than asking that rules preserve validity, as for example Garson (2013) does. One might argue, however, that there is a conceptual gain using the simpler notion of truth rather than the notion of validity. As far as our principles should also help us understand how speakers are able to learn the denotations of logical symbols, we do not need to assume that speakers master the concept of validity as such beforehand.

3.5.2 Some Consequences of (CCL)

In Section 3.4, we saw that the existential and universal quantifiers raise problems for Feferman's criterion when their standard meaning is not somehow presupposed. How does our (CCL) fare in this respect? Bonnay and Westerståhl (2016, 730) show that the only interpretations of $\forall$ consistent with classical consequence are principal filters, and recall that the maximal principal filter $\{M\}$ is the only principal filter which is invariant under permutation.[24] It follows from this that the standard interpretation of the universal quantifier is the only one satisfying *consistency* and *formality*, hence giving us *Carnap-categoricity*. As a consequence, $\forall$, in

[24] Permutation-invariance is the local analogue to bijection-invariance. In the following we sometimes use permutation-invariance rather than bijection-invariance for ease of exposition.

the context of standard first-order logical consequence and equipped with its standard interpretation, does qualify as logical on the basis of our combined criterion (CCL). Instead of demanding that the senses of candidate logical expressions uniquely pin down a (logical) meaning, (CCL) is more lenient in that it 'only' demands that the sense of logical expressions not conflate different logical meanings. Thus, the existential and universal quantifiers, for example, do not pin down a unique operator via their inference rules, and are therefore not categorical without further ado, but they do succeed in picking out a unique bijection-invariant operator, so that they are categorical under bijection-invariance, and thus qualify as logical according to (CCL).

By way of example, we now take a brief look beyond the standard first-order case to demonstrate how the criterion works in situations concerning more controversial candidates for logical quantifiers. We will here consider the quantifier $\mathcal{Q}_K$ with the intended interpretation "there exist uncountably many", $\mathcal{Q}_K(M) = \{A \subseteq M | |A| \geq \aleph_1$, as an example. This quantifier is of particular interest to debates on logicality due to the fact that Keisler (1970) proved FOL extended with this quantifier complete with respect to a small and natural set of axioms, and completeness has been considered one of the properties distinctive of logical,[25] as opposed to merely mathematical, axiom systems.

Let $\vdash_K$ be the consequence relation of FOL extended by the quantifier Q_K.[26,27] Now, $\mathcal{Q}_K$ satisfies *consistency* and *invariance* by design. In assessing the logicality of Q_K with respect to the consequence relation $\vdash_K$ and the interpretation $\mathcal{Q}_K$, we need to ask whether $\mathcal{Q}_K$ is the unique such interpretation, that is, whether the semantic value of Q_K is uniquely fixed by its axioms among its consistent bijection-invariant interpretations. In a first step we are then asking for the range of possible interpretations of Q_K that are consistent with $\vdash_K$. One can show that

THEOREM An interpretation $\mathcal{Q}_K^{\mathcal{M}}$ of the Keisler-quantifier is consistent with $\vdash_K$ iff

(i) for all $X \in \mathcal{Q}_K^{\mathcal{M}}$: $\omega \leq |X|$;
(ii) $\mathcal{Q}_K^{\mathcal{M}}$ is closed under supersets;[28]
(iii) if $\mathcal{Q}_K^{\mathcal{M}} \neq \emptyset$, then $min\{|A| | A \in \mathcal{Q}_K^{\mathcal{M}}\}$ is regular;

[25] See, e.g., Quine (1970) and Tharp (1975).
[26] K for "Keisler".
[27] See Keisler (1970) for details. Like Keisler, we work within ZFC.
[28] I.e., if $X \in \mathcal{Q}_K^{\mathcal{M}}$ and $X \subseteq Y$, then $Y \in \mathcal{Q}_K^{\mathcal{M}}$.

(iv) $\mathcal{Q}_K^{\mathcal{M}}$ is closed under equicardinal set[29]

It follows from this as an easy corollary that

COROLLARY An interpretation $\mathcal{Q}_K^{\mathcal{M}}$ of the Keisler-quantifier is consistent with $\vdash_K$ iff $\mathcal{Q}_K^{\mathcal{M}} = \{X \subseteq M | \aleph_\alpha \leq |X|\}$ for some regular cardinal $\aleph_\alpha$.

This includes of course the standard interpretation of Q_K, $\mathcal{Q}_K^{\mathcal{M}} = \{A \subseteq M | \aleph_1 \leq |A|\}$, but also many more. Hence, the class of interpretations of Q_K that are consistent with $\vdash_K$ proves to be much more encompassing, containing, for any regular cardinal $\aleph_\alpha$, interpretations of Q_K meaning "at least $\aleph_\alpha$ many".

This result is well-known: Keisler not only already proved the right-to-left direction of the above corollary,[30] but also made essential use of the fact that the logic of Q_K admits non-standard interpretations in his celebrated completeness proof. Nevertheless, in the context of the present enterprise the existence of these alternative interpretations calls into doubt the logicality of $\mathcal{Q}_K$ according to the criterion (CCL). For it is not hard to see that

PROPOSITION 5 For all regular cardinals $\aleph_\alpha$, $\mathcal{Q}_K^{\mathcal{M}} = \{A \subseteq M | \aleph_\alpha \leq |A|\}$ is invariant under permutation.

It thus follows that the constraint of bijection-invariance does not rule out the unintended interpretations, and therefore does not narrow down the space of possible meanings of Q_K consistent with $\vdash_K$ to a unique bijection-invariant interpretation. Restricting consistent interpretations of Q_K to those that are permutation-invariant does not remove the underdeterminacy of logical meaning as it did in the cases of $\forall$ and $\exists$. Due to the existence of many possible bijection-invariant interpretations of Q_K consistent with its axioms and inference rules, it therefore follows that $\mathcal{Q}_K$ is not *Carnap-categorical*, which provides grounds for counting $\forall$ and $\exists$, but not Q_K as belonging to the logical lexicon based on (CCL).

Let us wrap up. In order to deal with the underdetermination of semantic value by inferential behavior the criteria discussed and formulated in this paper already restrict the class of potential denotations of a candidate logical expression to those that constitute *logical* denotations – i.e., that

[29] I.e., for all $X, Y \subseteq M$, if $Y \in \mathcal{Q}_K^{\mathcal{M}}$ and $|X| = |Y|$, then $X \in \mathcal{Q}_K^{\mathcal{M}}$.
[30] See (Keisler 1970, Corollary 3.3.1).

are permutation-/bijection-invariant. Instead of demanding that the senses of candidate logical expressions uniquely pin down a (logical) meaning, the criterion treated here is more lenient in that it 'only' demands that the senses of logical expressions not conflate different logical meanings. Thus, $\forall$ and $\exists$ are deemed logical because their inferential role succeeds in uniquely determining a semantic value among all possible logical, i.e., invariant, extensions. Q_K is rejected as logical, on the other hand, not because its axioms fail to uniquely determine an extension, but because its axioms fail to pick out a unique *permutation-invariant* extension, or, as we would like to see it, because they fail to pick out a unique extension among all candidate *logical* extensions. The expression Q_K thus does not fail to be a logical expression because it is capable of picking out a non-logical extension, but because it fails to uniquely pick one out of a multitude of *logical* extensions.

3.5.3 Comparison and Outlook

In terms of setup, one important difference between (CCL) and Feferman's criterion is that our criterion talks directly about the inference rules for potential logical expression without having to take a detour through a meta-language – such as Feferman's $\mathcal{L}_2(Q)$ – in which inference rules can be described. An unalleviated burden of Feferman's approach is the vindication of the choice of $\mathcal{L}_2(Q)$ and its semantics for characterizing logicality, which is particularly pressing given that the choice of semantics is crucial in characterizing the extensional output of the criterion. Feferman (2015) proves that nothing is logical beyond the standard constants of the first-order quantificational calculus, which might be taken to provide indirect evidence for the adequacy of the criterion. However, Engström (2014) shows that this is essentially due to the choice of an extremely weak semantics which does not allow for a substantial use of second-order quantification. Without further modifying and strengthening the semantics, or imposing conditions on which subclasses of sentences of $\mathcal{L}_2(Q)$ can be considered adequate descriptions of first-order inference rules, Feferman's result reduces to the almost trivial result that nothing beyond what is already presupposed in $\mathcal{L}_2(Q)$ will qualify as logical.

Our criterion, on the other hand, offers an alternative circumventing the above described difficulties while at the same time providing a faithful implementation of the same kinds of constraints, on the basis of the same kinds of reasons, that matter to Feferman. But different from Feferman's criterion ours is able to provide a unified explanation, without apparent circularity, as to which expressions are logical. In particular, the logicality of the universal and existential quantifiers is explained by the very same criterion and on the very same basis on which the Keisler-quantifier "there exist uncountably many" is ruled out. This constitutes a crucial difference to Feferman's criterion which assumed the logicality of ∀ and ∃ in the meta-theory. Our criterion therefore succeeds more clearly in delivering a unified explanation regarding the logicality/non-logicality of expressions of a common grammatical category.

The precise scope of (CCL) remains an open problem. We originally conjectured that the only type-level 2 notions that qualify as logical according to (CCL) are the quantifiers of FOL and those notions definable in terms of them – that Feferman's and our criterion were conceptually distinct, but extensionally equivalent. Dag Westerståhl has since disproven this conjecture by showing that, under the assumption of the standard interpretation of ∀ and ∃ (yielded by *invariance*), the quantifier "there are infinitely many" will be Carnap-categorical with respect to a first-order consequence relation, and thus logical according to (CCL) as well (even if the resulting logic remains incomplete). This result can be further generalized to cover quantifiers such as "there are finitely many", and notions (generalized) definable from them (Speitel and Westerståhl 2019).

3.6 In Guise of a Conclusion: Two Open Questions

The current paper constitutes an attempt to show that, in trying to formulate a well-motivated criterion of logicality providing a mathematically precise and philosophically informative way to separate the logical from the non-logical expressions of a language, it is promising to take both inferential and semantic considerations into account. Not only does such an approach respect central tenets of logical practice, with its emphasis on both inference and truth, but it also promises a deeper understanding of the way notions characteristic of logic, such as its generality, formality, analyticity, etc., relate in their different proof- and model-theoretic guises.

We have outlined and argued for a particular combined criterion of logicality, (CCL), proposing mathematically precise implementations of *double life*, *formality*, and *analyticity* to tell apart the logical from the non-logical. In doing so, we criticized Feferman's related criterion for not fully realizing the motivations originally underlying his account. We hope that the criterion given in this paper might prove appealing to someone who shares with us the conviction that a natural notion of invariance captures a central aspect of what it means to be logical, but was dissatisfied with the fact that a criterion based solely on this notion appeared to "assimilate mathematics to logic".

As noted in the previous section, it is known that (CCL) does not vindicate first-order logic: there are non-first-order definable quantifiers which satisfy (CCL). This raises two questions. The first one, mentioned above, concerns the extent of logicality according to (CCL): what is the class of quantifiers such that, together with the corresponding extension of first-order consequence, they satisfy *invariance* and *Carnap-categoricity*? The second question concerns the possibility to pin down first-order logic by strengthening (CCL) with additional constraints: should we ask for completeness, or finite axiomatizability, and what happens if we do?

3.7 Appendix

This appendix contains central definitions and proof-sketches for the results mentioned in the main body of the text.

By *FOL* we denote standard first-order logic with its usual consequence relation $\vdash_{FOL}$. $\mathcal{L}_2(Q)$ is the second-order language of Feferman (2015), whose signature contains an n-ary second-order predicate symbol Q (i.e., $Q(X^{k_1}, \ldots, X^{k_n})$, with $X^{k_1}, \ldots, X^{k_n}$ k_i-ary predicate-letters, is a formula), with associated consequence relation $\vdash_2$. By $\mathcal{L}(Q_K)$ we denote first-order logic extended with a quantifier-symbol Q_K, and by $\vdash_K$ its consequence relation.[31]

A *(generalized) quantifier* of type $\langle k_1, \ldots, k_n \rangle$ is a class function $\mathcal{Q}$, s.t. for every domain M of a structure $\mathcal{M}$, $\mathcal{Q}(M) \subseteq \mathcal{P}(M^{k_1}) \times \ldots \times \mathcal{P}(M^{k_n})$.

Definitions and Results for Section 3.4 *Structures* for $\mathcal{L}_2(Q)$ are Henkin-style second-order structures without any restrictions on the permissible second-order domains of relations (see van Dalen (2008) and Feferman (2015) for details), including a set $\forall^{\mathcal{M}} \subseteq \mathcal{P}(M)$ and a set $\mathcal{Q}^{\mathcal{M}} \subseteq \mathcal{R}^{\mathcal{M}}_{k_1} \times \ldots \times \mathcal{R}^{\mathcal{M}}_{k_n}$ for Q of type $\langle k_1, \ldots, k_n \rangle$, serving as the interpretations of $\forall$ and Q, respectively. The satisfaction clauses for $\mathcal{L}_2(Q)$ are standard, except for:[32]

[31] See, e.g., Keisler (1970).

[32] Where $[\![\varphi(x_1, \ldots, x_n)]\!]^{\mathcal{M}}_{\sigma} = \{\langle a_1, \ldots, a_n \rangle \in M^n | \mathcal{M} \models \varphi(x_1, \ldots, x_n)[\sigma[a_1/x_1| \ldots |a_n/x_n]]\}$.

(i) $\mathcal{M} \models \forall x\varphi[\sigma]$ iff $[\![\varphi(x)]\!]^{\mathcal{M}}_{\sigma} \in \forall^{\mathcal{M}}$;
(ii) $\mathcal{M} \models Q(X^{k_1}, \ldots, X^{k_n})[\sigma]$ iff $\langle \sigma(X^{k_1}), \ldots, \sigma(X^{k_n})\rangle \in \mathcal{Q}^{\mathcal{M}}$.

An $\mathcal{L}_2(Q)$-*model* $\mathcal{M}$ is an $\mathcal{L}_2(Q)$-structure satisfying all instances of COMP (X^n not free in φ; φ may contain free variables other than $x_1, \ldots, x_n$):

$$(\text{COMP})\ \exists X^n \forall x_1 \ldots \forall x_n (X^n x_1 \ldots x_n \leftrightarrow \varphi(x_1, \ldots, x_n))$$

We call a structure/model *normal* if $\forall^{\mathcal{M}} = \{M\}$. $\mathcal{M}$ is *consistent with* a consequence relation $\vdash$ iff, whenever $\Gamma \vdash \varphi$, if $\mathcal{M} \models \Gamma$ then $\mathcal{M} \models \varphi$.

An $\mathcal{L}_2(Q)$-structure is called *general* if it is an $\mathcal{L}_2(Q)$-structure for which the interpretation of the first-order universal quantifier is any unary (generalized) quantifier for which standard first-order consequence is sound.[33] An $\mathcal{L}_2(Q)$-model is called *general* if it is consistent with $\vdash_2$. The reduct of a structure $\mathcal{M}$ to $\mathcal{L}$ is denoted by $\mathcal{M}|\mathcal{L}$. For a quantifier $\mathcal{Q}$ of type $\langle k_1, \ldots, k_n\rangle$ the value of $\mathcal{Q}$ on an $\mathcal{L}_2(Q)$-structure $\mathcal{M}$ is $\mathcal{Q}^{\mathcal{M}} = \mathcal{Q}(M) \cap \mathcal{R}^{\mathcal{M}}_{k_1} \times \ldots \times \mathcal{R}^{\mathcal{M}}_{k_n}$.

Proof of Proposition 1 Let $\mathcal{Q}(M) = \{M\}$. Then it is not hard to see that for an $\mathcal{L}_2$-structure $\mathcal{M}$, $\langle \mathcal{M}, \mathcal{Q}^{\mathcal{M}} = \mathcal{Q}(M) \cap \mathcal{R}^{\mathcal{M}}_1\rangle \models (U)$. Now suppose that, for two (normal) $\mathcal{L}_2(Q)$-structures $\mathcal{M}, \mathcal{N}$ with $\mathcal{M}|\mathcal{L}_2 = \mathcal{N}|\mathcal{L}_2$, we have that $\mathcal{M} \models (U)$ and $\mathcal{N} \models (U)$.

By unpacking the semantic clauses we see that an $\mathcal{L}_2(Q)$-structure $\mathcal{K} \models (U)$ iff $\mathcal{Q}^{\mathcal{K}} = \forall^{\mathcal{K}} \cap \mathcal{R}^{\mathcal{K}}_1$. If $M \in \mathcal{R}^{\mathcal{M}}_1 = \mathcal{R}^{\mathcal{N}}_1$, we immediately see that $\mathcal{Q}^{\mathcal{M}} = \mathcal{Q}^{\mathcal{N}} = \{M\}$. On the other hand, if $M \notin \mathcal{R}^{\mathcal{M}}_1 = \mathcal{R}^{\mathcal{N}}_1$, then $\mathcal{Q}^{\mathcal{M}} = \mathcal{Q}^{\mathcal{N}} = \emptyset$.

Proof of Proposition 2 Let $\mathcal{M}, \mathcal{N}$ be general structures, s.t. $M = N$, $\mathcal{R}^{\mathcal{M}}_i = \mathcal{R}^{\mathcal{N}}_i$ for all i, but $\forall^{\mathcal{M}} \cap \mathcal{R}^{\mathcal{M}}_1 \neq \forall^{\mathcal{N}} \cap \mathcal{R}^{\mathcal{N}}_1$. Set $\mathcal{Q}^{\mathcal{M}} = \forall^{\mathcal{M}} \cap \mathcal{R}^{\mathcal{M}}_1$ and $\mathcal{Q}^{\mathcal{N}} = \forall^{\mathcal{N}} \cap \mathcal{R}^{\mathcal{N}}_1$. By the above we know that $\mathcal{M} \models (U)$, $\mathcal{N} \models (U)$, and $\mathcal{M}|\mathcal{L}_2 = \mathcal{N}|\mathcal{L}_2$, yet $\mathcal{Q}^{\mathcal{M}} \neq \mathcal{Q}^{\mathcal{N}}$.

We say that an n-ary generalized quantifier $\mathcal{Q}$ is *definable* in *FOL* if there is a FOL-sentence $\varphi(P)$ (P an n-ary predicate constant), s.t., for all FOL-models $\mathcal{M}$:

$$\mathcal{M} \models \varphi(P) \text{ iff } P^{\mathcal{M}} \in \mathcal{Q}^{\mathcal{M}}$$

[33] See Bonnay and Westerståhl (2016) for a characterization of the shape of these structures for the first-order case.

Adapting a proof of Engström (2014) to our setting, we can establish the following Lemma:

LEMMA (Engström 2014, 68) If a generalized quantifier $\mathcal{Q}$ is implicitly definable over general structures, then it is definable in *FOL*.

Proof of Proposition 3 Suppose that $\mathcal{Q}(M) = \{M\}$ was implicitly definable over general structures. By the previous Lemma it follows that $\mathcal{Q}$ is definable in *FOL*, i.e., that there exists a sentence $\varphi(P)$ of *FOL*, s.t. $\mathcal{M} \models \varphi(P)$ iff $P^{\mathcal{M}} = M$. Since this holds in particular for all *FOL*-structures $\mathcal{M}$ for which $\forall^{\mathcal{M}}$ is normal, we see that for all normal *FOL*-structures $\mathcal{M}$, $\mathcal{M} \models \forall x Px \leftrightarrow \varphi(P)$, and therefore $\forall x Px \vdash_{FOL} \varphi(P)$.

By a theorem of Bonnay and Westerståhl (2016) and the above we know that if $\forall^{\mathcal{M}}$ is a principal filter, then, whenever $\mathcal{M} \models \forall x Px$ it follows that $\mathcal{M} \models \varphi(P)$. Let $\mathcal{M}'$ be a model in which $\forall^{\mathcal{M}}$ is a principal filter generated by a set $A \subset M'$ and $P^{\mathcal{M}'} = A$. Then, $\mathcal{M}' \models \forall x Px$ and thus also $\mathcal{M}' \models \varphi(P)$.

However, we then have a model, s.t. $\mathcal{M}' \models \varphi(P)$ yet $P^{\mathcal{M}'} \notin \mathcal{Q}^{\mathcal{M}'} = \{M'\}$, which is a contradiction to the assumption that $\mathcal{Q}(M) = \{M\}$ is definable in *FOL* and thus, by the above Lemma, $\mathcal{Q}(M) = \{M\}$ is not implicitly definable over general structures.

Proof of Proposition 4 Adapting a proof of Nour and Raffalli (2003) of the completeness of second-order logic with Henkin-semantics to our setting, we can prove a variant of Bonnay and Westerståhl's theorem w.r.t. $\vdash_2$, and give an argument analogous to the one given in the proof of Proposition 3.

We first show, following the structure of the proof in Bonnay and Westerståhl (2016), that a certain class of $\mathcal{L}_2(Q)$-models is consistent with $\vdash_2$ iff, for a model $\mathcal{M}$ of that class, $\forall^{\mathcal{M}} \cap \mathcal{R}_1^{\mathcal{M}}$ is a principal filter. Then, supposing that $\mathcal{Q}(M) = \{M\}$ is implicitly defined by a sentence $\Lambda(Q)$ of $\mathcal{L}_2(Q)$, we show, using the completeness of standard Henkin $\mathcal{L}_2(Q)$-models w.r.t. $\vdash_2$, that $\forall X^1(Q(X^1) \leftrightarrow \forall x X^1 x) \vdash_2 \Lambda(Q)$.

Similar to the proof of Proposition 3, we then show that there are models $\mathcal{M}$ – in which $\forall^{\mathcal{M}} \cap \mathcal{R}_1^{\mathcal{M}} \neq \{M\}$ is a principal filter – that satisfy $\forall X^1(Q(X^1) \leftrightarrow \forall x X^1 x)$. Thus, by the consistency of $\vdash_2$ w.r.t. these models, they also satisfy $\Lambda(Q)$, contrary to the assumption that $\mathcal{Q}(M) = \{M\}$ was implicitly defined by $\Lambda(Q)$.

Definitions and Results for Section 3.5 Let M be a set. A *permutation* π on M is a bijection $\pi : M \mapsto M$. A set X is *permutation-invariant* if, for all permutations π on M, $\pi[X] = X$. A quantifier $\mathcal{Q}$ is permutation-invariant if $\mathcal{Q}(M)$ is permutation-invariant for all M.
An $\mathcal{L}(Q_K)$-model is a tuple $\langle \mathcal{M}, \mathcal{Q}_K^{\mathcal{M}} \rangle$ where $\mathcal{M}$ is a FOL-model and $\mathcal{Q}_K^{\mathcal{M}} \subseteq \mathcal{P}(M)$. For an $\mathcal{L}(Q_K)$-model $\mathcal{M}$, we have that[34]

$$\mathcal{M} \models Q_K x \varphi(x)[\sigma] \text{ iff } [\![\varphi(x)]\!]_\sigma^{\mathcal{M}} \in \mathcal{Q}_K^{\mathcal{M}}$$

Proof-Sketch of the Theorem & Corollary The right-to-left direction of the Theorem and Corollary already follow from COROLLARY 3.3.1 in Keisler (1970). The less immediate parts of the left-to-right direction concern conditions (iii) and (iv) of the Theorem.

To prove condition (iii) one constructs, assuming $inf(\mathcal{Q}_K^{\mathcal{M}})$ to be singular,[35] a model in which $[\![\exists x \varphi(x, y)]\!] \in \mathcal{Q}_K^{\mathcal{M}}$ with $|[\![\exists x \varphi(x, y)]\!]| = inf(\mathcal{Q}_K^{\mathcal{M}})$, but $|[\![\exists y \varphi(x, y)]\!]| < inf(\mathcal{Q}_K^{\mathcal{M}})$ and $|[\![\varphi(a, y)]\!]| < inf(\mathcal{Q}_K^{\mathcal{M}})$ for all a, to obtain a countermodel to the axiom $Q_K y \exists x \varphi \rightarrow (\exists x Q_K y \varphi \vee Q_K x \exists y \varphi)$.

To prove condition (iv), one assumes that there are sets $X, Y \subseteq M$ with $|X| = |Y|$ and $X \in \mathcal{Q}_K^{\mathcal{M}}$, but $Y \notin \mathcal{Q}_K^{\mathcal{M}}$, and considers the graph of a bijection between X and Y in order to obtain a countermodel to the same axiom.

Proof of Proposition 5 Since $\mathcal{Q}_K^{\mathcal{M}}$ is closed under equicardinal sets it follows that for any permutation π, $\pi[\mathcal{Q}_K^{\mathcal{M}}] = \mathcal{Q}_K^{\mathcal{M}}$.

[34] For details of the language, axiomatization of the logic of $\mathcal{L}(Q_K)$, and completeness result, see Keisler (1970).

[35] Where, for $X \subseteq \mathcal{P}(M)$, $inf(X) = min\{|A| | A \in X\}$.

CHAPTER 4

Invariance without Extensionality

Beau Madison Mount

The project of analysing logicality in terms of invariance, inaugurated by a 1946 paper by Friedrich Mautner and a 1966 lecture of Tarski's (Mautner 1946; Tarski 1966/86), is familiar to philosophers and logicians. The tenability of the project in general and the details of the invariance criterion to be used have been the subject of extensive debate.[1] For the most part, this debate has been carried out in the framework of extensional type theory (or some fragment thereof, such as first-order logic with generalized quantifiers). Although some work exists concerning invariance criteria over Kripke models for modal languages (van Benthem and Bonnay 2008), there has been little discussion of how the invariance account of logicality should be applied to hyperintensional operations – operations too fine-grained to be captured by merely modal distinctions.

This choice of framework makes sense on the assumption that hyperintensionality is a fundamentally representational phenomenon, rather than an irreducible feature of reality; on such a picture, as Gil Sagi (2017) has noted, invariance criteria can be viewed as demarcating a 'sub-extensional' level of semantic values, even coarser-grained than those provided by standard extensional semantics. If, on the other hand, there exist fundamentally *worldly* hyperintensional distinctions, then an invariance-based account of logicality that can be applied without assuming extensionality should be developed.

In this paper, I set out a framework within which invariance criteria on hyperintensional operations can be assessed, using a non-extensional model theory developed by Reinhard Muskens (2007). It is fairly easily to find plausible necessary conditions on logicality, but sufficient conditions are more difficult. I tentatively defend an account on which the key notion is not the logicality of an individual operation, but the logicality of a family

[1] See, for example, van Benthem (1989); Sher (1991, 2003); McGee (1996); Feferman (1999, 2010); Casanovas (2007); Bonnay (2008, 2014); Paseau (2014); Dutilh Novaes (2014); Griffiths and Paseau (2016).

of operations: on my proposal, a family of operations is a candidate for logicality when it meets requirements of *presentational repleteness* and *closure*: every extension (or, strictly, 'quasi-extension', in a sense that will be spelled out) meeting certain standards of invariance has to be presented by some operation, and the operations must be closed under composition.

4.1 Operation-Logicality

Before beginning, it is necessary to say a bit about the target of the investigation. I am interested in providing an invariance-based account of what makes an *operation* logical, not what makes a *term* logical. Operation-logicality and term-logicality are obviously connected, but they should not be confused. Terms are syntactic objects in interpreted languages. Operations, in contrast, are worldly entities – propositions, functions, relations, higher-type functions and relations over other functions and relations, and so on. Extending terminology somewhat, we also count individuals as a degenerate case of operations.

For the purposes of this paper, I take it as given that reality has a type structure corresponding to that of an appropriately regimented language. I shall assume that standard relational type theory provides a good enough approximation to the ultimate type structure, whatever it is; it would not be difficult to adapt the formal machinery to allow for functional or transfinite types. I use a syntactic type framework $\mathscr{T}$ defined as follows: $\mathscr{T}$ contains a basic type e, the type of singular terms; if $\tau_1, \ldots, \tau_n \in \mathscr{T}$ for $n \in \mathbb{N}$, then $\langle \tau_1, \ldots, \tau_n \rangle \in \mathscr{T}$; nothing else is in $\mathscr{T}$. The type $\langle \tau_1, \ldots, \tau_n \rangle$ is to be understood as the type of n-ary predicates over types $\tau_1, \ldots, \tau_n$; we understand $\langle\rangle$, the $n = 0$ case, as the type of formulas. (Open formulas, like other open terms, are typed as a matter of convenience; our ultimate interest is in sentences.) To every syntactic type τ there corresponds a *metaphysical type*, which we denote by $\overline{\tau}$: every closed term of type τ has a type-$\overline{\tau}$ operation as its ultimate semantic value. 'Operations' of type $\overline{e}$ are simply individuals; those of type $\overline{\langle\rangle}$, propositions. I shall use 'entity' and 'operation' informally as dummy terms ranging over all metaphysical types, and in the case of type $\overline{\langle\rangle}$, I shall form a pseudo-name for the proposition expressed by the sentence ϕ using wide angle-brackets: $< \phi >$. Many of the theoretical issues I shall address already arise at the level of $\overline{\langle\rangle}$, $\overline{\langle\langle\rangle\rangle}$, and $\overline{\langle\langle\rangle, \langle\rangle\rangle}$. In my examples, I shall generally focus on these cases: extending a proposed constraint or analysis to the full type hierarchy is relatively straightforward.

Clearly, in order for a term in a given language to be logical, it must have as its ultimate semantic value an operation which is logical. But (at least on any one of a variety of standard metasemantic accounts) this will not always be sufficient for term-logicality, since in some settings the contribution of a term to complex expressions is not exhausted by its ultimate semantic value: intermediate semantic values of various kinds play a role.

For example, on a Fregean account, a term may have both a sense and a denotation; on a Kaplanian account, both a character and a content. Normally, ultimate semantic value is determined by a combination of intermediate semantic value and context (or other extralinguistic features). As Jack Woods (2016) has argued, a Kaplanian indexical may fail to be logical, even though its content – its ultimate semantic value – is a logical operation; on Woods's account, its character must also pass an appropriate (invariance-based) logicality test.

There is a long philosophical tradition according to which all hyper-intensionality – if not all intensionality whatsoever – is to be explained in terms of intermediate semantic values: the worldly operations that are the ultimate semantic values of terms remain coarse-grained. The most extreme version of the picture is Frege's. On his account, in type $\overline{\langle\rangle}$, there are only two entities that can serve as the ultimate semantic values of sentences: 'Puccini wrote *Tosca*', 'water is H_2O', '1729 is the smallest taxicab number', and '$\forall x\, x = x$' all have the same ultimate semantic value, the True.[2] (In our notation, <Puccini wrote *Tosca*> = <water is H_2O> = < 1729 is the smallest taxicab number> = $< \forall x\, x = x > = \mathfrak{True}$.)

Differences in the cognitive significance and embedding behaviour of these sentences are attributable solely to differences in sense. These differences may be relevant to determining when a *term* is logical, but they play no role in determining when the *operations* that serve as the ultimate semantic values of terms are logical. Similarly, on the Fregean picture, complex-type operations correspond one–one to functions-in-extension on lower types; thus there are, for example, only sixteen type-$\overline{\langle\langle\rangle, \langle\rangle\rangle}$ operations, each of which is counted as logical by standard invariance criteria. On a Fregean picture, although a difference in sense can explain why the type-$\langle\langle\rangle, \langle\rangle\rangle$ *term* '$\lambda pq \,.\, (p \wedge q)$' is logical whereas the term '$\lambda pq \,.\, (p \wedge q) \wedge$ water is H_2O' is not, both these terms denote the same type-$\overline{\langle\langle\rangle, \langle\rangle\rangle}$ *operation*, and that operation, like every other operation of the same type, is logical.

[2] The nth taxicab number is the least number that is the sum of two cubes in $n + 1$ different ways. See Hardy (1921: lvii–lviii).

There are alternative accounts which maintain a similar distinction between ultimate and intermediate semantic values, but allow a greater proliferation of the former: on Stalnaker's (1984) view, for instance, type-$\overline{\langle\rangle}$ entities are finer-grained but still not fine-grained enough to account for hyperintensional distinctions: 'water is H_2O' and '1729 is the smallest taxicab number' both have the class of all possible worlds (or something corresponding to it) as their ultimate semantic value.

For the purposes of this paper, I assume that accounts of this kind, however useful they may be for explaining some features of *term-logicality*, are not the whole story. Instead, I assume that at least some hyperintensionality is an irreducible feature of the world, and that any language capable of adequately representing the overall structure of reality will admit hyperintensional distinctions in the operations that serve as the ultimate semantic values of its terms. I shall not offer a philosophical defence of this position here; to my mind, the strongest arguments in its favour derive from considerations involving grounding operators (Rosen 2006; Fine 2012) and from intuitions about higher-type identity (Dorr 2016).

As far as possible, however, I shall avoid committing myself to a particular account of just how much hyperintensionality there is in the world. For instance, I take no stance on whether < 1729 is the smallest taxicab number> = <1729 is the smallest taxicab number ∧ 1729 is the smallest taxicab number>. This identity is a consequence, for example, of Booleanism – the doctrine that type-$\overline{\langle\rangle}$ entities and the standard connectives over them form a Boolean algebra (Dorr 2016: 67). Booleanism is convenient for building models, but I do not wish to assume it in the general case. We shall thus need a framework for languages over $\mathscr{T}$ that builds in neither Booleanism nor any similarly controversial principle about fineness of grain; the system of Muskens frames set out in § 4.3 is general enough for the task.

In principle, accounts of invariance for higher-order systems ought to be formulated in a higher-order metalanguage.[3] Unfortunately, however, higher-order model theory is not directly mathematically tractable. If we wish to know, for instance, how candidate invariance criteria are affected by the claim that at least two type-$\overline{\langle\alpha\rangle}$ entities exist corresponding to every function-in-extension from type-$\overline{\alpha}$ entities to type-$\overline{\langle\rangle}$ entities, we can formulate this hypothesis directly in higher-order terms; but in order to explore its consequences, we have to consider the fragments of the

[3] For general discussions about working 'natively' in higher-order languages, see Rayo and Yablo (2001) and Williamson (2003).

higher-order universe in which it holds, and in order to know whether there are any such fragments, we need to know about the *true* relationship of type-$\overline{\langle\alpha\rangle}$ entities to the corresponding extensional profiles – which is precisely the hypothesis in question.

For this reason, we are better off attacking the problem indirectly, by simulating higher-order models within set theory. Set theory is technically well understood, and we can choose the truth-in-a-model relation in such a way that there will always be well-behaved set models for the type-theoretic assumptions that interest us. The philosophical interest of this approach, of course, depends on a reflection principle: we assume that, at least for certain restricted languages, we can find a class of set models and a truth-in-a-model relation such that if a sentence in the language is true in every model in the class, then it holds on the intended (genuinely higher-order) interpretation. In the context of extensional higher-order logic, reflection principles such as this were first explicitly discussed by Kreisel (1967).[4] Their precise formulation and justification in a non-extensional context raises interesting philosophical issues, but these lie beyond the scope of the paper: here I shall merely follow the standard practice and formulate candidate invariance criteria in terms of the set-theoretical model theory, leaving their connexion to the intended (irreducibly higher-order) interpretation implicit.

First, I shall review the standard notions of invariance for operators in an extensional framework. We begin with the notion of an extensional frame – a typed collection of domains:

DEFINITION 1 *A* ***standard extensional frame (SE-frame)*** *is a collection* $\mathfrak{F} = \{D_\tau(\mathfrak{F}) : \tau \in \mathscr{T}\}$ *of sets where*

- $D_e(\mathfrak{F})$ *is nonempty, and* $\{0, 1\} \cap D_e(\mathfrak{F}) = \emptyset$,
- $D_{\langle\rangle}(\mathfrak{F}) = \{0, 1\}$, *and*
- $D_{\langle\tau_1,\ldots,\tau_n\rangle}(\mathfrak{F}) = {}^{D_{\tau_1}(\mathfrak{F})\times\cdots\times D_{\tau_n}(\mathfrak{F})}D_{\langle\rangle}(\mathfrak{F})$.

${}^{A}B$ *denotes the class of functions from A to B. We fix the convention that* 0 *represents* $\mathfrak{False}$ *and* 1 *represents* $\mathfrak{True}$. *The second constraint on* $D_e(\mathfrak{F})$ *ensures non-overlap of the* $D_\tau(\mathfrak{F})$*; it is not always included in presentations of type theory, but it makes no material difference to the model theory.*

We write D_τ *for* $D_\tau(\mathfrak{F})$ *when the frame is clear from context, and we write* $D(\mathfrak{F})$ *for* $\bigcup\mathfrak{F}$. *We write* $\mathfrak{Std}$ *for the class of all SE-frames, and we write* $\mathfrak{DomStd}$ *for* $\bigcup\{D(\mathfrak{F}) : \mathfrak{F} \in \mathfrak{Std}\}$.

[4] For an extensive discussion, see Shapiro (1987).

The frames are *standard* because, for any τ, every subset of D_τ is represented by its characteristic function in $D_{\langle\tau\rangle}$; *mutatis mutandis* for sets of tuples. Adding an interpretation function to a standard frame thus yields a standard model (as opposed to a Henkin model) of extensional higher-order logic.

For our purposes, however, we are interested not in interpretations of a fixed vocabulary within an individual model but in uniform ways of picking out appropriately typed entities from an arbitrary frame. We thus define an extensional operation as an appropriately typed class function; similarly, we define permutations and bijections as operations on frames.

DEFINITION 2 *A* ***standard extensional operation (SE-operation)*** *of type* τ *is a function* $f : \mathfrak{Std} \rightarrow \mathfrak{DomStd}$ *with the constraint that* $f(\mathfrak{F}) \in D_\tau(\mathfrak{F})$ *for all* $\mathfrak{F} \in \mathfrak{Std}$. *We write* $\mathfrak{SE}$ *for the class of SE-operations.*

DEFINITION 3 *Where* $\mathfrak{F}, \mathfrak{G} \in \mathfrak{Std}$, *a* ***base bijection*** *from* $\mathfrak{F}$ *to* $\mathfrak{G}$ *is a bijective function* $\pi : D_e(\mathfrak{F}) \rightarrow D_e(\mathfrak{G})$. *If* $\mathfrak{F} = \mathfrak{G}$, *then we term* π *a* ***base permutation***. *We write* $\mathfrak{Bij}_{Base}(\mathfrak{F}, \mathfrak{G})$ *for the class of all base bijections from* $\mathfrak{F}$ *to* $\mathfrak{G}$. *The* ***frame bijection*** $\hat{\pi} : D(\mathfrak{F}) \rightarrow D(\mathfrak{G})$ ***induced by a*** $\pi \in \mathfrak{Bij}(\mathfrak{F}, \mathfrak{G})$ *is defined as follows: for* $x \in D_e(\mathfrak{F})$, $\hat{\pi}(x) = \pi(x)$; *for* $x \in D_{\langle\rangle}(\mathfrak{F})$, $\hat{\pi}(x) = x$; *for* $x \in D_{\langle\tau_1,\ldots,\tau_n\rangle}(\mathfrak{F})$, $\hat{\pi}(x)$ *is the function* $\{\langle\hat{\pi}(y_1), \ldots, \hat{\pi}(y_n), \hat{\pi}(z)\rangle : \langle y_1, \ldots, y_n, z\rangle \in x\}$. *If* $\mathfrak{F} = \mathfrak{G}$, *then we term* $\hat{\pi}$ *a* ***frame permutation***. *We write* $\mathfrak{Bij}_{Frame}(\mathfrak{F}, \mathfrak{G})$ *for* $\{\hat{\pi} : \pi \in \mathfrak{Bij}_{Base}(\mathfrak{F}, \mathfrak{G})\}$.

DEFINITION 4 *An SE-operation* f *is* ***invariant***$_{perm}$ *if, for all* $\mathfrak{F} \in \mathfrak{Std}$ *and for all* $\pi \in \mathfrak{Bij}_{Base}(\mathfrak{F}, \mathfrak{F})$, $f(\mathfrak{F}) = \hat{\pi}(f(\mathfrak{F}))$. *An SE-operation* f *is* ***invariant***$_{bij}$ *if, for all* $\mathfrak{F}, \mathfrak{G} \in \mathfrak{Std}$ *such that* $|\mathfrak{F}| = |\mathfrak{G}|$ *and all* $\pi \in \mathfrak{Bij}_{Base}(\mathfrak{F}, \mathfrak{G})$, $f(\mathfrak{G}) = \hat{\pi}(f(\mathfrak{F}))$.

With these definitions in place, we can formulate the standard invariance-based logicality theses:

> TARSKI'S THESIS (TT) An SE-operation is logical just in case it is invariant$_{perm}$ (Tarski 1966/86: 149).
>
> MODIFIED TARSKI'S THESIS (MTT) An SE-operation is logical just in case it is invariant$_{bij}$ (McGee 1996: 576–8).

It is worth reiterating that TT and MTT are claims about the logicality of *operations*. MTT, for instance, must be distinguished from the Tarski–Sher thesis about the logicality of terms:

Tarski–Sher Thesis A term **t** in a purely extensional language is logical just in case the ultimate semantic value of **t** is an E-operation that is invariant$_{\text{bij}}$ (Sher 1991: 53–56).

For the purposes of this paper, I shall take the correctness of MTT as an account of operation-logicality, *when only extensional models are considered*, for granted. Those who believe that MTT overgenerates (Feferman 1999, 2010; Bonnay 2008) can modify the proposals that I shall develop reasonably easily to incorporate a more stringent criterion for extensional operation-logicality.

4.2 Muskens Frames, Hereditary Extensions, and an Initial Proposal

I now introduce a non-extensional framework based on that developed by Reinhard Muskens (2007). On Muskens's approach, a model is not a single typed collection of domains. Instead, in every model, for every type τ there are two domains, an *intensional domain* and an *extensional domain*. In type $\langle\rangle$, the elements of the intensional domain represent entities of type $\overline{\langle\rangle}$; those of the extensional domain represent $\mathfrak{True}$ and $\mathfrak{False}$. In every complex type $\langle \tau_1, \ldots, \tau_n \rangle$, the elements of the intensional domain represent genuine type-$\overline{\langle \tau_1, \ldots, \tau_n \rangle}$ entities; the elements of the extensional domain are functions-in-extension over entities in intensional domains of lower type. A typed extension function η associates every element of the intensional domain of a given type with a corresponding element of the extensional domain of that type.

The governing idea behind this framework is to provide a general way of representing higher-order entities using set-theoretical models while remaining as neutral as possible on the *nature* of those entities:

> The aim of this paper is not to add one more theory of intension to the proposals that have already been made, but is an investigation of their common underlying logic. The idea will be that the two-stage set-up is essentially *all* that is needed to obtain intensionality. For the purposes of logic it suffices to consider intensions as abstract objects; the question what intensions *are*, while philosophically important, can be abstracted from. (Muskens 2007, 101)

Indeed, most other non-extensional model theories for type theory – including higher-order Kripke models (Gallin 1975, 67–78; Williamson 2013, 222), impossible-worlds models (Rantala 1982a, 1982b), and applicative structures (Benzmüller, Brown, and Kohlhase 2004) – can be implemented as special cases of Muskens's framework. Muskens himself demonstrated how to apply the framework to a Fregean account of sense and denotation, with the elements of the various intensional domains representing intermediate semantic values, but it is equally natural to use the system to model worldly hyperintensionality, with the intensional domains representing fundamental elements of the world – the ultimate semantic values of terms – and the extensional domains playing a secondary role, tracking particular properties of the intensional domains.

As was the case in the standard extensional picture, we need not concern ourselves with the details of interpreting specific vocabulary on individual models; we can work with frames alone.

DEFINITION 5 *A* ***standard Muskens frame (M-frame)*** *is a triple* $\mathfrak{M} = \langle E_{\mathscr{T}}(\mathfrak{M}), I_{\mathscr{T}}(\mathfrak{M}), \eta \rangle$*, with the following requirements:*

- *(1) the* ***intension structure*** $I_{\mathscr{T}}(\mathfrak{M})$ *is a collection* $\{I_\tau(\mathfrak{M}) : \tau \in \mathscr{T}\}$ *of sets, where* $I_e(\mathfrak{M})$ *is nonempty,* $I_{\tau_1}(\mathfrak{M}) \cap I_{\tau_2}(\mathfrak{M}) = \emptyset$ *if* $\tau_1 \neq \tau_2$*; we write* $I(\mathfrak{M})$ *for* $\bigcup I_{\mathscr{T}}(\mathfrak{M})$*;*
- *(2) the* ***extension structure*** $E_{\mathscr{T}}(\mathfrak{M}) = \{E_\tau(\mathfrak{M}) : \tau \in \mathscr{T}\}$ *is defined as follows:* $E_{\langle\rangle}(\mathfrak{M}) = \{0, 1\}$*;* $E_e(\mathfrak{M}) = I_e(\mathfrak{M})$*; and* $E_{\langle \tau_1, \ldots, \tau_n \rangle}(\mathfrak{M}) = {}^{I_{\tau_1}(\mathfrak{M}) \times \cdots \times I_{\tau_n}(\mathfrak{M})} I_{\langle\rangle}(\mathfrak{M})$*; we write* $E(\mathfrak{M})$ *for* $\bigcup E_{\mathscr{T}}(\mathfrak{M})$*;*
- *(3) the* ***extension function*** $\eta : I(\mathfrak{M}) \rightarrow E(\mathfrak{M})$ *obeys the constraint that if* $x \in I_\tau(\mathfrak{M})$*,* $\eta(x) \in E_\tau(\mathfrak{M})$*;*
- *(4) for every* $\tau \in \mathscr{T}$ *and every* $y \in E_\tau(\mathfrak{M})$*, there exists some* $x \in I_\tau(\mathfrak{M})$ *such that* $\eta(x) = y$*.*

As before, we write I_τ for $I_\tau(\mathfrak{M})$ and E_τ for $E_\tau(\mathfrak{M})$ when the frame is clear from context. We write $\mathfrak{Msk}$ for the class of M-frames, and we write $\mathfrak{IntMsk}$ for $\bigcup\{I(\mathfrak{M}) : \mathfrak{M} \in \mathfrak{Msk}\}$ and $\mathfrak{ExtMsk}$ for $\bigcup\{E(\mathfrak{M}) : \mathfrak{M} \in \mathfrak{Msk}\}$. When $\eta(x) = y$, we say that x **presents** y.

I have modified Muskens's framework somewhat: I require the type-*e* intentional domain to be identical to the type-*e* extensional domain (reflecting the assumption that there is no worldly non-extensional behaviour among type-$\bar{e}$ entities); require a fullness condition analogous to the standardness requirement on extensional models (every subset of an E_τ must have a characteristic function presented by some $x \in I_\tau$, and so on); and impose non-overlap constraints for simplicity. The general idea behind my presentation is that intensions – whatever they are – must determine extensions, and a unique extension can be recovered from every intension: the η function performs this recovery. An intension, on this picture, 'carries' its extension with it – thus an M-frame fixes the relation between intensions and extensions, independently of the question of how terms in a particular language are to be interpreted (which must be specified to turn an M-frame into an Muskens *model*).

Just as we can associate SE-operations with class functions on SE-frames in the extensional framework, we can use class functions on M-frames for a similar purpose here:

DEFINITION 6 *A* ***Muskens operation (M-operation)*** *of type* τ *is a function* $f : \mathfrak{Msk} \rightarrow \mathfrak{IntMsk}$ *with the constraint that* $f(\mathfrak{M}) \in I_\tau(\mathfrak{M})$ *for all* $\mathfrak{M} \in \mathfrak{Msk}$*. We write* $\mathfrak{MO}$ *for the class of M-operations.*

If MTT represents the correct analysis of logicality in an extensional framework, it is natural to require that any logical M-operation *correspond*, in an informal sense, to invariant$_{\mathrm{bij}}$ SE-operation:

> EXTENSIONAL CORRESPONDENCE CONSTRAINT (ECC) If an M-operation f is logical, then it corresponds extensionally to a unique SE-operation g that is invariant$_{\mathrm{bij}}$.

To make ECC more precise, we need a characterization of the sense in which an M-operation corresponds to a unique SE-operation. At first, it might seem that we could simply inductively define an 'ultimate' extension $\widetilde{f}$ corresponding to every M-operation f using η; but some care is required: the most obvious approach fails in two kinds of cases to yield a unique SE-operation.

First, ultimate extensions will only be well defined for entities that meet a strong extensionality constraint. Where $f \in I_{\langle\tau\rangle}(\mathfrak{M})$, for example, $\eta(f)$ is a function not from $E_\tau(\mathfrak{M})$ to $E_{\langle\rangle}(\mathfrak{M})$ but from $I_\tau(\mathfrak{M})$ to $I_{\langle\rangle}(\mathfrak{M})$, so in order to define an ultimate extension for f, we need to replace the entities in the domain and range of $\eta(f)$ with the entities that they present. But it is possible that there exist $g_1, g_2 \in I_\tau(\mathfrak{M})$ such that $\eta(g_1) = \eta(g_2)$ but $f(g_1) \neq f(g_2)$. Consider, for instance, a model $\mathfrak{M}$ where $I_{\langle\rangle}$ contains three objects A, B, and C, $\eta(A) = \eta(B) = 1$, and $\eta(C) = 0$. There will exist an $f \in E_{\langle\langle\rangle\rangle}$ with $f(A) = A$ and $f(B) = C$, presented by some $g \in I_{\langle\langle\rangle\rangle}$. We cannot meaningfully ask what output f yields for the input whose extension is $\mathfrak{True}$, for there is more than one such entity and f discriminates among them. Thus f does not have an ultimate extension.

To deal with this, we introduce the notion of a *hereditary extension* – a *partial* function left undefined for entities in an M-frame that discriminate between extensionally equivalent inputs, or which have inputs that themselves discriminate between extensionally equivalent inputs, and so on.

DEFINITION 7 *We define the* ***extensional kernel*** $K_{\mathscr{T}}(\mathfrak{M})$ *of an M-frame* $\mathfrak{M}$ *as follows:* $K_{\mathscr{T}}(\mathfrak{M}) = \{K_\tau(\mathfrak{M}) : \tau \in \mathscr{T}\}$ *where* $K_e(\mathfrak{M}) = E_e(\mathfrak{M})$; $K_{\langle\rangle}(\mathfrak{M}) = E_{\langle\rangle}(\mathfrak{M})$; $K_{\langle\tau_1,\ldots,\tau_n\rangle}(\mathfrak{M}) = {}^{K_{\tau_1}(\mathfrak{M})\times\cdots\times K_{\tau_n}(\mathfrak{M})}K_{\langle\rangle}(\mathfrak{M})$.

In effect, the extensional kernel of an M-frame is the SE-frame that results when we quotient out all non-extensional features of the structure.

DEFINITION 8 *We define the* ***hereditary extension (h-extension)*** *partial function* $\tilde{\eta}_{\mathfrak{M}} : I(\mathfrak{M}) \dashrightarrow K(\mathfrak{M})$ *as follows (leaving off the subscript when no confusion will result):*

- *if* $x \in I_e \cup I_{\langle\rangle}$, $\tilde{\eta}(x) = \eta(x)$;
- *if* $x \in I_{\langle\tau_1,\ldots,\tau_n\rangle}$, $\tilde{\eta}(x)$ *is* $\{\langle\tilde{\eta}(y_1),\ldots,\tilde{\eta}(y_n),\tilde{\eta}(z)\rangle : \langle y_1,\ldots,y_n,z\rangle \in \eta(x)\}$
  - *if* $\tilde{\eta}(z)$ *and each* $\tilde{\eta}(y_i)$ *is defined and* $\{\langle\tilde{\eta}(y_1),\ldots,\tilde{\eta}(y_n),\tilde{\eta}(z)\rangle : \langle y_1,\ldots,y_n,z\rangle \in \eta(x)\}$ *is functional;*
  - $\tilde{\eta}(x)$ *is undefined otherwise.*

If $\tilde{\eta}(x) = y$, we say that x **h-presents** y.

The second complication arises because an M-operation, even if it meets the extensionality constraint on individual frames, might discriminate inappropriately among M-frames with the same extensional kernel. For instance, it may be the case that there exist two M-frames $\mathfrak{M}_1$ and $\mathfrak{M}_2$ and an M-operation f such that $\tilde{\eta}(f(\mathfrak{M}_1)) \neq \tilde{\eta}(f(\mathfrak{M}_2))$ even though $K(\mathfrak{M}_1) = K(\mathfrak{M}_2)$. For instance, let $\mathfrak{M}_1$ be chosen arbitrarily. In every M-frame $\mathfrak{M}$, there exist (not necessarily unique) elements $\top_{\mathfrak{M}} \in I_{\langle\rangle}(\mathfrak{M})$ and $\bot_{\mathfrak{M}} \in I_{\langle\rangle}(\mathfrak{M})$ such that $\eta(\top_{\mathfrak{M}}) = 1$ and $\eta(\bot_{\mathfrak{M}}) = 0$. Fix a $\top_{\mathfrak{M}}$ and $\bot_{\mathfrak{M}}$ for each $\mathfrak{M}$ and let f be the M-operation with $f(\mathfrak{M}_1) = \top_{\mathfrak{M}_1}$ and $f(\mathfrak{M}) = \bot_{\mathfrak{M}}$ for all $\mathfrak{M} \neq \mathfrak{M}_1$. Clearly, the h-extension of $f(\mathfrak{M})$ is defined for every $\mathfrak{M}$; but f singles out $\mathfrak{M}_1$, differentiating it from the proper class of M-frames with the same extensional kernel.

Thus, we need to use another partial function, leaving the extensional correlate of the M-operation undefined on any such M-frames.

DEFINITION 9 *For* $\mathfrak{M}_1, \mathfrak{M}_2 \in \mathfrak{MSk}$, *define* $\mathfrak{M}_1 \sim \mathfrak{M}_2$ *if* $K(\mathfrak{M}_1) = K(\mathfrak{M}_1)$; *let* $\frac{\mathfrak{MSk}}{\sim}$ *be the class of equivalence classes of M-frames under* $\sim$ *and let* $[\mathfrak{M}]_\sim$ *be the class in* $\frac{\mathfrak{MSk}}{\sim}$ *containing* $\mathfrak{M}$. *For* $f \in \mathfrak{MO}$, *we define the partial function* $\widetilde{f} : \mathfrak{Std} \dashrightarrow \mathfrak{DomStd}$, *which we term the* ***HE-correlate*** *of* f, *as follows.*

- *if, for every* $\mathfrak{M}_1, \mathfrak{M}_2 \in [\mathfrak{M}]_\sim$, $\tilde{\eta}(f(\mathfrak{M}_1)) = \tilde{\eta}(f(\mathfrak{M}_2))$, *then* $\widetilde{f}(K(\mathfrak{M})) = \tilde{\eta}(f(\mathfrak{M}))$;
- *otherwise,* $\widetilde{f}(K(\mathfrak{M}))$ *is undefined.*

DEFINITION 10 *An M-operation* f *is* ***invariant***$_{HE}$ *just in case* $\widetilde{f}$ *is total and invariant*$_{bij}$.

We can now give a more precise version of ECC:

(ECC1) An M-operation f is logical only if f is invariant$_{\text{HE}}$.

At this point it is natural to ask: can we strengthen ECC1 to a biconditional, giving us a full analysis of logicality over Muskens models? If so, given the extreme generality of the Muskens framework as a model for worldly hyperintensionality, we would seem to be in possession of an extension of MTT that is simple, compelling, and extremely powerful. We term the biconditional Proposal 1:

(P1) An M-operation f is logical if and only if f is invariant$_{\text{HE}}$.

4.3 Counterexamples to P1 and Two Alternative Proposals

Unfortunately, P1 overgenerates catastrophically. We can see this by considering a simple case involving only type-$\langle\langle\rangle\rangle$ M-operations. The pure identity SE-operation $\mathfrak{id}$ of type $\langle\langle\rangle\rangle$ is defined as follows: for every SE-frame $\mathfrak{F}$, for every $x \in D_{\langle\rangle}(\mathfrak{F})$, $\mathfrak{id}(x) = x$. Of course, $\mathfrak{id}$ is invariant$_{\text{bij}}$ and thus, by MTT, logical. Furthermore, it seems obvious that there should be at least one logical M-operation f of type $\langle\langle\rangle\rangle$ such that $\widetilde{f} = \mathfrak{id}$.

But, if P1 is true, we get not one hyperintensional M-operation with $\mathfrak{id}$ as its HE-correlate, but a plethora beyond number.

THEOREM 1 *There exists a proper class* $\mathscr{A}$ *of type-*$\langle\langle\rangle\rangle$ *M-operations such that, for each* $f \in \mathscr{A}$, $\widetilde{f} = \mathfrak{id}$.

Proof It suffices to show that, for every cardinal κ, there exists a set $\mathscr{A}_\kappa$ containing κ distinct type-$\langle\langle\rangle\rangle$ M-operations with $\mathfrak{id}$ as their HE-correlate. For any model $\mathfrak{M}$, we define $\mathsf{id}_t(\mathfrak{M})$ as the map $x \mapsto x$ for $x \in I_{\langle\rangle}(\mathfrak{M})$. For an arbitrary κ, we construct the M-frame $\mathfrak{N}_\kappa$ as follows: let $I_{\langle\rangle} = E_{\langle\rangle} = \{0, 1\}$; let $I_e = E_e = \{a\}$, for some arbitrary object a. We define $J = \{\langle\lambda, \mathsf{id}_t(\mathfrak{N})\rangle : \lambda < \kappa\}$. We set $I_{\langle\langle\rangle\rangle} = E_{\langle\langle\rangle\rangle} \cup J$. For every other type τ, we set $I_\tau = E_\tau$. We define η as follows: $\eta(x) = x$ unless $x \in J$, in which case $\eta(x) = \mathsf{id}_t(\mathfrak{N})$.

It is trivial to prove that there is at least one M-operation f with $\mathfrak{id}$ as its HE-correlate. For each M-model $\mathfrak{M}$, we well-order $S(\mathfrak{M}) = \{x \in I_{\langle\langle\rangle\rangle}(\mathfrak{M}) : \eta(x) = \mathfrak{id}_t(\mathfrak{M}) \wedge \tilde{\eta}(x) = 1\}$. ($S(\mathfrak{M})$ must be nonempty in virtue of the fullness requirement on M-frames.) Where α is an ordinal, let $\ell_\alpha(\mathfrak{M})$ be the αth element of $S(\mathfrak{M})$ under the ordering. We set $f(\mathfrak{M}) = \ell_0(\mathfrak{M})$. Now, for $\lambda < \kappa$, let f_λ be the M-operation that differs from f only in that $f_\kappa(N) = \ell_\lambda(\mathfrak{N})$. It is trivial to verify that $\widetilde{f_\lambda} = \mathfrak{id}$ for each λ; we thus set $\mathscr{A}_\kappa = \{\lambda < \kappa : \widetilde{f_\lambda}\}$.

As this construction evinces, M-operations can remain invariant$_{\text{HE}}$ whilst displaying precisely the kind of discrimination among entities that the Tarskian interpretation of logicality is supposed to rule out. How can an operation which singles out, on a completely arbitrary basis, one of the proper-class-many type-$\langle\langle\rangle\rangle$ operations with $\mathfrak{id}$ as their HE-correlate count as logical?

Of course, outside type e, some degree of discrimination is required. Even in the purely extensional case, invariance$_{\text{perm}}$ and invariance$_{\text{rm}}$ do not require that SE-operations treat 1 and 0 (i.e., the representatives of $\mathfrak{True}$ and $\mathfrak{False}$) interchangeably. Since $\mathfrak{True}$ and $\mathfrak{False}$, if such things there be, are *logical* entities par excellence, logical operations should be able to take their natures into account. Similarly, when we consider a framework that admits propositions beyond $\mathfrak{True}$ and $\mathfrak{False}$, but treats propositions in accordance with Booleanism, there are plausible grounds for singling out two propositions $\top$ and $\bot$. But M-frames do not come pre-programmed with Boolean structure. It seems reasonable to assume that, given *any* adequate model of the class of propositions, a $\top$ and $\bot$ can in principle be picked out; given any adequate model of the class of $\overline{\langle\langle\rangle\rangle}$ or $\overline{\langle\langle\rangle, \langle\rangle\rangle}$ entities, a negation operation or conjunction and disjunction operations can in principle be picked out, and so on; but ECC1 offers us no help in choosing which ones they are; P1 avoids the problem by including all reasonable candidates – with absurd results.

There is, however, a way to shift the problem in a more productive direction. The generality of the Muskens framework fails to offer us an obvious way to choose distinguished HE-invariant M-operations for the various types. Once such a choice has been made, however, we can meaningfully assess the properties of the resulting *family* of M-operations. If we are willing to accept (at least provisionally) that our search will lead only to stronger necessary conditions rather than sufficient conditions, we can try to improve on ECC1 by finding plausible properties of the class of logical M-operations as a whole.

We thus define:

DEFINITION 11 *A class* $\mathscr{C} \subseteq \mathfrak{MO}$ *is a* ***global invariance class*** *just in case: (1) every* $f \in \mathscr{C}$ *is invariant*$_{HE}$ *and (2) for each invariant*$_{bij}$ *SE-operation* g, *there exists a unique* $f \in \mathscr{C}$ *such that* $\widetilde{f} = g$.

The second clause, in essence, handles cases such as those in the previous example by imposing a uniqueness requirement by fiat. With this definition in place, we can formulate a thesis that goes beyond ECC1 but avoids the overgeneration problem with P1:

(P2) The logical M-operations form a global invariance class.

Unfortunately, however, P2 faces a potential undergeneration problem. P2 is based on the intuition that, although any number of entities in an M-frame may h-present a given SE-operation, *logical* entities – the ones which, so to speak, *logically* present an invariant$_{bij}$ SE-operation – should be (roughly speaking) sparse among the inhabitants of the M-frame. The easiest way to meet this intuition is to demand uniqueness for the distinguished M-operation presenting a given SE-operation. But there is another intuition that needs to be respected: the result of composing logical operations should itself be a logical operation. And nothing in P2 guarantees this.

For instance, let $\mathscr{C}$ be a global invariance class and let neg be the type-$\langle\langle\rangle\rangle$ SE-operation negation (i.e., the SE-operation defined by $\mathsf{neg}(0) = 1$ and $\mathsf{neg}(1) = 0$). Let pon be the type-$\langle\langle\langle\rangle\rangle\rangle$ SE-operation that *picks out negation* in the sense that $\mathsf{pon}(x) = 1$ if and only if $x = \mathsf{neg}$.

There must be an $f_1, f_2 \in \mathscr{C}$ such that $\widetilde{f_1} = \mathsf{pon}$ and $\widetilde{f_2} = \mathsf{neg}$. Fix an M-frame $\mathfrak{M}$. There exist an $f_1(\mathfrak{M}) \in I_{\langle\langle\langle\rangle\rangle\rangle}(\mathfrak{M})$ and an $f_2(\mathfrak{M}) \in I_{\langle\langle\rangle\rangle}(\mathfrak{M})$. We can of course apply $f_1(\mathfrak{M})$ to $f_2(\mathfrak{M})$ by way of the corresponding element in the extensional domain, yielding some $g = \eta(f_1(\mathfrak{M}))(f_2(\mathfrak{M})) \in I_{\langle\rangle}(\mathfrak{M})$. But there is no guarantee that there exists any $f_3 \in \mathscr{C}$ such that $f_3(\mathfrak{M}) = g$.

Clearly, this is a problem: logicality should be closed under composition. We thus specify a closure condition.

DEFINITION 12 *Say that a class* $\mathscr{C} \subseteq \mathfrak{MO}$ *is* ***closed under composition*** *just in case, for any* $f, g_1, \ldots, g_n \in \mathscr{C}$ *of types* $\langle\tau_1, \ldots, \tau_n\rangle, \tau_1, \ldots, \tau_n$ *respectively, there exists some* $h \in \mathscr{C}$ *such that* $\eta(h(\mathfrak{M})) = \eta(f(\mathfrak{M}))(g_1(\mathfrak{M}), \ldots, g_n(\mathfrak{M}))$.

Because the requirement that $\eta(h(\mathfrak{M})) = \eta(f(\mathfrak{M}))(g_1(\mathfrak{M}), \ldots, g_n(\mathfrak{M}))$ does not specify a unique h, we cannot simply define a closure operator that can be applied to classes of M-operations. We can, however, require that the closed class be minimal in the sense that it has no proper subclass properly extending an invariance class. This will not, in general, yield uniqueness, but it is a natural way of restricting the options. This leads to our third proposal:

DEFINITION 13 *A class* $\mathscr{C} \subseteq \mathfrak{MO}$ *is* ***charged*** *just in case (1) every* $f \in \mathscr{C}$ *is invariant*$_{\text{HE}}$*; (2)* $\mathscr{C}$ *is closed; and (3) there exists a* $\mathscr{B} \subseteq \mathscr{C}$ *such that* $\mathscr{B}$ *is a global invariance class. A class* $\mathscr{C} \subseteq \mathfrak{MO}$ *is* ***minimal charged*** *if it is charged and there is no* $\mathscr{D} \subsetneq \mathscr{C}$ *that is charged.*

(P3) The logical M-operations form a minimal charged class.

Unfortunately, P3 risks undergeneration as well. Plausibly, P3 suffices to satisfy our desiderata about how a family of M-operations should interact so that that their compositional behaviour *at the level of h-extensions* makes sense. But, intuitively, there are operations on operations which deserve to be considered logical for reasons not directly determined by their h-extensions. Consider, again, an arbitrary frame $\mathfrak{M}$. Now consider the map $x \mapsto x$ defined on elements of $I_{\langle\rangle}(\mathfrak{M})$, which we shall term $\mathfrak{ident}(\mathfrak{M})$. Unlike $\mathfrak{id}$, $\mathfrak{ident}(\mathfrak{M})$ is not the value of an SE-operation for $\mathfrak{M}$'s extensional kernel, since it does not act on elements of $0, 1$. It is also not the value of an M-operation for $\mathfrak{M}$, since it is located in $E_{\langle\langle\rangle\rangle}(\mathfrak{M})$, not $I_{\langle\langle\rangle\rangle}(\mathfrak{M})$. Nonetheless, it seems clear that $\mathfrak{ident}(\mathfrak{M})$ ought to be presented by some logical M-operation, since it is a paradigm case of an entity that disregards nonstructural differences among the entities on which it acts. But there is no guarantee that, if $\mathscr{C}$ is a class that satisfies P3, then there is some $f \in \mathscr{C}$ such that $\eta(f(\mathfrak{M})) = \mathfrak{ident}(\mathfrak{M})$.

4.4 A Final Proposal: Indicators and Repleteness

In order to expand candidate classes of logical operators to deal with this worry, we introduce a few new definitions.

DEFINITION 14 *We define* ***local relative invariance on*** ${}^{A_1 \times \cdots \times A_n}B$ *given some fixed* $C_1 \subseteq A_1, \ldots, C_n \subseteq A_n, D \subseteq B$ *as follows. We define a* C_i***-restricted*** A_i***-permutation*** *as a permutation* $\pi_i : A_i \to A_i$ *such that, for all* $x \in C_i$, $\pi_i(x) = x$. *We denote the class of* C_i*-restricted* A_i*-permutations by* $\mathfrak{LPerm}_{C_i}(A_i)$*;* mutatis mutandis *for* $\mathfrak{LPerm}_D(B)$.

An $x \in {}^{A_1 \times \cdots \times A_n}B$ *is locally invariant relative to* $C_1, \ldots, C_n, D$ *just in case, for every* $\pi_1 \in \mathfrak{LPerm}_{C_1}(A_1), \ldots, \pi_n \in \mathfrak{LPerm}_{C_n}(A_n), \pi_0 \in \mathfrak{LPerm}_D(B)$, $\pi_1 \cdots \pi_n \pi_0(x) = x$.

Local relative permutation-invariance permits us to select certain elements of an $I_{\tau_1}(\mathfrak{M}), \ldots, I_{\tau_1}(\mathfrak{M})$ and isolate the functions-in-extension in $E_{\langle \tau_1, \ldots, \tau_n \rangle}(\mathfrak{M})$ which are invariant under permutations *with those elements fixed.* We can thus proceed step by step within an M-frame, generating appropriate notions of invariance for higher and higher types: we require the functions-in-extension that display permutation invariance relative to entities of lower type that have already been classified as logical to be presented by specified entities in the appropriate intensional domain; these entities in turn are classified as logical and held fixed in an extensional domain of still higher type. We formalize this using what we term *indicators* of M-frames.

DEFINITION 15 *An **indicator** for an M-frame* $\mathfrak{M}$ *is a typed collection* $\{Q_\tau : \tau \in \mathscr{T}\}$ *with the following constraints.*

- $Q_e = \emptyset$;
- $Q_{\langle\rangle} = \{t, f\}$ *for some* $t, f \in I_{\langle\rangle}$ *where* $\eta(t) = 1$ *and* $\eta(f) = 0$;
- *for* $\tau = \langle \tau_1, \ldots, \tau_n \rangle$ *we define* $B_\tau = \{x \in I_\tau : x$ *is locally invariant relative to* $Q_{\tau_1}, \ldots, Q_{\tau_n}, Q_{\langle\rangle}\}$;
- $Q_\tau \subseteq I_\tau$ *such that for each* $y \in B_\tau$, *there exists a unique* $x \in Q_\tau$ *such that* $\eta(x) = y$. *We denote the class of all indicators for* $\mathfrak{M}$ *by* $\mathfrak{Ind}(\mathfrak{M})$; *we set* $\mathfrak{Ind} = \bigcup\{\mathfrak{Ind}(\mathfrak{M}) : \mathfrak{M} \in \mathfrak{Msk}\}$.

DEFINITION 16 *An **indicative framework** is a mapping* $\xi : \mathfrak{Msk} \to \mathfrak{Ind}$ *such that, for each* $\mathfrak{M}$, $\xi(\mathfrak{M}) \in \mathfrak{Ind}(\mathfrak{M})$. *We write* $\xi_\tau(\mathfrak{M})$ *for* $Q_\tau(\xi(\mathfrak{M}))$ *with* Q_τ *defined as above.*

DEFINITION 17 *A class* $\mathscr{C} \subseteq \mathfrak{MO}$ *is a **cover** of an indicative framework* ξ *just in case, for each* $\mathfrak{M} \in \mathfrak{Msk}$ *and each* $\tau \in \mathscr{T}$, *if* $x \in \xi_\tau(\mathfrak{M})$, *there exists some* $f \in \mathscr{C}$ *such that* $f(\mathfrak{M}) = x$.

DEFINITION 18 *A class* $\mathscr{C} \subseteq \mathfrak{MO}$ *is **replete** just in case*

- *there exists some indicative framework* ξ *such that* $\mathscr{C}$ *is a cover of* ξ,
- $\mathscr{C}$ *is closed;*
- *there exists some* $\mathscr{B} \subseteq \mathscr{C}$ *such that* $\mathscr{B}$ *is charged.*

A class $\mathscr{C} \subseteq \mathfrak{MO}$ is ***minimal replete*** *if it is replete and there is no $\mathscr{D} \subsetneq \mathscr{C}$ that is replete.*

With these definitions in place, we set out a final proposal.

(P4) The logical M-operations form a minimal replete class.

P4 is, as far as I can tell, the smallest natural modification of P3 that ensures that cases like that of $\mathfrak{ident}$ will not arise, by requiring, in every frame, for every type τ, presentation of every $x \in E_\tau$ that is invariant relative to the entities of lower type in the frame that have already been classified as logical by some logical $y \in I_\tau$. It remains only to show that at least one minimal replete class exists.

THEOREM 2 *There exists a minimal replete class.*

Proof (Sketch.) Let us say that an indicator Q_1 for an M-frame $\mathfrak{M}$ is minimal if there exists no indicator Q_2 for $\mathfrak{M}$ such that (1) for every $\tau \in \mathscr{T}$, $Q_{2\tau} \subseteq Q_{1\tau}$ and (2) for some $\tau \in \mathscr{T}$, $Q_{2\tau} \subsetneq Q_{1\tau}$. It can be shown, by construction, that minimal indicators exist for every M-frame. Choose an arbitrary minimal indicator $\xi(\mathfrak{M})$ for every $\mathfrak{M} \in \mathfrak{Msk}$; we fix ξ as our indicative framework. Because local relative invariance is a weaker condition than permutation (and thus bijection) invariance, if $h \in \mathfrak{EO}$ is invariant$_{\text{bij}}$, then there exists some $g \in \mathfrak{MO}$ such that $\tilde{g} = h$ and, for every $\mathfrak{M} \in \mathfrak{Msk}$, $g(\mathfrak{M}) = \xi(\mathfrak{M})$. Let us say that any such g is ξ-faithful. Let $\mathbb{G} = \{\mathscr{A} \subseteq \mathfrak{MO} : \mathscr{A}$ is a global invariance class and each $f \in \mathscr{A}$ is ξ-faithful$\}$; let $\mathbb{C} = \{\mathscr{B} \subseteq \mathfrak{MO} : \mathscr{B}$ is a cover of ξ and there exists an $\mathscr{A} \subseteq \mathscr{B}$ such that $\mathscr{A} \in \mathbb{G}\}$. Using global choice, we can well-order $\mathfrak{Msk}$, which induces a lexicographic well-ordering on the class $\{\{\mathfrak{M}_\tau : \tau \in \mathscr{T}\} : \mathfrak{M} \in \mathfrak{MO}\}$; making use of the fact that every $\xi(\mathfrak{M})$ is minimal, we can then use this to define a well-founded partial ordering $\prec$ on $\mathbb{C}$ such that, if $\mathscr{C}$ is minimal under $\prec$, then there is no $\mathscr{D} \in \mathbb{C}$ such that $\mathscr{D} \subsetneq \mathscr{C}$. Fix some such $\mathscr{C}$. By construction, there must be a $\mathscr{G} \in \mathscr{C}$ such that $\mathscr{G} \in \mathbb{G}$. We temporarily write $\mathscr{X}^\circ$ for the closure under composition of $\mathscr{X} \subseteq \mathfrak{MO}$. Since every $f \in \mathscr{G}$ is invariant$_{\text{HE}}$, and the result of composing invariant$_{\text{HE}}$ operations is invariant$_{\text{HE}}$, $\mathscr{G}^\circ$ is charged; since $\mathscr{G}^\circ \subseteq \mathscr{C}^\circ$, $\mathscr{C}^\circ$ is replete. It remains only to show that $\mathscr{C}^\circ$ is minimal replete. Assume that there exists some replete $\mathscr{E} \subsetneq \mathscr{C}^\circ$; since $\mathscr{X}_{1^\circ} \subsetneq \mathscr{X}_{2^\circ}$ only if there exists some $f \in \mathscr{X}_2$ such that $f \in \mathscr{X}_1$, we can construct from $\mathscr{E}$ a cover $\mathscr{H}$ for ξ such that $\mathscr{H} \subsetneq \mathscr{C}$. Contradiction.

Given the complexity inherent in working with M-frames, and our lack of experience refining intuitions on what should count as 'logicality' in this area, I do not wish to say that P4 is the last word. It is, however, the most plausible invariance-based account of logicality in a hyperintensional setting of which I am aware.

CHAPTER 5

There Might Be a Paradox of Logical Validity after All

Roy T Cook

5.1 The Paradox of (Non-Logical) Validity

A number of authors, including Whittle (2004), Field (2008), Shapiro (2010), and Beall (2013) have argued that validity is plagued by a paradox (or paradoxes) similar to the more well-known paradoxes that afflict notions such as truth, knowledge, or set. I argued in Cook (2014) that this paradox, if genuine, does not affect the notion of *logical* validity – that is, the notion explicated by Tarski in his important work on logical consequence:[1]

> Consider any class Δ of sentences and a sentence Φ which follows from the sentences of this class. From an intuitive standpoint it can never happen that both the class Δ consists only of true sentences and the sentence Φ is false. Moreover, since we are concerned here with the concept of logical, i.e. *formal*, consequence, and thus with a relation which is to be uniquely determined by the form of the sentences between which it holds... the relation cannot be affected by replacing the designations of the objects referred to in these sentences by the designations of any other objects. (Tarski 1983b, 414–15)

The paradox in question applies (if it applies at all) to something weaker, such as necessary preservation of truth (regardless of form). At least, that is what I argued in Cook (2014). Much of the subsequent literature on the paradox has taken up this insight. For example, Eduardo Barrio, Lucas Rosenblatt, and Diego Tajer write that:[2]

> It is worth remarking that the concept of validity we are discussing is not a purely logical concept, for it can be iterated (i.e., sentences about validity can themselves be valid). In fact, the purely logical notion of validity can actually be captured in any first-order arithmetical theory extending Robinson's arithmetic, as Ketland (2012) and Cook (2014) point out. (Barrio 2016, note 8)

[1] Ketland (2012) draws similar conclusions.

[2] See also: Murzi (2015), Shapiro (2013, 2015), Nicolai (2017), Murzi (2017).

In this essay I will present a new version of the paradox of validity – one that might be thought to be a genuine paradox of *logical* validity – at least, if one is willing to bite some philosophical and mathematical bullets. The remainder of this paper will be devoted to constructing and exploring this puzzle, and will proceed in three stages. In Section 5.2, I will rehearse (a version of) the argument against viewing the original version of the paradox as applying to logical validity. In Section 5.3 I will then construct the new version of the paradox, based on treating validity as a relation between properties or predicates rather than as a relation between sentences, and argue that there is a way that this puzzle might be viewed as applying to a genuinely logical notion of validity. Finally, in Section 5.4 I will discuss the general account of logicality suggested by the treatment of the paradox of validity in Section 5.3, and point out a few interesting aspects of the view that results.

A final note before moving on: I am not, by any means, endorsing the alternative account of logicality presented in Sections 5.3 and 5.4. On the contrary, I do not think the account developed there is particularly plausible. What I do think, however, is that the view is interesting, but more importantly, perhaps, it provides a nice lens through which to view the paradox of validity – a lens that helps us to bring in to focus exactly what we would have to accept in order to argue that the paradox in question is a paradox of *logical* validity.

5.2 The Original Paradox of Validity

Here I will use the formulation of the paradox found in Beall (2013) (similar comments apply to the variations found in Whittle (2004), Field (2008), Shapiro (2010), and elsewhere – see Ketland (2012) and Cook (2014) for details). Assume that we add to the language of Peano Arithmetic (PA) a predicate $\mathsf{Val}(x, y)$ that holds of the Gödel code $\ulcorner\Phi\urcorner$ of Φ and the Gödel code $\ulcorner\Psi\urcorner$ of Ψ (in that order) if and only if the argument whose sole premise is Φ and whose conclusion is Ψ is logically valid. What rules might we expect such an operator to obey? Beall and Murzi suggest that we should accept the following introduction and elimination rules for $\mathsf{Val}(x, y)$:

$$\begin{aligned}
\mathsf{V_I} :\ & \text{For any formulas } \Phi \text{ and } \Psi : \\
& \text{If} : \Phi \vdash \Psi \\
& \text{Then} : \varnothing \vdash \mathsf{Val}(\ulcorner\Phi\urcorner, \ulcorner\Psi\urcorner) \\
\mathsf{V_E} :\ & \text{For any formulas } \Phi \text{ and } \Psi : \\
& \varnothing \vdash \mathsf{Val}(\ulcorner\Phi\urcorner, \ulcorner\Psi\urcorner) \rightarrow (\Phi \rightarrow \Psi)
\end{aligned}$$

Put simply, V_I codifies the thought that, if we have a proof of Ψ from Φ, then the argument with Φ as premise and Ψ as conclusion is valid, and V_E codifies the very plausible thought that validity preserves truth.

Deriving the paradox is now straightforward: First, we apply the Gödelian diagonalization lemma to the predicate:

$$\mathsf{Val}(x, \ulcorner\bot\urcorner)$$

to obtain a sentence Π such that:

$$\Pi \leftrightarrow \mathsf{Val}(\ulcorner\Pi\urcorner, \ulcorner\bot\urcorner)$$

is a theorem. We can then, using arithmetic, V_I, and V_E, derive a paradox along lines similar to the reasoning underlying the Curry paradox:

1	$\quad\Pi$	Assumption for application of V_I.
2	$\quad\mathsf{Val}(\ulcorner\Pi\urcorner, \ulcorner\bot\urcorner)$	1, diagonalization.
3	$\quad\Pi \rightarrow \bot$	2, V_E.
4	$\quad\bot$	1, 3, *modus ponens.*
5	$\mathsf{Val}(\ulcorner\Pi\urcorner, \ulcorner\bot\urcorner)$	1 – 4, V_I.
6	$\Pi \rightarrow \bot$	5, V_E.
7	Π	5, diagonalization.
8	$\bot$	6, 7, *modus ponens.*

The problem, however, is that if $\mathsf{Val}(x, y)$ is meant to capture logical validity, then the argument given above is fallacious.

To see why, note that the equivalence between Π and $\mathsf{Val}(\ulcorner\Pi\urcorner, \ulcorner\bot\urcorner)$ is not a logical truth, but rather a truth of PA. Spelling out the reasoning above a bit more carefully, we should have noted that when we apply the Gödelian diagonalization lemma we obtain a Π such that:

$$\Pi \leftrightarrow \mathsf{Val}(\ulcorner\Pi\urcorner, \ulcorner\bot\urcorner)$$

is a theorem of PA, but not a theorem of first-order logic.[3]

[3] This is, of course, not some special property of PA, but applies to any theory that codes up syntax via (resursively) assigning an object (a number, in the case of PA) to each expression in the language in question. Since any such coding is arbitrary (we could have chosen a different coding), any coding-dependent theorem will depend on the particular coding being used – in particular, on which particular code is attached to which particular expression – and first-order logic is, in the relevant sense, insensitive to such distinctions.

In addition, the move from line 1 to 2 is not the only questionable step in the sub-proof terminating with the application of $\mathsf{V_I}$ at line 5. In addition, the application of $\mathsf{V_E}$ at line 3 is also of questionable legitimacy. After all, it is not *obvious* that $\mathsf{V_I}$ and $\mathsf{V_E}$ are, in fact, themselves logically valid rules, rather than being, for example, merely truth-preserving. And if $\mathsf{V_E}$ is not logically valid, then there is no guarantee that this sub-proof shows that line 4 follows from line 1 *as a matter of logic* (which presumably is what is needed in order for us to legitimately apply $\mathsf{V_I}$ at line 5).

Thus, the question is this: If $\mathsf{Val}(x, y)$ is meant to encode *logical validity*, then what resources are, and are not, allowed in a sub-proof of Ψ from Φ if we are to apply $\mathsf{V_I}$ to that sub-proof and conclude that $\mathsf{Val}(\ulcorner\Phi\urcorner, \ulcorner\Psi\urcorner)$ is true? There are three possible resources that we need to consider:

$$\mathsf{L} = \text{Pure First-Order Logic.}$$

$$\mathsf{PA} = \text{Peano Arithmetic.}$$

$$\mathsf{V} = \mathsf{V_I} + \mathsf{V_E}.$$

I take it to be obvious that the resources of pure first-order logic should be allowable in such sub-proofs – after all, if first-order logic doesn't preserve logical validity, then it is not clear that anything does. Additionally, and more substantially, I will assume in what follows that $\mathsf{V_I}$ and $\mathsf{V_E}$ stand or fall together – that is, either both of $\mathsf{V_I}$ and $\mathsf{V_E}$ are allowed in sub-proofs that can be terminated with an application of $\mathsf{V_I}$, or neither are.[4] This leaves us with four possibilities:

- L is logically valid, but $\mathsf{V_I}$, $\mathsf{V_E}$, and PA are not.
- L and PA are logically valid, but $\mathsf{V_I}$ and $\mathsf{V_E}$ are not.
- L, $\mathsf{V_I}$, and $\mathsf{V_E}$ are logically valid, but PA is not.
- L, $\mathsf{V_I}$, $\mathsf{V_E}$, and PA are all logically valid.

Of course, we have already ruled out the second and fourth option, since the axioms and rules of PA are not logically valid. Examining systems that allow the use of PA in such sub-proofs will turn out to be illuminating nevertheless. Given these four distinct possible answers to our question, we obtain four distinct versions of $\mathsf{V_I}$:

[4] Treating these rules separately, considering systems that allow $\mathsf{V_I}$ but not $\mathsf{V_E}$ to be allowed in such sub-proofs (or vice versa), would double the number of cases we need to consider, with no additional philosophical or mathematical insight.

$$\mathsf{V}_{\mathsf{I}}^{\mathsf{L}} : \text{For any formulas } \Phi \text{ and } \Psi : \\ \text{If} : \Phi \vdash_{\mathsf{L}} \Psi \\ \text{Then} : \varnothing \vdash_{\mathsf{L+PA+V}} \mathsf{Val}(\ulcorner \Phi \urcorner, \ulcorner \Psi \urcorner)$$

$$\mathsf{V}_{\mathsf{I}}^{\mathsf{L+PA}} : \text{For any formulas } \Phi \text{ and } \Psi : \\ \text{If} : \Phi \vdash_{\mathsf{L+PA}} \Psi \\ \text{Then} : \varnothing \vdash_{\mathsf{L+PA+V}} \mathsf{Val}(\ulcorner \Phi \urcorner, \ulcorner \Psi \urcorner)$$

$$\mathsf{V}_{\mathsf{I}}^{\mathsf{L+V}} : \text{For any formulas } \Phi \text{ and } \Psi : \\ \text{If} : \Phi \vdash_{\mathsf{L+V}} \Psi \\ \text{Then} : \varnothing \vdash_{\mathsf{L+PA+V}} \mathsf{Val}(\ulcorner \Phi \urcorner, \ulcorner \Psi \urcorner)$$

$$\mathsf{V}_{\mathsf{I}}^{\mathsf{L+PA+V}} : \text{For any formulas } \Phi \text{ and } \Psi : \\ \text{If} : \Phi \vdash_{\mathsf{L+PA+V}} \Psi \\ \text{Then} : \varnothing \vdash_{\mathsf{L+PA+V}} \mathsf{Val}(\ulcorner \Phi \urcorner, \ulcorner \Psi \urcorner)$$

We can sum up these rules as follows:

- $\mathsf{V}_{\mathsf{I}}^{\mathsf{L}}$ states that, if we have a sub-proof of Ψ from Φ that uses only the resources of first-order logic, then we can apply (this version of) V_{I} and conclude that $\mathsf{Val}(\ulcorner \Phi \urcorner, \ulcorner \Psi \urcorner)$ is true.
- $\mathsf{V}_{\mathsf{I}}^{\mathsf{L+PA}}$ states that, if we have a sub-proof of Ψ from Φ that uses only the resources of first-order logic and Peano Arithmetic, then we can apply (this version of) V_{I} and conclude that $\mathsf{Val}(\ulcorner \Phi \urcorner, \ulcorner \Psi \urcorner)$ is true.
- $\mathsf{V}_{\mathsf{I}}^{\mathsf{L+V}}$ states that, if we have a sub-proof of Ψ from Φ that uses only the resources of first-order logic, (this version of) V_{I}, and V_{E}, then we can apply (this version of) V_{I} and conclude that $\mathsf{Val}(\ulcorner \Phi \urcorner, \ulcorner \Psi \urcorner)$ is true.
- $\mathsf{V}_{\mathsf{I}}^{\mathsf{L+PA+V}}$ states that, if we have a sub-proof of Ψ from Φ that uses only the resources of first-order logic, Peano Arithmetic, (this version of) V_{I}, and V_{E}, then we can apply (this version of) V_{I} and conclude that $\mathsf{Val}(\ulcorner \Phi \urcorner, \ulcorner \Psi \urcorner)$ is true.

The following theorems (whose proofs can be found in Cook (2014)) settle the consistency question for the four systems in question:

THEOREM 5.2.1 The system that results from adding $\mathsf{V}_{\mathsf{I}}^{\mathsf{L+PA+V}}$ and V_{E} to PA is inconsistent.

THEOREM 5.2.2 The system that results from adding $\mathsf{V}_{\mathsf{I}}^{\mathsf{L+PA}}$ and V_{E} to PA is consistent.

Table 5.1 *Consistency of variants of* V_I

	PA allowed	PA disallowed
V_I, V_E allowed	Inconsistent	Inconsistent
V_I, V_E disallowed	Consistent	Consistent

COROLLARY 5.2.3 Ketland (2012): The system that results from adding V_I^L and V_E to PA is consistent.

THEOREM 5.2.4 The system that results from adding V_I^{L+V} and V_E to PA is inconsistent.

Thus, whether or not our theory of the logical validity predicate $\mathsf{Val}(x, y)$ is consistent co-varies with whether or not we allow the rules for the logical validity predicate themselves to appear in sub-proofs terminating with an application of V_I – that is, with whether or not we treat the rules for the validity predicate as being logically valid themselves (see Table 5.1). Whether or not we allow arithmetic within such sub-proofs turns out to be completely irrelevant to the *consistency* status of the resulting systems, however, strongly suggesting that the logical status of arithmetic and its use within such sub-proofs is orthogonal to a correct assessment of whether there truly is a paradox of logical validity.

Thus, we have two options: either we can conclude that the addition of a logical validity predicate to PA is paradoxical, in much the same way that the addition of an unrestricted truth predicate to PA is paradoxical, or we can conclude that the rules V_I and V_E are not logically valid, and hence cannot be applied in sub-proofs terminating in an application of V_I. In short, we need to decide whether V_I^{L+V} or V_I^L is the right introduction rule for the logical validity predicate.

The latter option is the correct one, or, at least, so I argued in Cook (2014). To see why, we need merely note that (on many common accounts of the nature of logical validity and logical vocabulary) an expression can only be logical if it is invariant under some class of transformations such as permutations or isomorphisms (see, e.g., Tarski (1966/86), Shapiro (1991), McGee (1996), and Bonnay (2008)).[5] We need not develop a fully fleshed-out account of this idea, nor need we engage with many of the

[5] For our purposes we can ignore the subtle differences between the views just cited.

subtle controversies surrounding how to do that developing, since for our purposes we merely need the following, rather simple case:[6]

> Perm : A first-level binary relation $R(x, y)$ is *permutation invariant* if and only if, for any model $\mathcal{M} = \langle \Delta, I \rangle$, any $\alpha, \beta \in \Delta$, and any permutation $\pi : \Delta \rightarrow \Delta$:
>
> $$\mathcal{M} \vDash R(\alpha, \beta) \text{ iff } \mathcal{M} \vDash R(\pi(\alpha), \pi(\beta))$$

If $\mathsf{V_I}$ and $\mathsf{V_E}$ are logically valid introduction and elimination rules for $\mathsf{Val}(x, y)$, then $\mathsf{Val}(x, y)$ must be a logical operator, and hence must be permutation invariant in the above sense. But it clearly is not:

THEOREM 5.2.5 If $R(x, y)$ is permutation invariant, then, for any model $\mathcal{M}$:

$$\mathcal{M} \vDash (\exists x)(\exists y)((x \neq y \wedge R(x, y)) \rightarrow (\forall z)(\forall w)(z \neq w \rightarrow R(z, w)))$$

Proof Assume that $\mathcal{M} = \langle \Delta, I \rangle$ is model where, for some $a, b \in \Delta$, we have:

$$\mathcal{M} \vDash a \neq b \wedge R(a, b)$$

Let $c, d \in \Delta$ be distinct – that is:

$$\mathcal{M} \vDash c \neq d$$

Then there is a permutation $\pi : \Delta \rightarrow \Delta$ such that $\pi(a) = c$ and $\pi(b) = d$. Since $R(x, y)$ is permutation invariant, this implies:

$$\mathcal{M} \vDash R(c, d)$$

Since c and d were arbitrary, we can conclude that:

$$\mathcal{M} \vDash (\exists x)(\exists y)((x \neq y \wedge R(x, y)) \rightarrow (\forall z)(\forall w)(z \neq w \rightarrow R(z, w)))$$

Theorem 5.2.5 clearly rules out $\mathsf{Val}(x, y)$ as a logical operator. If $\mathsf{Val}(x, y)$ were a logical operator, then it would be trivial:[7]

[6] In $\mathcal{M} = \langle \Delta, I \rangle$, Δ is the domain of the model, and I is the interpretation function.

- Clearly, given the intended meaning of $\mathsf{Val}(x, y)$, we should always have $\mathsf{Val}(\ulcorner\Phi\urcorner, \ulcorner\Phi\urcorner)$.
- Further, there are at least two distinct sentences Φ and Ψ such that $\mathsf{Val}(\ulcorner\Phi\urcorner, \ulcorner\Psi\urcorner)$. But then **Theorem 5.2.5** implies that, for any two distinct sentences Φ and Ψ, $\mathsf{Val}(\ulcorner\Phi\urcorner, \ulcorner\Psi\urcorner)$.

These two facts entail that $\mathsf{Val}(x, y)$ holds of (the codes of) any two sentences (distinct or not).

Thus, it appears as if we must either:

1. Give up the idea that logical validity is formal, and hence give up on the idea that logical operators must be permutation invariant.
2. Give up the idea that the validity predicate Val is a logical operator, and hence give up on the idea that the rules for Val are themselves logically valid.

Since giving up on the formality of logical validity would seem to be giving up on the intended and intuitive notion of logical validity altogether (robbing the claim that $\mathsf{V_I}$ and $\mathsf{V_E}$ are logically valid of most of its interest!), it seems that our only viable option is to abandon the idea that $\mathsf{V_I}$ and $\mathsf{V_E}$ are logically valid. Thus, the proper formulation of $\mathsf{V_I}$ must be its weakest formulation: $\mathsf{V_I^L}$, and as a result the purported paradox is dissolved. Or so it would seem.

5.3 The New Paradox of Validity

There is a possible third way out of the puzzle rehearsed above – we can formulate a different, more lenient criterion of logicality that retains the idea that logicality is intimately tied to permutation invariance yet judges the validity predicate $\mathsf{Val}(y, x)$ (or some similar, and similarly paradox-prone, validity predicate) to be logical.[8]

[7] I am assuming here, as is standard, that we are using a coding where distinct sentences receive distinct codes.

[8] It is worth noting that this move is not unprecedented. For example, Woods (2014) endorses a (different) liberalization of the permutation invariance criterion – one that classifies Hilbert's ϵ operator, Russell's definite description operator, and (some of) the term-forming operators in neo-logicist abstraction principles as logical.

To motivate such an account, we'll begin by considering a new version of the paradox.[9] Assume that we add to the language of Peano Arithmetic (PA) a predicate $\mathsf{VPred}(x, y)$ that holds of the Gödel code $\ulcorner\Phi(x)\urcorner$ of a *unary predicate* $\Phi(x)$ and the Gödel code $\ulcorner\Psi(y)\urcorner$ of a *unary predicate* $\Psi(y)$ (in that order) if and only if the argument whose sole premise is $\Phi(z)$ and whose conclusion is $\Psi(z)$ is logically valid (note the agreement of variable in the final clause). What rules might we expect such an operator to obey?

If we ignore the points made in the previous section, and unreflectively adopt the introduction and elimination rules for $\mathsf{VPred}(x, y)$ analogous to those initially suggested for $\mathsf{Val}(x, y)$:

$$\begin{aligned} \mathsf{VPred}_{\mathsf{I}} : & \text{ For any unary predicates } \Phi(x) \text{ and } \Psi(y) : \\ & \text{ If} : \Phi(z) \vdash \Psi(z) \\ & \text{ Then} : \varnothing \vdash \mathsf{VPred}(\ulcorner\Phi(x)\urcorner, \ulcorner\Psi(y)\urcorner) \\ \mathsf{VPred}_{\mathsf{E}} : & \text{ For any formulas } \Phi(x) \text{ and } \Psi(y) : \\ & \varnothing \vdash \mathsf{VPred}(\ulcorner\Phi(x)\urcorner, \ulcorner\Psi(y)\urcorner) \rightarrow (\Phi(z) \rightarrow \Psi(z)) \end{aligned}$$

then we can, along similar lines, once again derive a paradox. Let $f_{\neq}(y)$ be the (recursive) function that maps each number n to $\ulcorner w \neq w\urcorner$ (for some arbitrary variable w). Now, apply the predicate version of the Gödelian diagonalization lemma to the predicate:[10]

$$\mathsf{VPred}(x, f_{\neq}(y))$$

to obtain a predicate $\Pi^*(y)$ such that:

$$(\forall y)(\Pi^*(y) \leftrightarrow \mathsf{VPred}(\ulcorner\Pi^*(x)\urcorner, f_{\neq}(y)))$$

is a theorem. We can then, using arithmetic, $\mathsf{VPred}_{\mathsf{I}}$, and $\mathsf{VPred}_{\mathsf{E}}$, derive the requisite contradiction:[11]

[9] As we shall see in Section 5.4 below, we can apply the strategy developed here to the original paradox formulated in terms of $\mathsf{Val}(x, y)$. Motivating the strategy, however, is a bit more natural if we use the new formulation of the puzzle given here.

[10] The predicate version of the Gödelian diagonalization lemma states that, for any binary predicate $\Phi(x, y)$ there is a unary predicate $\Psi(x)$ such that:

$$(\forall y)(\Psi(y) \leftrightarrow \Phi(\ulcorner\Psi(x)\urcorner, y))$$

is a theorem of PA.

[11] Note that:

$$\vdash_{\mathsf{PA}} (\forall z)(f_{\neq}(z) = \ulcorner w \neq w\urcorner)$$

1	$\Pi^*(z)$	Assumption for application of $\mathsf{V}\mathit{Pred}_{\mathsf{I}}$.
2	$\mathsf{VPred}(\ulcorner\Pi^*(x)\urcorner, f_{\neq}(z))$	1, diagonalization.
3	$\mathsf{VPred}(\ulcorner\Pi^*(x)\urcorner, \ulcorner w \neq w\urcorner)$	2, df. of $f_{\neq}$.
4	$\Pi^*(z) \rightarrow z \neq z$	3, $\mathsf{VPred}_{\mathsf{E}}$.
5	$z \neq z$	1, 4, *modus ponens*.
6	$\mathsf{VPred}(\ulcorner\Pi^*(x)\urcorner, \ulcorner w \neq w\urcorner)$	1 – 5, $\mathsf{VPred}_{\mathsf{I}}$.
7	$\Pi^*(z) \rightarrow z \neq z$	6, $\mathsf{VPred}_{\mathsf{E}}$.
8	$\mathsf{VPred}(\ulcorner\Pi^*(x)\urcorner, f_{\neq}(z))$	6, df. of $f_{\neq}$.
9	$\Pi^*(z)$	8, diagonalization.
10	$z \neq z$	7, 9, *modus ponens*.

Thus, our new predicate version of the validity predicate $\mathsf{VPred}(x, y)$ allows us to derive a contradiction along lines similar to those used in the original Beall-Murzi paradox constructed in terms of $\mathsf{Val}(x, y)$.[12] So what? After all, presumably this predicate fails the permutation invariance test codified in **Perm** above, and hence is not a genuine logical operator. And if it is not a logical operator, then the rules $\mathsf{VPred}_{\mathsf{I}}$ and $\mathsf{VPred}_{\mathsf{E}}$ (in full generality) are not logical introduction and elimination rules (respectively), and hence we have no reason to think they are *logically* valid rules at all. As a result, along lines similar to those outlined in the previous section, we should object to the application of $\mathsf{VPred}_{\mathsf{I}}$ at line 6 in the proof just given, since the sub-proof upon which this inference depends (lines 1–5) contains inferences that are not purely logical. Aren't we right back where we started?

In one sense, yes. If the right notion of invariance to apply, in order to determine whether $\mathsf{VPred}(x, y)$ is a logical operator, is the notion codified as **Perm** above, then we can easily show that, if $\mathsf{VPred}(x, y)$ were a logical operator, then it must be trivial (along lines similar to the argument given for $\mathsf{Val}(x, y)$ in the previous section – details are left to the reader). But – and this is the rub – there might be some wiggle room with regard to whether **Perm** is the right notion of permutation invariance to apply to $\mathsf{VPred}(x, y)$ in the first place.

Strictly speaking, $\mathsf{VPred}(x, y)$ is a first-level predicate that holds of objects – in particular, of numbers that serve as codes of predicates. But,

[12] Further, the trick used in Cook (2014) to eliminate the dependence on arithmetic in the sub-proof from lines 1 to 5 can be applied here as well. Details are left to the reader.

putting things a bit loosely, $\mathsf{VPred}(x, y)$ is intended to codify a relation that holds, not between numbers understood as codes of predicates, but between the predicates so coded or, perhaps better, between the concepts denoted by those predicates.[13] If this is right, then one might be forgiven for thinking that applying a notion of permutation invariance that operates on the codes of predicates, rather than some modified notion that applies directly to the concepts those predicates denote, is a mistake. To put the point bluntly, $\mathsf{VPred}(x, y)$ is not meant to capture a logically salient relation that holds between natural numbers, but is instead intended to capture a logical relation that holds between the concepts denoted by the predicates whose codes occur in true instances of $\mathsf{VPred}(x, y)$.[14] Hence, the appropriate test of logicality to apply to $\mathsf{VPred}(x, y)$ should, in some sense, apply not to the codes, but to the concepts.

What would such a modified notion of permutation invariance look like? Here is one way we might formulate such a criterion:

> PredPerm: A first-level binary relation $R(x, y)$ in $\mathcal{L}$ is *predicate permutation invariant* if and only if there is a second-level binary relation $\mathsf{R}(X, Y)$ (holding of unary concepts) such that:
>
> 1. $\mathsf{R}(X, Y)$ is definable in some $\mathcal{L}^*$ such that $\mathcal{L} \subseteq \mathcal{L}^*$ and $\mathcal{L}^*$ contains the logical necessity operator $\Box$.
> 2. $\mathsf{R}(X, Y)$ is permutation invariant.[15]
> 3. For any $\mathcal{L}^*$ model $\mathcal{M}$ (which, again, is both second-order and intensional) and any predicates $\Phi(x), \Psi(x) \in \mathcal{L}$:
>
> $$\mathcal{M} \vDash (\forall X)(\forall Y)[(\Box(\forall z)(X(z) \leftrightarrow \Phi(z))) \wedge (\Box(\forall x)(Y(z) \leftrightarrow \Psi(z))) \\ \rightarrow \Box(R(\ulcorner \Phi(x) \urcorner, \ulcorner \Psi(x) \urcorner) \leftrightarrow \mathsf{R}(X, Y))]$$

[13] I use "concept" in the Fregean sense. Readers uncomfortable with this usage are free to uniformly substitute "property" in what follows. Importantly, on this conception concepts are instantiated intensionally, in terms of their extensions at all possible worlds. In symbols:

$$(\forall X)(\forall Y)[X = Y \leftrightarrow \Box(\forall z)(X(z) \leftrightarrow Y(z))]$$

[14] And it's a good thing that $\mathsf{VPred}(x, y)$ isn't intended to capture some special relationship between particular numbers themselves, since the relevant relationship changes with any change in the particular coding we use!

[15] A second-level *intensional* relation $\mathsf{R}(X, Y)$ (where X and Y are unary concept variables) is permutation invariant if and only if, for any model $\mathcal{M}$ of $\mathcal{L}^*$ (which, recall, includes $\Box$, interpreted as the logical necessity operator):

$$\mathcal{M} \vDash (\forall \pi)(\mathsf{Perm}(\pi) \rightarrow \Box(\forall X)(\forall Y)(R(X, Y) \leftrightarrow R(\pi(X), \pi(Y))))$$

where $\mathsf{Perm}(\pi)$ abbreviates the (purely logical, assuming that $\Box$ is logical) claim that π is a permutation on the domain of each world, and, for any X, $\pi(X)$ is the concept whose extension at each world is the image of the extension of X under π at that world.

More informally: A first-level binary relation $R(x, y)$ in a language $\mathcal{L}$ is predicate permutation invariant if and only if we can expand $\mathcal{L}$ in such a way that the new language $\mathcal{L}^*$ contains a second-level relation $\mathsf{R}(X, Y)$ that is permutation invariant (in a straightforward extension of the usual sense – see footnote 15), and where it is logically necessary that $\mathsf{R}(X, Y)$ holds of the concepts denoted by two predicates if and only if $R(x, y)$ holds of (the Gödel codes of) those predicates (in the same order).[16] Note that, since we are individuating concepts intensionally, the necessity operators are required in the antecedent to insure the concepts in question are in fact the concepts denoted by the relevant predicates (i.e., are *necessarily* pairwise co-extensional with the predicates), rather than merely being co-extensional with the predicates in the actual world (and similar comments apply to the consequent). Note further that determining whether a relation definable in a first-order language is predicate permutation invariant requires a detour through (at least) a corresponding second-order language containing a logical necessity operator $\Box$.[17]

So far, so good. But is $\mathsf{VPred}(x, y)$ predicate permutation invariant? In order to answer this question, we will have to be a bit more specific about which particular account of permutation invariance we adopt. In particular, we will need to determine whether (and which) modal operators are permutation invariant. Fortunately for us, there already exist standard accounts of how to extend the notion of permutation invariance to modal operators – via permuting the domains of possible worlds, see van Benthem (1989) and Scott (1970) – and according to these accounts the modal operators in **S5** are permutation invariant, and hence logical. As a result, **S5** can be coherently (and fruitfully) understood as the logic of logical truth ($\Box$) and logical consistency ($\Diamond$) – see also Burgess (1999). The logical operators in other modal logics – in particular, the modal operators in **S4** and the provability logics **GL** and **GLS** – fail to be permutation invariant in the relevant sense, however.[18]

Before giving the argument that $\mathsf{VPred}(x, y)$ is predicate permutation invariant, it is worth noting that, if we read the occurrences of $\Box$ in

[16] Note that we are assuming that the same coding of formulas is in play at each possible world in the model. Thus, although coding is context dependent, it is not intensional. Thanks are owed to Gil Sagi for helping to clarify this issue.

[17] See Shapiro (1991) for a good treatment of classical second-order logic.

[18] In order to show that $\mathsf{VPred}(x, y)$ is predicate permutation invariant, we need to use the logical truth *operator* ($\Box$), rather than the logical truth predicate, since the former, but not the latter, is logical on the traditional account of logicality-as-permutation-invariance (or, a bit more carefully, $\Box$ is logical according to the extension of the traditional account to modal operators discussed above). The logical truth operator is predicate permutation invariant, however – see below.

the third clause of our definition of predicate permutation invariance as instances of the logical necessity operator, then this criterion is itself expressible in terms of purely logical resources (on the traditional reading of "purely logical").

The argument that $\mathsf{VPred}(x,y)$ is predicate permutation invariant is now straightforward: Let $\mathcal{L}$ be the language of Peano Arithmetic supplemented with $\mathsf{VPred}(x,y)$. Now consider $\mathcal{L}^*$, where $\mathcal{L}^*$ is $\mathcal{L}$ supplemented with second-order quantification and a unary modal operator $\Box$ interpreted as the ($\mathsf{S5}$) logical truth operator. Then the second-level relation witnessing the predicate permutation invariance of $\mathsf{VPred}(x,y)$ is just:

$$\Box(\forall z)(X(z) \rightarrow Y(z))$$

Thus, if predicate permutation invariance is sufficient for the logicality of expressions that codify relations holding between concepts indirectly via Gödel coding (and if we accept the standard means for extending the standard account of permutation invariance to modal notions), then $\mathsf{VPred}(x,y)$ is logical. But, if $\mathsf{VPred}(x,y)$ is logical, then we have no grounds for objecting to the use of the elimination rule $\mathsf{VPred_E}$ within the sub-proof terminating in an application of the introduction rule $\mathsf{VPred_I}$. As a result, we have a genuine paradox of logical validity after all.

5.4 Some Brief Notes on the General Account

The previous section sketches a strategy for arguing that (a version of) the paradox of validity is, indeed, a paradox of logical validity. The claim that the paradox applies to the (or a) notion of *logical* validity is only as plausible as the claim that predicate permutation invariant first-level relations are in fact logical. As I already noted in the introduction, I am not going to attempt to defend that claim here. Instead, however, we will conclude with some general observations regarding what such an account might look like.

First, it is worth noting that the account can easily be generalized to provide an account of logicality in terms of permutation invariance – one somewhat more lenient than the standard account(s). We will not try to flesh out such account in full detail – I will leave that task to anyone (if there is anyone) who thinks the account might be *correct*. But a quick sketch can be given. We can generalize the account given for binary first-level relations as follows:

> ExpressionPerm: A first-level n-ary relation $R(x_1, \dots x_n)$ in $\mathcal{L}$ is *expression permutation invariant* if and only if there is a function $f(x)$

(in the metatheory) that maps each variable x_i ($1 \leq i \leq n$) to a expression-type (in $\mathcal{L}$), and there is a higher-order n-ary relation $\mathsf{R}(\alpha^{f(x_1)}, \dots \alpha^{f(x_n)})$ (where each $\alpha_i^{f(x_i)}$ ($1 \leq i \leq n$) is a variable ranging over the (intensional) argument type denoted by an expression of type $f(x_1)$) such that:

1. $\mathsf{R}(\alpha_1^{f(x_1)}, \dots \alpha_n^{f(x_n)})$ is definable in some $\mathcal{L}^*$ such that $\mathcal{L} \subseteq \mathcal{L}^*$ and $\mathcal{L}^*$ contains the logically necessary operator $\Box$.
2. $\mathsf{R}(\alpha_1^{f(x_1)}, \dots \alpha_n^{f(x_n)})$ is permutation invariant (on some appropriate generalization of permutation invariance to intentional notions – see footnote 15).
3. For any $\mathcal{L}^*$ model $\mathcal{M}$ and any expressions $\eta_1, \dots \eta_n \in \mathcal{L}$ where, for each i ($1 \leq i \leq n$), η_i is of type $f(x_i)$:[19]

$$\mathcal{M} \vDash (\forall \alpha^{f(x_1)} \dots (\forall \alpha^{f(x_n)})[(\Box(\alpha^{f(x_1)} \equiv_{f(x_i)} \eta_1) \wedge \dots \Box(\alpha^{f(x_n)} \equiv_{f(x_n)} \eta_n)) \rightarrow \Box(R(\ulcorner \eta_1 \urcorner, \dots \ulcorner \eta_n \urcorner) \leftrightarrow \mathsf{R}(\alpha_1^{f(x_1)}, \dots \alpha_n^{f(x_n)}))]$$

In short, the intuitive idea is that an expression in $\mathcal{L}$, where the intention, presumably, is that this expression holds of codes of $\mathcal{L}$ expressions, is expression permutation invariant if and only if it expresses, indirectly via Gödel coding, a relationship that can also be expressed directly in terms of a permutation invariant expression (where the latter might involve higher-order resources as well as possibly other recourses not present in $\mathcal{L}$, such as the logical necessity operator $\Box$).

We can now give a straightforward statement of the new notion of logicality that results:[20]

> **Logicality:** An expression Φ is *logical* if and only if it is either permutation invariant or expression permutation invariant.

The first thing to notice is that, if we allow propositional variables (i.e., 0-ary concept variables) in the second-order language that can be used in the expanded language $\mathcal{L}^*$ (and there is no reason why we shouldn't), then this new account of logicality also judges as logical the original

[19] I use $\equiv_{f(x_n)}$ here as a general place-holder for the relevant equivalence relation expressing "sameness" for the type in question: identity for terms, coextensionality for predicates, etc.

[20] Note that every formula that is permutation invariant in the traditional sense will also be expression permutation invariant (assuming we allow n^{th}-order resources for arbitrary $n \in \mathbb{N}$), since we can obtain the required witnessing formula merely by replacing the open objectual variables in the expression with higher-order variables ranging over the appropriate types.

sentential validity predicate $\mathsf{Val}(x, y)$ used in the original Beall-Murzi version of the paradox.[21] The witnessing formula is:

$$\Box(X \rightarrow Y)$$

where the modal operator is again necessity in $\mathsf{S5}$, understood as logical necessity.[22]

In addition, this account provides, amongst other things, one way of drawing a distinction between formulas of Peano Arithmetic (or any extension of any language that contains the resources for the standard Gödelian constructions) that are, in some sense, *about* the natural numbers, and those sentences of Peano Arithmetic that are, although formulated in terms of arithmetical notions, nevertheless in some sense *not* directly *about* the numbers – the expression permutation invariant formulas.[23]

That being said, it is worth noting that the present account draws this border rather differently from previous accounts – such as the account found in Isaacson (1987, 1992, and 1994) – that attempt to divide those sentences that are *about* the natural numbers from those, such as the undecidable Gödel sentence, that involve (on such accounts) non-arithmetical syntactic or semantic notions such as provability (in this case) or truth. Such accounts typically draw the line (or attempt to do so) in such a way that independent sentences in general, and the Gödel sentence in particular, end up on the non-arithmetical side. On the account sketched (but not endorsed) here, however, the Gödel sentence:

$$G \leftrightarrow \mathsf{Bew}(\ulcorner G \urcorner)$$

does not end up being non-arithmetical (i.e., logical, in the extended sense of logical developed here), since the provability predicate $\mathsf{Bew}(x)$, unlike $\mathsf{Val}(x, y)$ or $\mathsf{Val}_{\mathsf{Pred}}(x, y)$, is not expression permutation invariant.

[21] Importantly, propositional variables must be understood here intensionally – that is, as something akin to functions from worlds to truth values.

[22] Further, the logical truth operator is expression permutation invariant, since witnessed by:

$$\Box(X).$$

[23] Note that the arithmetically definable predicate holding between two formulas if and only if the second is a first-order consequence (i.e., a consequence in the system not containing $\mathsf{Val}_{\mathsf{I}}$ or $\mathsf{Val}_{\mathsf{E}}$) of the first:

$$\mathsf{Val}_{\mathsf{FOL}}(x, y)$$

is a *purely arithmetical* expression permutation invariant formula since the witnessing statement is definable in second-order logic (with propositional variables) plus the logical necessity operator $\Box$. Hence, it is a purely arithmetical formula that is, in the informal sense described above, not directly *about* the numbers.

At first glance, it might seem like we could argue that the provability predicate $\mathsf{Bew}(x)$ was logical on the account being considered here, since we might think it is witnessed in the relevant sense by a higher-order predicate of the form:

$$\Box_{\mathsf{G}}(X)$$

where $\Box_{\mathsf{G}}$ is the necessity operator from a provability logic like **GL** or **GLS** (see Boolos (1993) for details). But, although this formula gets the truth conditions for a witnessing formula of the relevant sort correct (i.e., it satisfies the third clause of the definition of expression permutation invariance), we need only recall the fact – already mentioned in the previous section – that these modal operators (and hence this complex construction built from one or another of these modal operators) are not permutation invariant. As a result, the provability predicate is not logical, even on the more lenient account of logicality being considered here.[24]

One concrete consequence of this observation is that the distinction presented here, between those expressions in the language of arithmetic that are expression permutation invariant, and hence (in some sense) not directly about the numbers, and those expressions that are not expression permutation invariant, and hence (in some sense) are directly about the numbers, will be of little help in arguing that particular undecidable sentences, such as the famous Paris-Harrington result, are (unlike the original, provability-involving Gödel sentence) in some sense purely arithmetical. Even if the Paris-Harrington sentence does fail to be expression permutation invariant, and hence is directly about the numbers in that sense, this

[24] A bit more technical detail: For the provability predicate $\mathsf{Bew}(x) \in \mathcal{L}_{\mathsf{PA}}$ to be expression permutation invariant, there must be some language $\mathcal{L}^*$ that extends $\mathcal{L}_{\mathsf{PA}}$ and expression $\Phi(X) \in \mathcal{L}^*$ such that $\mathsf{Bew}(x)$ and $\Phi(X)$ agree on all sentences in $\mathcal{L}_{\mathsf{PA}}$ (but not necessarily on all sentences in $\mathcal{L}^*$) in the appropriate way – that is, for any sentence $\Psi \in \mathcal{L}_{\mathsf{PA}}$

$$\mathsf{Bew}(\ulcorner \Psi \urcorner) \leftrightarrow \Phi(\Psi)$$

In addition, $\Phi(X)$ must be permutation invariant in the standard sense. One might wonder whether, despite the fact that provability operator $\Box_{\mathsf{G}}$ fails to be permutation invariant, we might nevertheless find some other expression, in some expressively more powerful language, that could nevertheless witness the expression permutation invariance of $\mathsf{Bew}(x)$. The answer is "no", since the failure of permutation invariance for $\Box_{\mathsf{G}}$ can be witnessed using sentences from $\mathcal{L}_{\mathsf{PA}}$, hence, any formula in any language that agrees with $\mathsf{Bew}(x)$ in the relevant sense – and hence simply agrees with $\Box_{\mathsf{G}}$ – on the sentences of $\mathcal{L}_{\mathsf{PA}}$ will also fail to be permutation invariant: this failure can be witnessed by the same sentences from $\mathcal{L}_{\mathsf{PA}}$.

It might turn out, however, that a potential witnessing formula that "agrees" with an expression in $\mathcal{L}_{\mathsf{PA}}$ might fail to be permutation invariant, but nevertheless be permutation invariant "locally" – that is, when restricted to the appropriate expressions in $\mathcal{L}_{\mathsf{PA}}$. In such cases, there is the possibility of a different formula being a genuine witnessing formula.

would not distinguish it from the Gödel sentence, which also fails to be expression permutation invariant.

Thus, despite the fact that the account of logicality being considered here is much more lenient than any account that requires an expression to itself be permutation invariant if it is to be logical (rather than either being permutation invariant or corresponding to a witness statement in a higher-order language that is permutation invariant), it nevertheless, despite possible first impressions, does not allow in *any* expression that seems to express something *about* sentences, terms, predicates, or other expressions. The distinction made by the account being considered here is not the distinction between those sentences that are simply about the domain of quantification (including the natural numbers, or whatever objects are used to code up expressions) and those sentences that are in some sense only indirectly about the domain and are instead best understood as being about the expressions coded up by elements of that domain. Instead, the account sketched above only allows those sentences that are, in some sense, about expressions to count as logical if they correspond to permutation invariant notions (in the traditional sense of permutation invariance) in a principled manner. Of course, corresponding to a logical notion and actually being a logical notion are different things. Nevertheless, this heads off at least one easy objection to the account being considered here – that the distinction is too permissive in that it allows in *any* expression that is, in some sense, about expressions.[25]

[25] Thanks are owed to Jack Woods, Gil Sagi, and the audience at the 2018 Semantic Conception of Logic Workshop at the Munich Center for Mathematical Philosophy for helpful feedback on this material.

PART II

Critiques and Applications of the Semantic Approach

CHAPTER 6

Semantic Perspectives in Logic

Johan van Benthem

6.1 Introduction

The erudite and still highly relevant paper Beth (1963) presents logic as consisting of three strands, historically entangled and complementary: linguistic definability (semantics, if you will), proof, and algorithm (i.e., computation). With this perspective in mind, in this survey paper, I weave a story demonstrating the broad scope of semantic themes in logic. The starting point is model-theoretic invariance, but gradually, other key themes come in, such as consequence relations, preservation theorems, and the role of games and agency. There will be no new technical results, and just a minimum of details: these are found in the references. My emphasis is on broad integrating themes in logic, along a path that the reader may find unusual, with occasionally different vistas from those found in standard texts.

6.2 Semantic Invariance and Definability

A common view, reflecting the standard textbook order of presentation, is that a logical language is uninterpreted syntax that stands in need of semantic interpretation in order to yield meaningful assertions. In that sense, syntax is prior to semantics. But historical reality may well have been the other way around. What came first in evolution was meaningful communication and description of reality, and human languages evolved to provide a vehicle for this. And logical languages arose out of reflection on human language.

I thank the editors Gil Sagi and Jack Woods for their comments.

6.2.1 Invariants, Languages, and Logics

The world-to-language perspective goes back to the influential view of Helmholtz 1878. Reality has stable structure, and its patterns reveal themselves to the human observer as *invariants* under suitable *transformations*. In particular, Helmholtz thought that the basic geometric notions in daily life and in mathematics reveal themselves as invariants for natural transformations in visual perspective arising from basic human motion, viz. walking in a straight line and turning around. In mathematics, going from observer movements to translations and rotations of geometrical space, or to more abstract transformations in algebra, this is reflected in Klein's 'Erlanger Program': a mathematical theory needs to start from structures plus transformations setting its 'invariance level'.

The next step is this: invariant structure is important, so a *language* will emerge to express and communicate information about it. A perceptive exposition is in Weyl (1963), who observes that mathematical languages define invariants for matching transformations over relevant structures, while also drawing attention to what he calls the much harder converse question whether, given those transformations, the language is rich enough to define all invariants. By now, invariance thinking is ubiquitous, in physics, computer science, ecology, psychology (Suppes 2002), and philosophy (Barwise and Perry 1983).

Alternatives There are also other views of how language may have arisen, more in terms of human abilities to pick up information from the world, or the process structure of acts of communication. We will encounter such views later on in this article.

Invariance in logic Invariance has been around in logic for a long time, with invariance for isomorphisms acting as a constraint on properly 'logical' notions (Mostowski 1957, Lindström 1966). It became a well-known philosophical concern only as late as the 1980s, when Tarski 1983a and others independently emphasized the general power of this idea. Some authors even think that invariance, suitably conceived, is all there is to 'logicality' (Sher 1991). It is not my aim to recount this foundational debate. van Benthem (2002) discusses strengths and weaknesses of invariance as a criterion for logicality, and surveys literature, and alternative approaches. Here, I will start with links between logical languages and invariance, broadening the canvas to include other themes as we go along.

Permutation invariance Much of the philosophical and linguistic literature chooses for its transformations *permutations* π of some fixed domain D of objects. For instance, the identity relation is invariant for permutations, as $x = y$ iff $\pi(x) = \pi(y)$. Together with non-identity, the universal relation and the empty relation, we get the only four permutation-invariant relations between objects. In other linguistic categories, say, the quantifier "some" is permutation-invariant, since some*(A)* holds for a set of objects A (that is, A is non-empty) iff some($\pi[A]$) holds for any permutation image $\pi[A]$ of the set A. The only thing that matters here is the cardinality of the set A, and this also holds for many other quantifiers, such as "all", "one", or "most": after all, these are expressions of 'quantity' only.

Permutation invariants occur in many different kinds of expression in natural language, when we match linguistic categories with a type system in Montague's style, with base domains of objects and truth values, and operations of product and implication. Läuchli (1970) proved the intriguing result that the types whose domains always contain permutation-invariant objects correspond precisely, when read as propositional formulas with conjunction and implication, to the validities of intuitionistic logic. For a discussion of permutation invariance in finite type theory, see van Benthem (1989), occurring in a seminal issue of the *Notre Dame Journal of Formal Logic* that collected many approaches to the nature of logical constants co-existing at the time: semantic, but also proof-theoretic.[1]

Isomorphisms and automorphisms A more general view, in line with the cited history, is that of invariance between any two structures. Permutations are a special case of *bijections* between structures: these preserve only identity and non-identity of objects, measuring cardinality. But there are other basic connections between structures, such as *isomorphisms* ('automorphisms', when inside the same structure), bijections that also preserve all relevant atomic predicates and operations,[2] – as well as other less demanding notions of similarity. The cross-model perspective on invariance and definability is standard in model theory, and it is how we will mostly phrase things henceforth.

[1] Further results, including connections to notions of invariance appropriate to the lambda calculus due to Statman (1982) and Plotkin (1980), can be found in van Benthem (1991).

[2] Eventually, the functional character of bijections or isomorphisms is not all-important; any suitably defined similarity relation between models can induce invariance.

We now continue with the theme of general isomorphism invariance in logic. For a start, the basic system of first-order logic satisfies the following property:

Fact If F is an isomorphism from a model $\boldsymbol{M}$ to model $\boldsymbol{N}$, then, for any first-order formula ϕ and assignment of objects $\boldsymbol{d}$ to the free variables $\boldsymbol{x}$ of ϕ, $\boldsymbol{M}, \boldsymbol{d} \models \phi$ iff $\boldsymbol{N}, F(\boldsymbol{d}) \models \phi$.

Many basic results about definability in first-order logic involve invariances of this sort. For instance, Beth's Definability Theorem can be viewed in this light – but for present purposes, we rather state Svenonius' Theorem, in a version from van Benthem (1982). This result gives a precise sense, in first-order logic, to the invariance-related intuition that being able to fix denotations up to isomorphism leads to definability.

Theorem A predicate P is fixed up to isomorphism by theory $T(P, \boldsymbol{Q})$, meaning that, in all models for $T(P, \boldsymbol{Q})$, any $\boldsymbol{Q}$-automorphism is automatically a P-automorphism, iff P is explicitly definable in T up to finite disjunction: $\vee_i \forall \boldsymbol{x}(P\boldsymbol{x} \leftrightarrow \delta_i(\boldsymbol{Q}, \boldsymbol{x}))$ is a semantic consequence of T for some finite set of formulas $\delta_i(\boldsymbol{Q}, \boldsymbol{x})$).

We have concentrated on first-order logic here, since it is still a hothouse for developing logical ideas, but isomorphism invariance is so widespread in logic that it is a defining characteristic of logical systems in Abstract Model Theory (Barwise and Feferman 1985).

6.2.2 Combining Invariance and Inference

Isomorphism invariance is an abstract criterion, but it gets more bite when combined with other phenomena. For an example, we return to bijection invariance, and consider binary quantifiers with their usual pattern of occurrence in natural language, the type

$Q\,A\,B$ standing for binary relations between sets of objects.

This setting was studied in van Benthem (1984), over finite domains of objects. On any given domain of size n, isomorphism invariance tells us that the denotation of the quantifier is fixed by the set of 4-tuples (n_1, n_2, n_3, n_4) with $n_1 + n_2 + n_3 + n_4 = n$, where the four numbers refer to the cardinalities of the zones $A \cap B, A \cap -B, B \cap -A, -B \cap -A$.[3]

[3] This reduces the size of the set of all generalized quantifiers on n objects by an exponential.

Next, we combine this with one more widely accepted semantic restriction, satisfied by all the usual quantifiers studied in logic and linguistics:

Conservativity $\quad Q\,A\,B$ *iff* $Q\,A\,(B \cap A)$

Conservativity makes the A-domain of a quantifier paramount, as setting the scene of objects that the statement is about. This further restriction leads to a view of permutation-invariant quantifiers as sets of pairs of numbers (adding up to the size of A), visualized geometrically as a set of points $(|A \cap B|, |A - B|)$ in the so-called 'Tree of Numbers'

(0,0)

(1,0) (0,1)

(2,0) (1,1) (0,2)

...

Conservativity is a basic Boolean property, and what it shows is how quantifiers display inferential behavior from the start. Many quantifiers in natural language satisfy even further Boolean inference properties such as the following:

Monotonicity $\quad Q\,A\,B,\ B \subseteq B'$ *imply* $Q\,A\,B'$

This property may be called 'upward right'. In total, there are four basic monotonicity properties: right or left, upward or downward. One of each occurs in the Square of Opposition: 'all', 'some', 'no', 'not all'. These are doubly monotone in the sense of supporting monotonicity inferences in both arguments, making them 'inference-rich'.

In the Tree of Numbers, monotonicity properties acquire a direct geometrical meaning. For instance, upward right monotonicity means that on horizontal rows in the tree, once a position is accepted, so is everything to its right, while upward left monotonicity says that once a point is accepted, so is the whole downward subtree generated from it.

Now we can classify all possible quantifiers that are permutation-invariant, conservative, and that support a rich set of inferences. Here is a sample result (van Benthem 1986a):

Theorem All doubly monotone isomorphism invariant quantifiers are first-order definable.

These first-order quantifiers exhibit a geometrical pattern of finite unions of convex sets, as can be seen from the above facts, which typically differs from the tree pattern for "most AB" ($|A \cap B| > |A - B|$) whose boundary follows a zigzag line through the middle of the tree.

The motivation for results like this in the 1980s was the question to which extent natural language can define all 'natural' or 'useful' counting expressions. Part of this utility is richness in inferences, where monotonicity reflects a sort of stability: the quantifier still holds when we make changes in one or both of its arguments. The entanglement of invariance with inference is a theme to which we shall return several times in what follows.

6.2.3 Potential Isomorphism

However, isomorphism is not the only game in town, and there are alternatives, even in the heartland of logic. First-order logic is also invariant for a much less demanding invariance called potential isomorphism. This is a family $\boldsymbol{F}$ of finite partial isomorphisms F between two models satisfying the Back and Forth properties, saying that, given any object a in one of the models, there is an object b in the other model such that the extended function $F \cup (a, b)$ also belong to the family $\boldsymbol{F}$. Isomorphisms induce potential isomorphisms by taking all finite submaps, but the converse does not hold. van Benthem and Bonnay (2008) identify the abstract form of the Back and Forth properties as a natural diagrammatic form of "commutation of an invariance relation with object expansion".

Theorem Potential isomorphism is the smallest relation between models that commutes with object expansions.

They then prove a very general result, whose technical details do not matter here, that for any binary relation E and equivalence relation S over some class of objects, S commutes with E iff the inverse E^{-1} preserves S-invariance. As a consequence, potential isomorphism is the smallest similarity relation S between models that respects truth values of atoms while object projection is S-invariant. This analysis applies to modal logic and bisimulation, a topic introduced below, where these abstract formulations become more concrete.

But the fit is still not precise: invariance for potential isomorphism also holds for strong extensions of the first-order language. It only becomes a precise match for *infinitary first-order logic* L_ω^∞, which has conjunctions and disjunctions over arbitrary sets of formulas.[4]

[4] The proof uses some basic model-theoretic definability techniques involving 'Scott sentences' describing types of objects occurring in models up to any given ordinal depth of recursion.

Theorem Two models have a potential isomorphism between them iff they satisfy the same sentences of L^{∞}_{ω}.

For more on the foundational importance of potential isomorphism, as a set-theoretically 'absolute' version of isomorphism, compare with Barwise and Feferman 1985.

6.2.4 *Fixed-Point Logic and Computation*

Another important extension of first-order logic, incomparable in that it can define well-foundedness of binary orders (something that L^{∞}_{ω} cannot), is first-order fixed-point logic *LFP*(*FO*), having operators for defining smallest fixed points for formulas

$\mu P, x \cdot \phi(P, Q, x)$ where P occurs only positively in ϕ,

In any model $\boldsymbol{M}$, $\mu P, x \cdot \phi(P, Q, x)$ defines the smallest predicate P of an arity indicated by the tuple of variables $\boldsymbol{x}$, such that the equivalence $Pd \leftrightarrow \phi(P, Q, d)$ holds for all tuples of objects $\boldsymbol{d}$ in $\boldsymbol{M}$. Such smallest predicates exist by the Tarski-Knaster theorem on fixed points for monotonic maps on complete partial orders, compare with Ebbinghaus and Flum 2005. Likewise, this logic has definitions $\nu P, x \cdot \phi(P, Q, x)$ for greatest fixed points.

This system is a natural extension of first-order logic, with still countable syntax, that encodes the fundamental theory of induction and recursion, the staples of computation. For instance, *LFP*(*FO*) defines transitive closure, as well as the mentioned well-foundedness of orderings, that supports reasoning by induction. It can also define semantics for programming and computation, where it provides an interesting abstraction. Fixed-point logic makes no assumptions about the data one computes over, disentangling the mix of recursion and coding in natural numbers that is characteristic of Recursion Theory.

One might zoom in more closely on fixed-point logic by tightening invariance for potential isomorphism. But this may also be a point where notions beyond invariance come into play. As mentioned, induction and recursion are basic structures of computation, the third ingredient in Beth's historical view of logic that we endorsed, and fixed-point formulas can be seen as recipes for algorithms. What is characteristic of systems like *LFP*(*FO*) may be the interplay of two factors: semantic invariance, and algorithmic structure with intuitions of its own.[5] The computational aspect of logic ties in well with our earlier concerns.

6.2.5 Discussion

More permissive views of logicality In the above setting, we thought of the logical constants as invariants for bijections: rough isomorphisms respecting only identity of objects. However, invariance for *L*-isomorphisms suggests a more liberal view that is not all-or-nothing. We allow the atomic predicates in the language *L* as parameters (invariant by definition), and we ask which complex predicates are then also invariant, or perhaps better: which constructions (sometimes called 'logical glue') maintain invariance. Then, even in the realm of logical expressions, some parameters in *L* may have a special status.

Consider 'mass quantifiers' in natural language, such as "all the wine", "some wine", "most wine". Now there is no discrete base domain, we are rather in a mereological setting of continuous objects (say, bits of water) ordered by inclusion. In this case, the definition of the quantifiers may refer essentially to this inclusion structure, not just to identity of objects. Thus, mass quantifiers fail the above test of invariance under bijections. But they are still natural, they support Boolean inferences just like the standard 'count quantifiers' (Peters and Westerståhl 2006), and they are invariant under inclusion isomorphisms. In other words, the appropriate invariance level for even logical expressions may differ.[6]

Whence the base structures? The case of mass versus count quantification also shows how invariance analysis depends on a prior choice of structures. Working in a standard set-theoretic universe suggests an underlying domain of primitive objects out of which everything is constructed 'upward' by set-forming operations. Mereology, on the other hand, suggests a different view: a universe where we can only analyze things 'downward' into smaller components, without a guarantee that we will

[5] The lack of a purely semantic fit also shows in the lack of an abstract model-theoretic Lindström theorem characterizing LFP(FO), compare with van Benthem, ten Cate and Väänänen (2009).

[6] A similar point holds for logical operators in intuitionistic, rather than classical logic on models of information stages ordered by inclusion. Permutation invariance is not the right notion there, invariance for inclusion-isomorphisms is – or even stronger criteria to be discussed below.

hit smallest objects. Building up versus analyzing represent very different views of what logical semantics is about, and the choice between these two perspectives is not made for us by invariance thinking.

Still, an irreducible role for one's conceptual choice of base objects and patterns does not invalidate our story: it makes our discussion all the more one of semantic analysis.

Function words in natural language The picture emerging here fits natural language. The choice between logical words and others is not all-or-nothing. In addition to expressions like Booleans and quantifiers, there are many 'function words' with a logic-like behavior, such as modals ("can", "may", "must"), prepositions ("in", "out"", "of", "with", "to"), or other functional items such as comparatives or indexicals. These expressions are not completely free in their interpretation, like nouns or verbs, as they encode facts about the way we use language to structure and convey information. In line with this role, functional linguistic expressions come with their own inferential behavior, and their own notions of invariance referring to semantic structure that is appropriate to them. Thus, from an invariance perspective, logicality, both in mathematical and in ordinary linguistic settings, is a widespread phenomenon that can come in degrees, or levels.

6.2.6 Conclusion

Invariance under structure transformations is a pervasive aspect of semantical analysis for logical languages, and it supports sophisticated notions and technical results. Also, it makes us acknowledge a wide variety of 'logicality' across natural language, depending on the invariance level. Finally, and significantly, invariance is naturally entangled with other basic logical notions. One of these, as we saw in the pilot case of monotonicity, is inference and proof. A second entangled notion is computation, the third strand in Beth's view of logic, which emerged when we looked at fixed-point logics for induction and recursion.

Coda: syntax Both proof and computation depend crucially on syntax, the code one works with. While invariance might suggest a priority for structure over syntax, once a language is there, other intuitions become available and can operate freely, sometimes without semantic counterparts. This entanglement of themes will return in what follows.

6.3 Plurality of Zoom Levels

6.3.1 Plurality and Zoom

Plurality While invariance has been proposed as a unique underpinning for a core of logic, scientific practice seems very different. When analyzing reality, many different levels of structure can play a role. Consider mathematics: it has a wide array of legitimate theories of space, ranging from affine and metric geometry to topology. These different theories represent different 'zoom levels' for looking at reality. Each of these levels comes with its own logical language: richer if the structural similarity relation is more detailed (making invariance more demanding), poorer if the similarity relation is coarser.

Looking down, or up This observation fits with a fact about logical analysis. Many people see the task of logic as providing ever more detail, formalizing each small step in reasoning and each feature of structure. The paradigm for this would be total formalization of informal mathematics into machine-checkable languages, with nothing taken for granted.

But proofs in this style may be as unreadable as machine code, and an equally legitimate form of logical analysis does the exact opposite. One looks at a reasoning practice, abstracts away from details, and looks for global patterns representing some high level of reasoning that may even bring to light patterns undiscovered so far. For a sample of the contrast, compare the extremely detailed first-order language of Tarski's 'elementary geometry' with the highly general modal logic of the interior operation in topology (van Benthem and Bezhanishvili 2008). We will discuss two ways in which this plurality may arise, and point at some open problems that emerge when we look at this generally.

6.3.2 Logics for Graphs

Let us fix one similarity type, annotated directed graphs $(W, R, \mathbf{P})$ with a set of points W, a binary relation R, and a set of unary predicates $\mathbf{P}$ of points. If we choose isomorphism as our invariance, then languages appropriate to studying graphs are first-order logic and its extensions – and these are a good fit with much reasoning in Graph Theory. However, there are also other natural ways of looking at graphs, where we identify more structure.

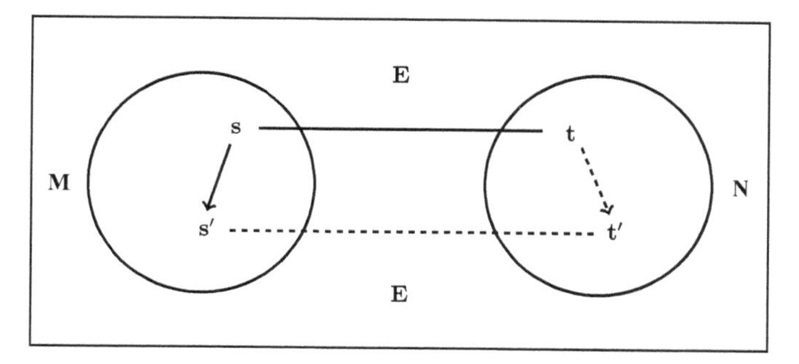

Figure 6.1 Diagram for the forward clause of bisimulation

One such invariance is *bisimulation*, where we are only interested in local properties of points as well as the structure of accessibility at each point, i.e., its arrows to other points (Figure 6.1).[7] A bisimulation E between two annotated graphs $\boldsymbol{M}$, $\boldsymbol{N}$ is a binary relation between points in $\boldsymbol{M}$ and $\boldsymbol{N}$ matching only points with the same local properties from $\boldsymbol{P}$, and satisfying the following Back and Forth properties: (a) if $s\ E\ t$, and $s\ R\ s'$ in $\boldsymbol{M}$, then there exists a point t' with $t\ R\ t'$ in $\boldsymbol{N}$ such that $s'\ E\ t'$, (b) the same clause in the direction from $\boldsymbol{N}$ to $\boldsymbol{M}$.

It is easy to find non-isomorphic bisimilar graphs, and indeed bisimulation is a rougher invariance level, proposed independently in modal logic, set theory, and computer science. Finding natural structural similarity relations is often a matter of having some intuition in mind – and for bisimulation, a powerful intuition is one of equivalence of processes, with points in graphs as states of a process, and arrows as possible state transitions.

The modal landscape The language that fits bisimulation is modal logic, and a rich theory has sprung up at this level that we cannot survey here: compare with Blackburn et al. (2001); van Benthem (2010) on the mathematics of modal logic. Our point here is just that modal logic is a good illustration of how invariance thinking ties up with language design.

[7] Notice how points in graphs derive their identity from two sources: their local properties, plus their connections with other points. This recursive nature, which can be made precise in terms of our earlier fixed-point logics, provides a much more sophisticated view of these structures than what might be suggested by standard discussions of 'possible worlds' as isolated entities.

For a start, there is a translation from the modal language into first-order logic, sending universal modalities $\Box p$ to guarded quantifier formulas $\forall y\,(Rxy \rightarrow Py)$, and existential $\Diamond p$ to $\exists y\,(Rxy \wedge Py)$. So, we can view the logic of a coarser level as a fragment of the logic of a richer level in a precise manner. Moreover, there is a certain dynamics of design. There is a whole landscape of languages in between modal logic and first-order logic, which can be viewed as arising in two ways. Either we extend the syntax of modal logic by certain 'hybrid' expressive devices available in first-order logic, or we devise new notions of simulation in between bisimulation and isomorphism, and create matching languages.

Zoom levels This setting also explains why working at different zoom levels can be useful, when we bring in the computational perspective mentioned several times already. The first-order language, though richer than modal logic, and natural in its own way, comes at a cost. Its theory is undecidable by Church's Theorem – and at a more domestic level, its syntax is more complicated. Modal logic has a variable-free notation that makes checking for truth provably easier, and allows for perspicuous notation of basic proof patterns without variable management. This demonstrates a much more general point that is often under-appreciated. In logic, as in acting, saying less is sometimes saying more.

Zooming out: from actions to powers Invariance thinking also suggests coarser levels with weaker modal languages. Here is an illustration for extensive games, or processes of interactive computation. Consider a finite game tree with transitions as moves, and atomic properties of nodes recording turns for players and pay-offs at end nodes. In Game Theory, one is often interested only in players' powers for controlling the outcomes of a game, not in local moves. We can identify different game trees where players have the same powers using a natural notion of 'power bisimulation', van Benthem (2014).

For a concrete illustration, consider the following two game trees:

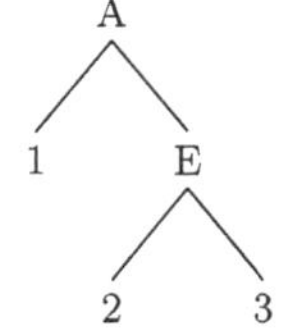

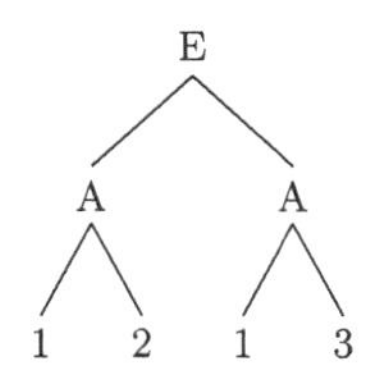

In the game to the left, player E has two strategies, going Left and Right. If she plays Left, she forces the game to end in the set {1, 2}, where the outcome *1* must be included since player A might have chosen Left at the start. This set {1, 2} is called a power for player E. Likewise, if E plays her strategy Right, she exercises the power {1, 3}. Now these powers are exactly the same in the game to the right, even though E's moves are different there. An analogous, slightly more complex, argument shows that powers for player A are the same in both games – bearing in mind that A has 4 strategies in the game to the right.

Naturally, there exists a language matching this invariance, less expressive than the modal logic of game trees. To be aligned with the power structure, it has 'forcing modalities' $\{i\}\phi$ saying that player i has a strategy for playing the game such that, under any counterplay by the other players, only end nodes result that satisfy ϕ. The logical theory of forcing has many similarities with modal logic – be it that $\{i\}\phi$ only has upward monotonicity $\Box\phi \rightarrow \Box(\phi \vee \psi)$ as a base law (plus laws linking powers of different players). The aggregation law $(\Box\phi \wedge \Box\psi) \rightarrow \Box(\phi \wedge \psi)$ of basic modal logic is invalid for powers, as is easy to see.[8]

The search for natural invariances and logical languages that fit games is still ongoing.[9] We have mentioned this illustration to show how invariance thinking is a live topic today, while also, games will make a brief appearance in this paper later on.

6.3.3 Interlevel Connections

Translation While we advocated variety of simulations and matching languages, logic abhors chaos. To see a deeper unity, one must also study connections between levels. We already mentioned translations between logics, and these can be related to invariances. For instance, the Modal Invariance Theorem (van Benthem 1977) says the following:

> A first-order formula ϕ in the signature $(R, \mathbf{P})$ is equivalent to the translation of a modal formula iff ϕ is invariant for bisimulations.

[8] Powers suggest a transition from graph models to 'neighborhood models' for modal logic, compare with van Benthem (2014). We do not elaborate this here, as it might confuse the reader with neighborhood models as a richer, rather than a poorer, semantic level beyond modal logic to be introduced below.

[9] For a new 'instantial bisimulation', in between modal and power bisimulation, that fits with game-theoretic equilibria involving all players, see van Benthem, Enqvist and Bezhanishvili (2018).

Many such characterization results exist for modal logics. But there are also other ways of relating different levels and their languages, using the apparatus of Category Theory, compare with the work on modal logic and co-algebra surveyed in Venema 2012.

Digression: weaker or stronger? While we have suggested that invariance perspectives are rewarding, we did not offer a complete theory. For instance, we pointed out how logical languages at different invariance levels can be related, but did not offer a definitive view on precisely how. One of the vexing (but also intriguing and wonderful) things about logic is that one can look at the same topic in different ways in tandem. For instance, is the modal language really more specialized and weaker than that of first-order logic, or is it more general? The latter view is developed in generalized modal semantics for first-order logic (Andréka, van Benthem and Németi 1998), and also, via an abstract correspondence between potential isomorphism and bisimulation, in van Benthem and Bonnay (2008).[10]

6.3.4 Neighborhood Structures

Generalizing a similarity type Searching for new invariances may also make us change the very similarity type of the models we work with. For example, let us generalize annotated graphs to *neighborhood models* (W, N, P) with $N\,s\,X$ a relation between points s in W and sets of points X (cf. the textbook Pacuit 2017).[11]

Neighborhoods, evidence and plausibility For a concrete case, consider the treatment in van Benthem and Pacuit (2011) of agents' evidence for their beliefs. Starting at the ordered graph level, belief is a modality using a reflexive transitive plausibility ordering $\leq$ of worlds (points in the graph), with $B\phi$ saying that ϕ is true in all the most plausible worlds, while conditional belief $B^{\psi}\phi$ refers to truth in the most plausible ψ–worlds. Indeed, plausibility models support a standard modal language with even further doxastic notions.

Now, let us add semantic structure explaining how the plausibility order $\leq$ came about. We give models a family $\boldsymbol{E}$ of subsets encoding the evidence received at the current stage, and stipulate that $s \leq t$ iff for each set E in

[10] Generalized semantics (cf. Andréka, van Benthem, Bezhanishvili and Németi 2014) is related to our themes here. But we still do not see exactly how things fit, and omit further discussion.

[11] This move is natural, since neighborhood models are close to parametrized 'hypergraphs'.

E, if $s \in E$, then $t \in E$. Thus, each evidence model ***M*** induces a plausibility graph *ord(**M**)*. A language matching this richer structure contains not just analogues for the old modalities at the plausibility level, but also new ones, such as $\Box\phi$ saying there exists an evidence set E 'supporting' ϕ in the sense that each point $s \in E$ makes ϕ true. In fact, several languages make sense, depending on the similarity relation chosen for evidence structures (van Benthem, Bezhanishvili, Enqvist and Yu 2016). Now comparison between levels is more delicate: there are not just different invariances on the same structures, but maps transforming structures at one level into those at another.

Without further details of this ongoing work, here is the point we want to make. Generalization of similarity types is another active force in semantics, going beyond invariance inside one type. Having said this, type change will be a side theme in this paper.

6.3.5 Conclusion

Finding natural zoom levels in semantics is an ongoing process. We gave illustrations from modal logic, driven by intuitions about space, processes, or information. But this variety was not just a free-for-all. There are systematic connections between different zoom levels and their invariances – though open problems abound concerning the total picture, and maintaining the unity of logic lies partly in getting clear on these.

6.4 Preservation, Generalized Consequence, and Model Change

6.4.1 Semantic Consequence

A semantic analysis of meaning and truth also offers an account of consequence. A standard notion of consequence says that $\Sigma \models \phi$ iff for all models making all formulas in Σ true, ϕ is also true – though logical systems can differ in what they take as the relevant models. Thus, consequence gets defined in terms of truth in a model, which can be analyzed in a recursive manner. This is not an arbitrary choice since it matches with other perspectives. Completeness theorems say that, for many logical systems, there is an extensional equivalence (qua transitions from sets of formulas to formulas) between semantic consequence and syntactic derivability, reflecting independent proof-theoretic intuitions.

Of course, consequence is not yet the same as actual inference and reasoning, but at least, it adds a new perspective that tends in this direction – in line with what we observed earlier about the tandem of invariance and inference for quantifiers.

6.4.2 *Interpolation and Invariance*

Consequence mixes well with our earlier topic of invariance. We give one illustration. Consider Craig's Interpolation Theorem for first-order logic.

Theorem For all first-order formulas ϕ, ψ with $\phi \models \psi$ there exists a formula α whose non-logical vocabulary is contained in that both ϕ and ψ such that $\phi \models \alpha \models \psi$.

Now, intuitively, there is a semantic surplus to the existence of an interpolant involving only part of the vocabulary of the antecedent and the consequent of an inference. What the existence of an α as above guarantees is the following 'transfer property':

> Let $\boldsymbol{M} \models \phi$ and let there be an $L_\phi \cap L_\psi$-potential isomorphism between $\boldsymbol{M}$ and any other model $\boldsymbol{N}$: then $\boldsymbol{N} \models \psi$.

Let us say that, in this case, 'ϕ entails ψ along $L_\phi \cap L_\psi$-potential isomorphism'. The following analysis comes from Barwise and van Benthem (1999). Here is a new version of the first-order Interpolation Theorem highlighting this special behavior:

Theorem The following are equivalent for all first-order formulas ϕ, ψ: (a) there is an L-interpolant for ϕ, ψ, (b) 'ϕ entails ψ along L–potential isomorphism'.[12]

Meta-logic Standard Interpolation fails for natural extensions of first-order logic such as the earlier infinitary logic L^∞_Ω. However, the preceding invariance version of Interpolation can be shown to hold for L^∞_Ω. Thus, invariance versions of well-known meta-properties, equivalent for first-order logic, may have better prospects of extending to other logics. This point is seldom appreciated. What counts as a crucial property of a logical system may depend on historical accidents of formulation. Received views of holding or failing of meta-properties across systems in the landscape of logics should be examined critically.

[12] One can also rework most standard preservation theorems in this interpolation style. This reanalysis also carries over to modal logic. For instance, a valid consequence between two modal formulas ϕ, ψ has a modal L-interpolant iff ϕ entails ψ along L-bisimulation.

6.4.3 *Preservation and Generalized Consequence*

Preservation theorems The same thinking applies to preservation theorems, to some the most attractive results that started Model Theory in the 1950s. A key example is the Los–Tarski Theorem saying that a first-order formula is preserved under submodels iff it is definable in a syntax starting from literals (atoms and their negations) using only conjunction, disjunction, and universal quantifiers. Another such result, relevant to the earlier monotonicity inferences, is Lyndon's Theorem: a first-order formula is upward monotonic in the predicate P iff it is equivalent to a formula in which P has only positive syntactic occurrences. These results embody what logicians like: syntactic form determines semantic behavior. In line with our discussion of invariance, we could also reverse this to: useful semantic transfer across models will lead to the emergence of matching syntax.[13]

Entailment along a relation From our current perspective, preservation theorems are really about a generalized notion of consequence, allowing for transfer in that, when the premises hold in one model, the conclusion holds in some other model. Standard consequence is the special case where we stay, a bit timidly, inside the same model.

A general way of reasoning about entailment along any relation R is in a modal format

$$\phi \to [R]\psi$$

The general properties of this style of reasoning come out in a calculus having different relations. For instance, with ; as relational composition, and v as converse, we have:

(a) $\phi \to [R]\psi$ and $\psi \to [S]\alpha$ imply $\phi \to [R;S]\alpha$,
(b) $\phi \to [R]\psi$ implies $\neg\psi \to [R^v]\neg\phi$

van Benthem (1996, 1998) give complete logics for the 'universal Horn fragment' of propositional dynamic logic (PDL), which describes the structural properties of such generalized reasoning in its most abstract form. The claim that real inference usually involves jumps across situations was already emphasized in Barwise and Perry (1983).

[13] In addition, restrictions to special syntax, say, 'universal' or 'positive', can make inferences much more perspicuous and efficiently computable than in first-order logic (van Benthem 1986b).

6.4.4 Excursion: 'Alternative Logics'?

By making the jumps in transfer inferences explicit as modalities, implication stays classical, supporting all the usual properties of inclusion, except for Reflexivity ($\phi \rightarrow [R]\phi$ clearly fails in general). Still, our setting supports variations that break classical principles. New consequence notions arise if models come with an order of relevance or importance. If we then say that a conclusion holds if it is true in all most important models for the premises (so premises influence which models are relevant), we get non-classical *non-monotonic logics*, widely studied since the 1980s. But classical logic lies close by. There is a clear analogy between this way of thinking and the earlier plausibility models for belief. Instead of insisting on non-monotonic consequence $\phi \Longrightarrow \psi$ we might just as well add a formula $B^{\phi}\psi$ with an explicit modality for belief to classical logic (van Benthem 2011).

6.4.5 Dynamic Logics of Model Change

Transfer or transformation of information across models occurs more often in the recent literature. In particular, logics of information change analyze how knowledge, belief, or yet other attitudes of agents change as new information comes in. We cannot go into the details of such 'dynamic epistemic logics' (see Baltag and Smets 2006; van Benthem, van Eijck and Kooi 2006; van Ditmarsch and Kooi 2007; van Benthem 2011). But relevant to us is this: update with new information is seen as definable model change.

In particular, an event $!\phi$ of getting the 'hard information' that proposition ϕ is the case restricts a current model $\boldsymbol{M}$ to a definable submodel $M|\phi$. This is a semantic 'relativization' of a model to a definable subdomain. If we now want to axiomatize a matching logic with explicit dynamic modalities $[!\phi]$ for informational actions, and static modalities K and B for agents' knowledge and beliefs, we are not only interested in transfer implications, but in equivalences telling us how the updated model relates to the original model. A good example are the following two valid 'recursion laws' for semantic relativization:

$$[!\phi]K\psi \leftrightarrow (\phi \rightarrow [!\phi]\psi)$$
$$[!\phi]B\psi \leftrightarrow (\phi \rightarrow B^{\phi}[!\phi]\psi)$$

More general logics of update also axiomatize operations that modify plausibility order (say, $\Uparrow\phi$ puts all ϕ–points in the current model on top of all $\neg\phi$–points, while retaining the old order inside these zones), or transform models in yet other definable ways.

Technically, these logical systems formalize the theory of cross-model relations matching definable transformations. This is more special than the scenarios considered before, where entailment could be along any relation, say, arbitrary submodels, definable or not. For definable transformations, often the static base logic has enough expressive power to supply all needed recursion laws (cf. van Benthem and Ikegami (2008) on "product closure" of logics). Indeed, dynamic-epistemic logics tend to be decidable if their static base logic is, whereas logics of consequence along arbitrary relations can be much more complex.[14]

Still, this simplicity of logics for model change is fragile. Löding and Rohde (2003), a study of "sabotage games" where one player can delete arrows from a graph, while the other tries to travel to some goal region (cf. van Benthem 2014), showed how the modal logic of removing arbitrary links from relations is undecidable. Aucher and Grossi (2018) show how even simpler 'stepwise' versions of dynamic-epistemic logics, where just some point lacking the property ϕ (or some link failing to pass the relevant update recipe) gets removed, may become undecidable – even when the base logic is quite simple.

6.4.6 Zoom Levels and Tracking

The general point here is sweeping. The universe of models for semantics is criss-crossed by links of many sorts. In inference, we are often interested in what holds for one model in terms of what we know about another model linked to it. And to describe this transfer, logics of model change seem a good medium. Moreover, there are still strong connections with our earlier semantic themes. In particular, update operations should respect whatever invariance relation we chose for our static models. But there is more. Updates can take place at various zoom levels, and in that case, there is a significant issue of when updates at a coarser level faithfully 'track' updates at a finer level. For more on this connection between model change and zoom levels, compare with van Benthem (2016); Ciná (2017).

[14] For instance, first-order logic with an added modality over model extensions no longer has a recursively axiomatizable set of validities – but a proof would take us too far for a survey paper.

6.4.7 Conclusion

We have shown how consequence is a natural companion to semantic invariance and definability. Generalized notions of consequence make sense then, that merge with model-theoretic preservation results and with dynamic logics of model transformations.[15]

6.5 Games and Agents

While our topics so far were within the realm of model theory as description of the world, in this final section, we explore a more agent-oriented perspective.

6.5.1 Logic Games

Model comparison games A well-known technique for analyzing fine-structure of similarity and invariance are 'model comparison games', due to Ehrenfeucht and Fraïssé, played between two models $\boldsymbol{M}$ and $\boldsymbol{N}$. Such games and their strategies are a natural bridge between logical syntax and semantic structure. They work as follows. A player $\boldsymbol{S}$ ('Spoiler') who claims dissimilarity of $\boldsymbol{M}$ and $\boldsymbol{N}$ chooses one of the models, and picks an object d in its domain. The counter-player $\boldsymbol{D}$ ('Duplicator') then chooses an object e in the other model, and the pair (d, e) is added to the current list of matched objects. After k rounds, the object matching is inspected. If it is a partial isomorphism, $\boldsymbol{D}$ wins; otherwise, $\boldsymbol{S}$ does. For concrete examples of the game and its uses, see Doets 1996; van Benthem 2014.

It can be shown that this game is adequate in the following sense.

Theorem For all models $\boldsymbol{M}$, $\boldsymbol{N}$, and all $k \in N$ the following two assertions are equivalent:
(a) $\boldsymbol{D}$ has a winning strategy in the k-round game,
(b) $\boldsymbol{M}$, $\boldsymbol{N}$ agree on all first-order sentences up to quantifier depth k.

In fact, the correlation with syntax is much tighter still.

[15] There are also independent more syntactic proof-theoretic intuitions concerning consequence – but these are not the topic of this paper.

Theorem There is an explicit correspondence between
(a) winning strategies for ***S*** in the k-round comparison game for ***M***, ***N***,
(b) first-order sentences ϕ of quantifier depth k with $\mathbf{M} \models \phi$, not $\mathbf{N} \models \phi$.

Likewise, ***D***'s winning strategies can be made explicit in terms of 'towers' of partial isomorphisms. In the game over infinitely many rounds, the winning strategies for ***D*** start resembling potential isomorphisms between the models (if any).

Other logic games The same sort of strategic analysis provides fine-structure to other logical notions such as truth, proof, or model construction (Väänänen 2011). All these games show interesting connections. For instance, the above match of first-order formulas and strategies for the Spoiler ***S*** provides a precise correspondence between modal comparison games and evaluation games for first-order formulas in single models.

6.5.2 Introducing Agents

The use of games in logic signals something more. Games are played by agents, and what comes to the fore here is the role of agents dealing with truth, similarity, or consequence.[16] This brings in the other face of logic: not as description of the world, in either physical or mathematical structure, but as analyzing *structured activity* such as communication or argumentation – having to do with information flow, and interaction between agents.

Games are a concrete focus for studying agency, and for modern *computation* (one of our basic themes) as interactive agency between computers and humans in social networks. van Benthem (2014) develops various links between game solution procedures, game-theoretic equilibria, and the fixed-point logics that we discussed earlier.

Logical constants once more A game-theoretic stance may affect our understanding of logic itself. In this perspective, logical constants are not so much most general invariants of reality, as in our earlier discussion, but the most general structures found in games. Conjunction and disjunction

[16] This agency theme also underlies the dynamic-epistemic logics in the preceding section.

reflect choices different players can make in games, negation correlates with role switch, and quantifiers involve sequential composition of games. Even further logical constants, beyond the classical repertoire, arise with further natural operations on games such as parallel composition, or infinite iteration. The consequences of such a shift for semantics as traditionally understood remain to be fully explored.

6.5.3 Agent Diversity

A last illustration concerns earlier topics that seemed settled. Consider similarity between models. So far, this was an absolute notion, with a Yes/No answer in any given case. But with agents that use or inhabit these models, similarity need not be objective, but may depend on the views and abilities of the agents doing the comparing. To make this more precise, we need to delve more deeply into the nature of agents and how they work.

Bounded agents and automata What capacities are required for tasks associated with standard logical notions? Clearly, real agents operate under bounds on what they can infer, observe, or remember. Accordingly, computational logic and parts of game theory model agents by *automata* of various sorts. As just one illustration, in 'pebble versions' of model comparison games, players have finite memories given by a fixed number of pebbles at their disposal that they can use to mark objects when drawing samples from models and checking for partial isomorphism. This leads to more fine structure than we had earlier, winning strategies now match the syntax of finite-variable fragments of first-order logic. For other uses of automata in logic, compare with Graedel, Wolfgang and Wilke (2002).

From fragments to agents What might agents do in the heartland of logic? Normally, we think in terms of logical systems, complete machineries for definition and inference that can be highly complex. If we think of users, we do so implicitly, looking for fragments of complex systems, with lower complexity of model-checking or inference. Alternatively, however, a bounded agent using a full logical system will only be able to use part of that system correctly. But then we can rethink fragments in terms of abilities of automata with computational limitations. One instance are the above pebble games: adequacy results in terms of finite-variable fragments determine for which part of full predicate logic such agents perform model comparison correctly. A proof-theoretic example of this agent view would be simple automata doing a parity count to determine positive or negative occurrences, and working correctly only with the monotonicity subcalculus of full first-order logic.

Invariance and agent types A second way in which agents may enter semantics concerns invariance. When we ask whether two given structures are 'the same', there is a hidden parameter: 'the same for *whom*'? Agents with restricted powers of inspection or memory distinguish fewer games than idealized ones. And differences also arise with preferences. To see this, consider again games. One might view two extensive games as equivalent when they have the same Backward Induction solution (van Benthem 2014). But then, earlier judgments of invariance can change since players can have different preferences.

For instance, consider our earlier example of power-equivalent games in Section 3.2, but now annotated with preferences. Here a node marked *(1, 0)* indicates that player *A* assigns utility *1* and player *E* assigns *0*, and similarly for all other end nodes.

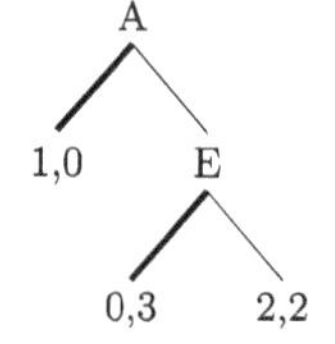

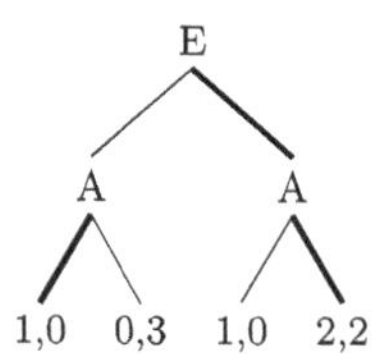

Backward Induction works as follows. In the game to the left, player ***E*** would go Left at her turn since this yields the best outcome for her among her available actions: *3* versus *2*. This is indicated by the bold-face line. Foreseeing this, player *A* gets *1* if he moves Left, and only *0* when he moves Right, so he moves Left. Thus the game ends in *(1, 0)*. But the same analysis in the game to the right yields the moves marked by the bold-face lines there, and the outcome is the node labeled *(2, 2)*. Thus earlier judgments of game equivalence change when players with preferences reason in a rationality-based manner.

Now rationality is a strong assumption about a type of player choosing best actions for oneself given one's preferences and beliefs. Other players can behave differently, making 'agent types' a parameter to be set explicitly before we analyze a game – and in the same vein, our discussion of invariance might also need an additional parameter when things are viewed from an agent perspective. When are two structures the same for whom?

Diversity once more Here is one more step. The usual assumption in logic is that agents have the same powers and styles of behavior. Say, in logic games, players may have different information, roles and pay-offs, but their styles of observation and reasoning are exactly the same. But in reality, agents can differ in all these respects (Liu 2009), and life even pits very different players against each other, such as humans versus machines. This suggests a new object of study. What about taking this diversity seriously, and introducing more realistic scenarios? What are solutions to logic games when players are different, say in model comparison or dialogue? What happens to "resource logics" (Restall 2000) when we model interplay of agents with different reasoning resources? Bits and pieces of this line of thinking occur in the literature (cf. the players with different 'sights' in Grossi and Turrini 2012), but no general theory exists.[17]

Relocating logic Does logic now dissolve in a concert, or cacophony, of agent voices? I do not think so. What we give up is one unique style of rationality for all agents. But the role of logic shifts to something more subtle. With agent types as a parameter, much structure remains, and logic then regulates the rational interaction of different kinds of agents.

6.5.4 Conclusion

An agent perspective using games as its vehicle is a natural supplement to our semantic considerations, bridging between structure and syntax in new ways.[18] But if we take this viewpoint seriously, it may also be much more radical then it looks at first sight, and it may start affecting our very understanding of the basic logical notions.

6.6 Conclusion

This short piece has given a broad picture of the semantic strand of logic, following the main themes of similarity, invariance, consequence, and a

[17] Uniformity assumptions about agents are also pervasive in philosophy, and van Benthem (2014) discusses, e.g., moral agents that are different in how much moral reasoning they are capable of.

[18] Games can also be studied from proof-theoretic and category-theoretic viewpoints, so we are not claiming an exclusive tie with semantics here.

bit of games and agency to spice things up. Admittedly, our treatment was sketchy – and in particular, all that we have said about games and agency was meant as an appetizer only, not as a serious introduction. Even so, we hope to have given an impression of the richness of our themes, and of the way they integrate across logic, crossing between different systems and subfields.[19]

[19] Still, we had to leave out many topics. Thus, we have underplayed the role of alternative logics arising on an agent conception of invariance or inference. One reason for this de-emphasis is the 'voracity' of classical logic that so far has been able to translate most alternative logics faithfully.

CHAPTER 7

Overgeneration in the Higher Infinite

Salvatore Florio and Luca Incurvati

George Boolos once remarked that first-order Zermelo-Fraenkel set theory is 'a satisfactory if not ideal theory of infinite numbers' (Boolos 1971, 229). An ideal theory, he added, 'would decide the continuum hypothesis, at least'. Despite its success in capturing facts about the cardinality of infinite collections, ZF, even when augmented with the Axiom of Choice, fails to settle many natural questions about which sizes there are.[1] In particular, ZFC cannot decide whether a number of interesting cardinality notions expressible in its language are instantiated.

In a first-order setting, these notions are expressed by using the non-logical symbol for the membership relation. However, the language of second-order logic affords us the resources to characterize analogues of these notions without resorting to non-logical vocabulary. One can then formulate counterparts of the cardinality questions left open by ZFC in the pure language of second-order logic.

On a Tarskian account, also known as *semantic* account, logical validity is identified with truth in every model. Given this account, a special relation appears to exist between certain cardinality statements in set theory and the logical validity of their counterparts in second-order logic. More specifically, there are cardinality statements that are true if and only if their second-order counterpart is true in every set-theoretic model. A number of authors have claimed this *entanglement* of second-order logic with mathematics to be problematic (Etchemendy 1990, Gómez-Torrente 1996, Gómez-Torrente 1998, Hanson 1997, Hanson 1999, Priest 1995, Ray 1996, Etchemendy 2008). Others have recently defended the opposite view (Parsons 2013, Paseau 2014, Griffiths and Paseau 2016). In previous work (Florio and Incurvati 2019), we have exhibited a conflict between the entanglement and the view that logic should be neutral. This can be seen as vindicating the claim that the entanglement is problematic and provides the basis

[1] This, of course, holds *modulo* relevant consistency assumptions.

for a plausible reconstruction of John Etchemendy's (1990) *Overgeneration Argument.*

Etchemendy concluded that the entanglement undermines the semantic account of logical validity. Other writers have retained the semantic account, taking the entanglement to cast doubt on the view that second-order logic is pure logic (Parsons 2013, Koellner 2010). Both reactions are premature in that they assume a specific implementation of the semantic account, one in which models are set-theoretic constructions. After all, the conflict between the entanglement and the neutrality of logic might be due to this assumption.

Indeed, earlier results of ours suggest that adopting a different semantics for second-order logic might provide an alternative way of resisting our reconstruction of the Overgeneration Argument. In particular, if we embrace higher-order resources in the metatheory and take models to be higher-order entities, the paradigmatic example of the conflict either disappears or is no longer problematic. This example involves the Continuum Hypothesis (CH), which states that every set has cardinality aleph-1 if and only if it has the cardinality of the continuum.

In this article, we extend this approach to other well-known cases of apparent entanglement involving the existence of certain small large cardinals, i.e., cardinals whose existence is consistent with the Axiom of Constructibility. We consider what happens to these cases of entanglement when second-order logic is given a higher-order semantics. Our findings are consistent with the optimistic outlook suggested by our earlier results. The adoption of a higher-order semantics for second-order logic helps avoid the conflict between the entanglement and the neutrality of logic.

7.1 Overgeneration

Using a set-theoretic semantics, i.e., a semantics that construes models as sets, it is provable in ZFC that CH is true if and only if the following sentence of second-order logic is valid:[2]

$$\forall X(\textsc{aleph-1}(X) \leftrightarrow \textsc{continuum}(X)) \qquad (\textbf{CH2})$$

The cardinality notions involved in **CH2** can be defined by pure formulas of second-order logic (see, e.g., Shapiro 1991, 101–5). As our discussion will make clear, what matters is the provability in a given theory of the biconditional stating that CH is true if and only if **CH2** is valid. We

[2] As usual, we indicate second-order variables by means of upper-case letters.

will refer to this provability as the *entanglement of second-order logic with CH*. Note that one can also formulate a sentence **NCH2** of pure second-order logic which is entangled with the negation of CH when ZFC is used as the background theory and the semantics is construed set-theoretically (Shapiro 1991, 105).

In our discussion, it will be important to distinguish between a sentence of a formal language and the English (or mathematical English) sentence it formalizes. To this end, we shall adopt the following typographical convention: boldface indicates that a sentence belongs to a formal language, whereas its informal counterpart is indicated using roman characters. So, for example, CH2 stands for the English sentence that **CH2** formalizes.[3] If we read the second-order quantifier as ranging over properties, CH2 states that every property is ALEPH-1 if and only if it is CONTINUUM, where being ALEPH-1 and being CONTINUUM are characterized in terms of properties and their relations. Note that CH2 is different from CH, which states that every *set* is aleph-1 if and only if it is continuum, where these notions of size are characterized using the membership relation. Another point related to the distinction between formal and informal sentences: we use 'validity' to denote the property a formal sentence has just in case it is true in every model. Validity is meant to correspond to the notion of *logical truth*, which applies to informal sentences.

As noted above, the claim that the entanglement of second-order logic with CH is problematic has recently been disputed. Against this, we have identified two arguments in support of the claim, both of which take the entanglement to be problematic insofar as it is in tension with the neutrality of second-order logic (Florio and Incurvati 2019). Their difference lies in the way in which they articulate the notion of neutrality. In both cases, the intended conclusion is that second-order logic fails to be *sound with respect to logical truth*, i.e., there are second-order validities whose informal counterpart is not a logical truth. In this sense, the arguments purport to show that second-order logic *overgenerates*.

[3] We are assuming that for every formal sentence there exists a unique informal counterpart, and that for every informal sentence we consider there exists a unique formalization. The uniqueness assumption is made for convenience but is not strictly needed for the purposes of the paper. What is needed is, roughly put, that the formalization operation and its inverse respect the informal relation of logical equivalence. The existence assumption is a presupposition of the debate. It may be justified by appealing to natural language sentences that seem to require variable binding of predicate positions (Higginbotham 1998, 3). Alternatively, it may be justified by appealing to a second-order translation of sentences involving plural quantification (Boolos 1984, Boolos 1985) or non-nominal quantification of the kind explored by Rayo and Yablo (2001).

The first argument is based on the idea that logic is *dialectically* neutral: it should be able to serve as a neutral arbiter in disputes. The argument uses the entanglement of second-order logic with CH, soundness of second-order logic with respect to logical truth and *completeness of second-order logic with respect to logical truth*, i.e., the thesis that there are no logical truths whose second-order formalization is not valid. Let us spell out the argument. If second-order logic is to serve as a neutral arbiter in a legitimate dispute, settling the question of whether a statement is a logical truth should not close that dispute. But now consider a dispute over CH in the context of ZFC. Suppose, on the one hand, that we agree that CH2 is a logical truth. By completeness, we have that **CH2** is a higher-order validity. Using the entanglement of second-order logic with CH, we conclude that CH is true. Hence, if we agree that CH2 is a logical truth, we also have to agree that CH is true. Suppose, on the other hand, that we agree that CH2 is *not* a logical truth. By soundness, we have that **CH2** is not a higher-order validity. Using the entanglement of second-order logic with CH again, we conclude that CH is false. Hence, if we agree that CH2 is not a logical truth, we also have to agree that CH is false. Therefore, settling the question of whether CH2 is a logical truth settles the question of whether CH is true, thereby violating dialectical neutrality.[4] The proponent of the Overgeneration Argument concludes that second-order logic is not sound with respect to logical truth.[5]

The second argument is based on the idea that logic is *informationally* neutral: it should not be a source of new information. This idea licenses the following principle of *informational neutrality*:

(IN) if a theory T does not informationally contain p and p is a consequence of T together with q, then q is not a logical truth.

For had q been a logical truth, it could not have brought about such an increase in informational content. With the principle (IN) in mind, we

[4] There is an alternative version of the argument which involves NCH2. This version uses the entanglement of second-order logic with the negation of CH together with soundness and completeness of second-order logic with respect to logical truth. Below we will see that there are statements which are not entangled with second-order logic whereas their negation is.

[5] To save soundness, the defender of second-order logic might be tempted to deny that second-order logic is *complete* with respect to logical truth. In fact, the argument uses only a particular instance of completeness, namely the one which states that if CH2 is a logical truth, then **CH2** is valid. Hence, denying completeness is acceptable only if one can point to features of CH2 which are not captured by the formalization. Compare with the case of propositional logic: it is usually agreed that this logic is not complete with respect to logical truth, but this is acceptable only because we can point to features of certain natural language sentences (e.g., the presence of predicates and quantifiers) which are not captured by a propositional language.

can now turn to the argument. The argument is more complex than the first one, and we divide it into to steps.

The first step is as follows. Suppose that CH is true. By the entanglement of second-order logic with CH, it follows that **CH2** is valid. By soundness, we conclude that CH2 is a logical truth.

The second step relies on this key assumption: ZFC together with the thesis that second-order logic is complete with respect to logical truth does not informationally contain CH. However, CH can be derived from ZFC plus completeness if we assume that CH2 is a logical truth. Using the key assumption, we derive from (IN) that it is not a logical truth that CH2 is a logical truth. On the supposition that logical truth can be iterated (i.e., if φ is a logical truth, then it is a logical truth that φ is a logical truth), we conclude that CH2 is not a logical truth.

Therefore, the conclusion of the second step of the argument contradicts the conclusion of the first step. The argument supposes that CH is true but a parallel argument can be formulated using the assumption that CH is false. This parallel argument relies on the entanglement of second-order order logic with the negation of CH. The proponent of the Overgeneration Argument concludes, once again, that second-order logic is not sound with respect to logical truth.

Both arguments make essential use of the entanglement of second-order logic with CH. The second argument, in addition, makes use of the entanglement of second-order order logic with the negation of CH. Both cases of entanglement, recall, are provable in ZFC when second-order logic is given a set-theoretic semantics. But this is just *one* implementation of the semantic conception of logic. So the question arises of what happens when second-order logic is given a different semantics.

7.2 Higher-Order Semantics

Higher-order semantics is an alternative way of providing a semantics for second-order logic which has gained popularity. On this semantics, one adopts higher-order resources in the metatheory and uses second-order entities, rather than sets, as the values of object-language second-order variables (Boolos 1984; Boolos 1985, Rayo and Uzquiano 1999, Rayo 2002, Rayo and Williamson 2003, Yi 2005, Yi 2006, McKay 2006, Oliver and Smiley 2013). This is compatible with various interpretations of the second-order resources. For example, one may interpret these resources in terms of properties, plurals, or Fregean concepts. Alternatively, one may take the second-order resources to be *sui generis* and not reducible to more familiar linguistic constructions (Williamson 2003, 459). For convenience, we will read the second-order quantifier as ranging over properties.

Higher-order semantics allows us to define validity for second-order logic without ascending beyond second-order resources (see McGee 1997). Let us focus on a language whose sole non-logical symbol is the membership predicate of set theory ($\in$). For such a language, the definition of validity amounts to truth under any reinterpretation of $\in$ in any second-order domain. More formally, let $\varphi[E/\ \in]$ be the result of replacing all occurrences of $\in$ in φ with E and the superscript indicates the restriction of the formula's quantifiers to U. Then we say that φ is a *higher-order validity* if for every non-empty property U and every binary relation E, $\varphi[E/\ \in]^U$ holds. For instance, $\exists x \exists y\ x \in y$ is a higher-order validity just in case the following statement is true: for every non-empty property U and every binary relation E, there are x and y having U such that Exy. But the latter statement can be refuted in pure second-order logic and hence $\exists x \exists y\ x \in y$ is not a higher-order validity.

If we are concerned only with arguments with finitely many premises, entailment can be defined in terms of validity. In particular, we can say that $\gamma_1, \ldots, \gamma_n$ *entail* φ just in case $(\gamma_1 \wedge \ldots \wedge \ldots \gamma_n) \rightarrow \varphi$ is a higher-order validity.

A key issue concerns the resources used in the metatheory. As we will see, the strength of the resources allowed has an impact on whether second-order logic is entangled with mathematics. In terms of logical resources, we will work with an axiomatization of second-order logic obtained by adding to first-order logic the rules for the second-order quantifiers, full second-order Comprehension and the second-order principle of Property Choice, stating that there is a choice function corresponding to every relation. As for the non-logical resources, we will consider two options. For the first option, we need the notion of a second-order closure of a theory **T**. This is denoted by $\mathbf{T}^*$ and defined as the set of sentences derivable from **T** in the axiomatization of second-order logic just described. The first option is then to use the informal counterpart of $\mathbf{ZFC}^*$. The second option is to use ZFC2, the informal counterpart of the theory obtained by replacing **ZFC**'s Replacement Schema with the corresponding second-order axiom. Since this theory has finitely many axioms, we can take it to be axiomatized by their conjunction. Thus, unlike the first option, the second option involves a theory whose non-logical axioms are formulated using distinctively higher-order resources.

It is worth noting that second-order logic with higher-order semantics retains its expressive power. For example, Shapiro (in unpublished work) and Väänänen and Wang (2015) show that an internal version of Zermelo's (1930) categoricity theorem is provable in pure second-order logic: there is a quasi-isomorphism between any two higher-order models of **ZFC2**.[6] This implies that **ZFC2** semantically decides any statement concerning the hierarchy up to V_κ, where κ is the first inaccessible. Since **CH** concerns the first few infinite levels of the hierarchy, it follows that **ZFC2** semantically decides **CH**. That is, it is provable in pure second-order logic that **ZFC2** entails either **CH** or it entails ¬**CH**.

7.3 The Continuum Hypothesis

We are now in a position to answer the question left open: 'What happens to the entanglement of second-order logic with CH when a higher-order semantics is adopted?'. As we have shown in Florio and Incurvati (2019), the answer turns on the metatheory used. In particular, assume ZFC and suppose that **ZFC** has an ω-model, i.e., a model which is correct about the natural numbers. Then we have:

FIRST NEGATIVE THEOREM **ZFC*** does not prove that if **CH2**, then **CH**.

SECOND NEGATIVE THEOREM **ZFC*** does not prove that if **CH**, then **CH2**.

In addition, the following lemma is provable in pure second-order logic.

EQUIVALENCE LEMMA **CH2** is true if and only if **CH2** is a higher-order validity.

Combining these results, one can conclude that **ZFC*** proves neither direction of the following biconditional:

CH if and only if **CH2** is a higher-order validity.

Does the informal counterpart of this result hold? That is, can we prove in ZFC* that CH if and only if CH2 is a logical truth? We think not. As an analogy, consider the case of standard independence results. The independence of **CH** from **ZFC** (Gödel 1939, Cohen 1963) is usually taken to show

[6] A function F is a *quasi-isomorphism* between two higher-order models $\langle U_1, E_1 \rangle$ and $\langle U_2, E_2 \rangle$ if F is a bijection between U_1 and a subproperty of U_2 or between U_2 and a subproperty of U_1 such that F preserves the relations E_1 and E_2.

that CH is independent of ZFC. For, presumably, any purported informal proof of CH or its negation from ZFC could be turned into a corresponding formal proof, contradicting the Gödel-Cohen result. Similarly, suppose one could give an informal proof in ZFC* that CH if and only if CH2 is a logical truth. Such an informal proof could then be turned into a formal proof in **ZFC*** that **CH** if and only if **CH2** is a higher-order validity. This point extends to the relation between formal and informal statements in the case of small large cardinals considered below.

While the entanglement of second-order logic with CH is not provable in ZFC*, matters are different if we strengthen the set-theoretic background. In particular, **ZFC2** proves:

FIRST POSITIVE THEOREM If **CH2**, then **CH**.

SECOND POSITIVE THEOREM If **CH**, then **CH2**.

Thus, given the Equivalence Lemma, we have that **ZFC2** proves that **CH** if and only if **CH2** is a higher-order validity. It follows that the informal counterpart of this biconditional may be established in ZFC2.

Recall that the argument from dialectical neutrality and the argument from informational neutrality make essential use of the entanglement of second-order logic with CH. Thus, if we use ZFC* as our background theory, both arguments are blocked. What about the case in which we use ZFC2? Luckily for the defender of second-order logic, the arguments can be blocked in that case too. The issue turns on a property of **ZFC2** mentioned above, namely that it semantically decides **CH**.

Consider the argument from dialectical neutrality first. The idea was that second-order logic could not serve as a neutral arbiter in a dispute over CH in the context of ZFC. That is because, in the presence of the entanglement, settling whether CH2 is a *logical* truth settles whether CH is true. One might reformulate the argument so that it applies to ZFC2, given that this theory proves the entanglement.

However, note that the original version of the argument assumed that we need to allow for a dispute over CH in the context of ZFC. This assumption is made plausible by the fact that **ZFC**, if consistent, entails neither **CH** nor **¬CH**. The alternative version of the argument would replace this assumption with the new assumption that we need to allow for the same dispute in the context of ZFC2. But the fact that **ZFC2** semantically decides **CH** gives us reason to reject the new assumption. For the fact that either **ZFC2** entails **CH** or it entails its negation means that one of

ZFC2 → CH and **ZFC2 → ¬CH** is a higher-order validity. Therefore, in the context of ZFC2, it is not possible to agree on the logic while disagreeing on CH. Once it is agreed whether **ZFC2 → CH** or **ZFC2 → ¬CH** is a higher-order validity, the dispute over CH can be settled by an appeal to soundness of second-order logic with respect to logical truth followed by a simple *modus ponens*. Hence, the alternative version of the argument is undermined.[7]

Similar considerations apply to the argument from informational neutrality. The argument relied on the assumption that neither CH nor its negation is implicitly contained in ZFC. A reformulation of the argument would therefore rest on the assumption that neither CH nor its negation is implicitly contained in ZFC2. But, again, the assumption is undermined by the fact that **ZFC2** semantically decides **CH**.

The defender of second-order logic has thus reasons to switch to higher-order semantics. By adopting such a semantics, she can respond to both arguments against the soundness of second-order logic with respect to logical truth, no matter whether she chooses ZFC* or ZFC2 as her background theory. The viability of this response turns on two facts. First, the entanglement of second-order logic with CH is not provable in ZFC*. Second, even on a higher-order semantics, **ZFC2** semantically decides any statement concerning sets occurring below V_κ (with κ the first inaccessible) and hence **CH**.

However, there are statements other than CH whose truth is equivalent over ZFC to the set-theoretic validity of their second-order counterparts. The entanglement of these statements with second-order logic has also been deemed problematic. What's more, some of these statements concern V_κ or higher levels of the hierarchy. Therefore, despite Zermelo's categoricity theorem and its internal version mentioned in Section 7.2, these statements are not semantically decided by **ZFC2** in either set-theoretic or higher-order semantics (modulo the appropriate consistency assumptions). Can the move to higher-order semantics help with these cases of entanglement beyond categoricity? In what follows, we focus on one notable set of examples concerning the higher infinite, namely statements asserting the existence of certain small large cardinals.[8]

[7] It may be observed that the alternative version of the argument from dialectical neutrality uses the entanglement of second-order logic with the *negation* of CH. However, analogues of the results described in this section can be proved for the case of **NCH2** and **¬CH** (Florio and Incurvati 2019). Thus, the alternative version of the argument is also blocked when moving to a higher-order semantics.

[8] To see that **ZFC2** does not semantically decide these statements given the appropriate consistency assumption, suppose that a small large cardinal κ exists. Then, one can obtain higher-order

7.4 Inaccessible Cardinals

As usual, we say that a cardinal is inaccessible if it is uncountable, regular and strong limit. We use IC to denote the assertion that there is such a cardinal. As is well known, ZFC proves that κ is inaccessible if and only if V_κ is a set-theoretic model of **ZFC2**. This fact suggests the standard second-order counterpart of **IC**, namely the pure statement that there is a domain and a binary relation that satisfies **ZFC2**. In symbols:

$$\exists U \exists R\, \mathbf{ZFC2}[R/\in]^U \qquad (\mathbf{IC2})$$

Now, there is no entanglement of second-order logic with IC in set-theoretic semantics (as well as in higher-order semantics). Indeed, **IC2** is false in any model with a finite domain and hence the semantics easily refutes the validity of **IC2**. Therefore, **IC** does not imply that **IC2** is valid. However, the other direction of the entanglement holds so we have what we might call *semi-entanglement*. For the fact that **IC2** is not valid trivially implies that

if **IC2** is valid, then **IC**.

Things change if we consider the *negation* of **IC**. In particular, let **NIC2** be the negation of **IC2**. Then, given a set-theoretic semantics, it is provable in **ZFC** that

$\neg$**IC** if and only if **NIC2** is valid.

The argument from informational neutrality as stated requires both directions of the entanglement for a statement *and* its negation. So it is not applicable in the present case. The argument from dialectical neutrality, on the other hand, only needs that a statement *or* its negation be entangled with second-order logic. Thus it can be formulated using the entanglement of the negation of **IC**.

Does this case of entanglement carry over to higher-order semantics? The answer is negative, no matter whether the background theory is **ZFC*** or **ZFC2**. Let us begin with **ZFC2**. The following two theorems hold.

THEOREM 3 ***ZFC2*** *proves that if* **NIC2** *is a higher-order validity, then* $\neg$**IC**.

THEOREM 4 ***ZFC2*** *does not prove that if* $\neg$**IC**, *then* **NIC2** *is a higher-order validity.*

models of **ZFC2** which disagree over the existence of κ by truncating at κ. The existence of the corresponding set-theoretic models can be established by assuming that there is an inaccessible above κ.

In both cases, the key observation is that, by a simple existential introduction, **ZFC2** proves that **IC2** holds and hence that **NIC2** is *not* a higher-order validity. This statement is the negation of the antecedent of the conditional in Theorem 3. Thus the theorem follows trivially. For Theorem 4, suppose towards a contradiction that **ZFC2** proved that if ¬**IC**, then **NIC2** is a higher-order validity. Since **ZFC2** proves that **NIC2** is not a higher-order validity, it would follow that **ZFC2** proved **IC** and hence its own consistency. But this cannot be for standard consistency reasons.

Let us now turn to **ZFC***. To start with, note that **ZFC*** is a subtheory of **ZFC2**. Therefore, Theorem 4 yields immediately:

THEOREM 5 ***ZFC*** does not prove that if* ¬**IC**, *then* **NIC2** *is a higher-order validity.*

Unlike **ZFC2**, **ZFC*** does not prove the other direction of the conditional in Theorem 5. This follows from the next two results, namely Lemma 6 and Theorem 7.

Lemma 6 is a version of the Equivalence Lemma above and is provable in pure second-order logic.

LEMMA 6 **NIC2** *if and only if* **NIC2** *is a higher-order validity.*

The right-to-left direction is straightforward. For the other direction, assume **NIC2** and suppose towards a contradiction that **NIC2** is not a higher-order validity. This means that there is a non-empty U^* and there are U and R such that $\left(\mathbf{ZFC2}[R/\in]^U\right)^{U^*}$. By second-order Comprehension we have a property Z which something has if and only it has both U and U^*. It is easy to verify that $\mathbf{ZFC2}[R/\in]^Z$, which contradicts **NIC2**.

The proof of Theorem 7 uses the assumption that there is an ω-model of the theory consisting of **ZFC** and the assertion that there are two inaccessibles. Note that the theorem is an underivability result. So, in the course of the proof, it is permissible to give the higher-order metalanguage a set-theoretic interpretation and use countermodels constructed according to the set-theoretic semantics.

THEOREM 7 ***ZFC*** does not prove that if* **NIC2**, *then* ¬**IC**.

Proof Let us abbreviate the statement that there are two inaccessibles in the constructible hierarchy **L** as $\mathbf{2IC^L}$. First, we use the consistency assumption to construct an ω-model m of **ZFC** $+ \neg$**IC** $+$ **2IC**$^{\mathbf{L}}$. Start from an ω-model with at least two inaccessibles and obtain by truncation a model with exactly two inaccessibles. Using forcing, make the two inaccessible cardinals of the model only weakly inaccessible (thus destroying their strong inaccessibility). This gives the desired model, since weakly inaccessible cardinals are strongly inaccessibles in **L**.

Next, we use m to establish the consistency of the theory **ZFC*** $+$ **NIC2** $+$ **IC**. Let κ be the second cardinal of m which is inaccessible in **L**. From the perspective of m, L_κ is a model of **ZFC** $+$ **IC**. Hence, again from the perspective of m, $\langle L_\kappa, \mathcal{P}(L_\kappa)\rangle$ is a model of **ZFC***$+$**IC**. Now since m satisfies **ZFC**, it thinks that every set-theoretic model satisfies **NIC2** if and only if $\neg$**IC** is true. But m is a model of $\neg$**IC** and therefore it thinks that $\langle L_\kappa, \mathcal{P}(L_\kappa)\rangle$ is also a model of **NIC2**. Hence m thinks that the theory **ZFC*** $+$ **NIC2** $+$ **IC** is consistent. Since m is an ω-model, it is correct about consistency facts. Therefore the theory **ZFC*** $+$ **NIC2** $+$ **IC** really is consistent.

The situation, therefore, is the following. If the background theory is ZFC*, neither direction of the entanglement of second-order logic with $\neg$IC is provable in higher-order semantics. If we strengthen the background theory to ZFC2, one direction of the entanglement becomes provable. Either way, the argument from dialectical neutrality is blocked. However, one might worry that the provability of semi-entanglement for the case of **IC** and $\neg$**IC** still poses a threat to the neutrality of logic. We address this issue in Section 7.6.

Of course, there are natural strengthenings of ZFC2 which do prove both directions of the entanglement of second-order logic with $\neg$IC in higher-order semantics. For instance, the result of adding to ZFC2 the assertion that there inaccessible cardinals trivially proves that if there are no inaccessibles, then **NIC2** is a higher-order validity. However, the dispute over the existence of an inaccessible is closed by such a theory quite independently of the entanglement. Therefore, its availability poses no problems for the dialectical neutrality of second-order logic. The same holds for other theories which imply the existence of inaccessibles, e.g., the theory obtained by adding a form of second-order reflection to ZFC2.

7.5 Mahlos and Weakly Compacts

The fact that **ZFC2** proves exactly one direction of the entanglement of $\neg$**IC** with second-order logic is a direct consequence of the definition of **NIC2**. That is, **ZFC2** proves that there is an inaccessible property, contradicting the higher-order validity of **NIC2**. Thus one may wonder what the situation is with respect to other small large cardinals, such as Mahlos and weakly compacts. For as well as not proving the existence of Mahlo and weakly compact *sets*, **ZFC2** does not prove the existence of Mahlo and weakly compact *properties*. As it turns out, perfect analogues of the results obtained for ¬**IC** can be obtained for the case of Mahlos and weakly compacts. Let us consider these cases.

Recall that a cardinal κ is *Mahlo* if every normal function on κ has an inaccessible fixed point.[9] We denote the set-theoretic statement that there is a Mahlo cardinal with M. Now, it can be shown in ZFC that κ is Mahlo if and only if V_κ is a set-theoretic model of **ZFCM2**, i.e., **ZFC2** plus the following axiom:

$$\forall F(\mathbf{normal}(F) \rightarrow \exists\kappa(\mathbf{Inacc}(\kappa)\ \&\ F(\kappa,\kappa)))$$

Here **normal**(F) and **Inacc**(κ) formalize the statements that F is a normal functional property and κ is an inaccessible set.

As before, this suggests the standard second-order counterpart of **M**, namely the pure statement that there is a domain and a binary relation that satisfies **ZFCM2**:

$$\exists U \exists R\ \mathbf{ZFCM2}[R/\in]^U \tag{M2}$$

Again, there is only semi-entanglement of second-order logic with M in set-theoretic semantics (as well as in higher-order semantics). On the other hand, if we let **NM2** be the negation of **M2**, the following is provable in **ZFC** given the set-theoretic semantics:

¬**M** if and only if **NM2** is valid.

We now switch to higher-order semantics. Again, we first examine what happens when the metatheory is **ZFC2**. In that case, we have:

THEOREM 8 ***ZFC2*** *proves that if* **NM2** *is a higher-order validity, then* ¬**M**.

THEOREM 9 ***ZFC2*** *does not prove that if* ¬**M**, *then* **NM2** *is a higher-order validity.*

[9] A function on the ordinals is normal if it is strictly increasing and continuous. Since in the definition of a Mahlo cardinal one restricts attention to normal functions on a set of ordinals κ, the notion of being Mahlo is first-order definable. Note that the definition of a Mahlo cardinal given above is equivalent over the ZFC axioms to the following definition, which is perhaps more standard: a cardinal κ is Mahlo if it is regular and the set of inaccessibles below κ is stationary in κ. A set $a \subseteq \kappa$ is stationary in κ if $a \cap c \neq \emptyset$ for each closed unbounded subset c of κ.

For Theorem 8, we reason in **ZFC2** and show that if there is a Mahlo cardinal, then **NM2** is not a higher-order validity. Let κ be Mahlo. It can be verified that the axioms of **ZFCM2** hold when restricted to V_κ. So the property of being a set in V_κ and the relation of membership restricted to V_κ provide witnesses to the existential quantifiers of **M2**. So **M2** is true and hence **NM2** is not a higher-order validity.

For Theorem 9, we reason about provability in **ZFC2** and establish that the relevant conditional is unprovable by providing a set-theoretic countermodel. We make the consistency assumption that there is a set-theoretic model of **ZFCM2**. This model must be a V_κ with κ a Mahlo cardinal. Now truncate this model at the first Mahlo cardinal μ. (If there are none, let μ simply be κ.) Clearly, V_μ is a model of $\neg$**M**. On the other hand, since μ is Mahlo, V_μ is a model of **ZFCM2** and hence of **M2**. Therefore V_μ is a model of the statement that **NM2** is not a higher-order validity.

The perfect parallel between the case of inaccessibles and that of Mahlos holds even if we weaken the background theory to **ZFC***. Obviously, the direction of the entanglement unprovable in **ZFC2** is also unprovable in **ZFC***:

THEOREM 10 ***ZFC**** *does not prove that if* $\neg$**M**, *then* **NM2** *is a higher-order validity.*

Moreover, the following two results can be established by adapting the proofs of Lemma 6 and Theorem 7.

LEMMA 11 **NM2** *if and only if* **NM2** *is a higher-order validity.*

THEOREM 12 ***ZFC**** *does not prove that if* **NM2**, *then* $\neg$**M**.

Thus, **ZFC*** also fails to prove the other direction of the entanglement.

How about weakly compact cardinals? One can formulate a second-order statement which, when added to **ZFC2**, yields a theory whose set-theoretic models are all and only the V_κ with κ a weakly compact cardinal (Hellman 1989).

Proceeding as before, one can then find a pure second-order statement (**WKC2**) whose set-theoretic validity implies the assertion that there is a weakly compact (**WKC**). As in the previous cases, **ZFC** proves that the negation of this statement (**NWKC2**) is true in every set-theoretic model if and only if there are no weakly compacts ($\neg$**WKC**). When moving to a higher-order semantics, one can prove analogues of the theorems obtained above. This means that we have no entanglement when the background theory is **ZFC***and only semi-entanglement when the background theory is **ZFC2**. These results can be established using the techniques employed in the case of Mahlo cardinals.

7.6 Semi-entanglement

We now want to reassess the arguments from dialectical and information neutrality in the light of the results proved in the previous section. As mentioned, the argument from informational neutrality requires the entanglement for a statement and its negation. Therefore it does not apply in the case of small large cardinals, even on a set-theoretic semantics. In contrast, the argument from dialectical neutrality requires the entanglement of a statement or its negation. So it continues to apply with respect to small large cardinals. In these cases, higher-order semantics provides a clear way out: none of the relevant statements which are entangled in set-theoretic semantics continues to be entangled in higher-order semantics. So the new instances of the argument from dialectical neutrality are blocked.

If the background theory is ZFC2 rather than ZFC*, we still have semi-entanglement. Thus one might wonder whether semi-entanglement suffices to run a *modified* version of the arguments we have articulated. In the case of informational neutrality, it is hard to see how the argument could be modified so as to rely only on semi-entanglement. The case of dialectical neutrality requires closer scrutiny.

Let S be any of the statements involving small large cardinals which are entangled with second-order logic in set-theoretic semantics (i.e., ¬IC, ¬M and ¬WKC) and let S2 be its second-order counterpart. Then the argument from dialectical neutrality concerning S goes as follows. By logic, either S2 is a logical truth or it isn't. If it is, using completeness of second-order logic with respect to logical truth and one direction of the entanglement, we derive that S is true. It follows that S is settled. If it isn't a logical truth, using soundness of second-order logic with respect to logical truth and the other direction of the entanglement, we derive that S is false. It follows, again, that S is settled. Therefore, no matter what logical status S2 has, S is settled. This means that second-order logic cannot be a neutral arbiter in a dispute over S.

Since in the context of higher-order semantics we only have semi-entanglement when the background theory is **ZFC2**, it is not the case that S is settled independently of the logical status of S2. For each of S and ¬S is consistent with the assumption that S2 is not a logical truth. So the conclusion that second-order logic cannot be a neutral arbiter in a dispute over S is not available in general. However, one may point out that the conclusion *is* available if it is assumed that S2 is a logical truth. For, in the presence of semi-entanglement, this assumption together with completeness of second-order logic with respect to logical truth entail that S is true. So it follows that S is settled.

To start with, note that this modified version of the argument leads at best to denying completeness, not soundness. Thus it cannot be considered an *overgeneration* argument. Rather, the denial of completeness results in the claim that second-order logic *undergenerates*: there are logical truths whose second-order formalization is not valid.

In any case, the question we should ask is: what reasons do we have to assume that S2 is true and hence a logical truth? To answer this question, it is useful to consider first another case, namely that of CH2 and NCH2. As Michael Potter remarks:

> [W]e do not seem to have any intuitions about whether these second-order principles [i.e., CH2 and NCH2] that could settle the continuum hypothesis are themselves true or false. So this observation [that there is entanglement] does not seem especially likely to be a route to an argument that will actually settle the continuum hypothesis one way or the other. (Potter 2004, 271)

The same could be said about S2. Of course, one might deny this and insist that either we have no intuitions about whether S2 is true or we have intuitions that it is false. But suppose that we did think that S2 is true. Presumably this is because we have certain views about the extent of the set-theoretic hierarchy. For example, it is because we think that the hierarchy does not have Mahlo height that we think that there is no Mahlo property and hence that NM2 is true. Clearly, this is also a reason for thinking that the hierarchy contains no Mahlo set: if there was such a set, one could obtain a Mahlo property by simply considering the property of being a member of that set. Therefore the reason to accept NM2 *ipso facto* closes the dispute over the existence of a Mahlo.

Note that this response does not carry over to the situation in which we have full entanglement. For not every reason to deny the existence of a Mahlo is *ipso facto* a reason to accept NM2. For instance, one might think that the universe is V_κ with κ the first Mahlo cardinal. In that case, one would hold that the hierarchy has Mahlo height without thereby thinking that there is a Mahlo set.

So, in all these cases of semi-entanglement, if we have reasons to take the second-order statement to be true, we also have reasons to take its set-theoretic counterpart to be true independently of semi-entanglement. Thus there appears to be no conflict between the relevant cases of semi-entanglement and dialectical neutrality.

7.7 Conclusion

The semantic account of logic identifies validity with truth in every model. When models are construed set-theoretically, this account gives rise to cases of entanglement which are in conflict with neutrality of second-order logic. This conflict is captured by the arguments from dialectical neutrality and informational neutrality. These arguments make use of the assumption that second-order logic is dialectically neutral and the assumption that second-order logic is informationally neutral. Thus, one could block them by rejecting those assumptions.

We have shown that the defender of second-order logic has an alternative way of blocking the arguments, namely by adopting a higher semantics. This semantics embodies a different implementation of the semantic account of logic in that it takes models to be higher-order constructions rather than sets. Of course, one might challenge the legitimacy of higher-order semantics. But this would require a separate argument targeting the use of second-order resources in semantic theorizing. Pending such an argument, the defender of second-order logic may resort to higher-order semantics in response to the Overgeneration Argument.

In this article, we have focused on some cases of entanglement in the higher infinite. We have shown that these cases of entanglement, previously thought to be problematic, disappear when a higher-order semantics is adopted. In the absence of entanglement, both of our reconstructions of the Overgeneration Argument are blocked. Moreover, a modified version of the argument from dialectical neutrality based on semi-entanglement fails to establish its intended conclusion.

This provides additional evidence that the proponent of the semantic account can hold on to the neutrality of second-order logic by employing a higher-order semantics. When one takes into account models that are not set-sized, denying the existence of small large cardinals does not translate into a commitment to the validity of certain second-order sentences.

There remains one well-known case of entanglement in set-theoretic semantics which we have not considered. This is the one involving the Generalized Continuum Hypothesis, and we plan to explore it in future work. Together with the results obtained in this article, this new investigation will provide a clearer picture of the extent to which higher-order semantics can help vindicate the neutrality of second-order logic.[10]

[10] This work has received funding from a Leverhulme Research Fellowship held by Salvatore Florio and from the European Research Council (ERC) under the European Union's Horizon 2020 research and innovation programme (grant agreement No 758540) within Luca Incurvati's project *From the Expression of Disagreement to New Foundations for Expressivist Semantics*. For comments and discussion, we would like to thank Michael Glanzberg, Mario Gómez-Torrente, Graham Leach-Krouse, Beau Madison Mount, Ian Rumfitt, Gil Sagi, Kevin Scharp, Gila Sher, Jack Woods, and Elia Zardini. Earlier versions of this material were presented at the workshop *The Semantic Conception of Logic* in Munich and the conference *Disagreement within Philosophy* in Bonn. We are grateful to the members of these audiences for their valuable feedback.

CHAPTER 8

Propositional Logics of Truth by Logical Form

A.C. Paseau and Owen Griffiths

The propositional 'logic' of truth by logical form consists of the general propositional principles satisfied by the operator 'It is true by logical form that'. We investigate which formulas of modal propositional logic (MPL) are true when $\Box$ is read in this way. An example is any instance of the T-schema $\Box\phi \rightarrow \phi$, since if it is true by logical form that A then it's the case that A, where ϕ is read as A. This chapter explains why there are different ways of understanding the question and sets out the respective answers.

8.1 Introduction

8.1.1 The Question

The language of modal propositional logic MPL is that of classical propositional logic augmented with the operator $\Box$. (Throughout, we think of MPL primarily as a language.) The specification of MPL's formulas adds to the usual propositional clauses the condition that if ϕ is a formula then so is $\Box\phi$. MPL-formulas are uninterpreted, and their interpretations – or *readings* as we shall generally call them – will be natural-language sentences.[1]

Since we are interested in truth by logical form, $\Box$ will be read as 'it is true by logical form that', sentence letters as declarative sentences, and connective symbols as their natural-language truth-functional counterparts ($\wedge$ as 'and', $\vee$ as 'or', etc.).[2] For example, if we read p as 'It's raining' and q as 'It's sunny' the resulting reading of $p \vee \Box q$ is 'It's raining or it's true

[1] As opposed to interpretations that assign T/F to formulas.

[2] Throughout, we speak of a 'reading' ambiguously, as both a function from (uninterpreted) formal-language sentences to meaningful sentences, as well as its particular values; context should make plain which is intended. We also generally dispense with quotation marks, corner quotes and similar niceties.

by logical form that it's sunny'. Our question is: which **MPL**-formulas are true under *any* such readings?

When $\Box$ is read as 'It is metaphysically necessary that', the analogous question is: which propositional modal logic is the right one for metaphysical necessity? When metaphysicians affirm or deny that this logic is S5, what they mean is that S5's theorem set consists of all and only the **MPL**-formulas that are true however sentence letters are read, with the reading of $\Box$ constrained to be 'It is metaphysically true that'.[3] Williamson (2013, ch. 3), for example, calls an **MPL**-sentence $\phi = \phi(p_1, \cdots, p_n)$ such that $\forall p_1 \cdots \forall p_n \phi$ is true on $\Box$'s intended interpretation *metaphysically universal*, and the set made up of such sentences **MU**. Our question is the parallel one about truth by logical form, resulting in a set of **MPL**-sentences we label **TLF**. Analogous questions have of course been asked about epistemic logic, deontic logic, etc.

Why be interested in **TLF**? One use to which **TLF** might be put is to assess the validity of arguments involving 'It is true by logical form that'. If ϕ is true by logical form, then every sentence with the same form is true. In this way, the truth of every substitution instance of sentence S appears *necessary* for S's logical truth. A substitutional account of logical truth would also take this to be *sufficient* and equate the notions of logical truth and truth by logical form.[4] At any rate, if an argument is valid by logical form, then every argument with the same form is truth-preserving. Compare, for example, the argument scheme

$$\frac{\begin{array}{c} A \\ \text{It is true by logical form that } (A \text{ iff } B) \end{array}}{\text{It is true by logical form that } B}$$

with

$$\frac{\begin{array}{c} \text{It is true by logical form that } A \\ \text{It is true by logical form that } (A \text{ iff } B) \end{array}}{\text{It is true by logical form that } B.}$$

[3] And connective symbols are read as their meaningful counterparts.

[4] There are well-known problems for substitutional accounts, such as whether they can be extended to logical consequence and whether they can apply to expressively impoverished languages. Dogramaci (2015, §4) and Halbach (2018) have tried to address such issues.

The former has invalid instances, whereas the latter does not. These facts are of the sort **TLF** aims to capture:[5] the formula $(p \wedge \Box(p \leftrightarrow q)) \rightarrow \Box q$ should not be a member of **TLF**, whereas any instance of $(\Box\phi \wedge \Box(\phi \leftrightarrow \psi)) \rightarrow \Box\psi$ ought to be. To say that any instances of the latter scheme are valid is to say that, for any instantiating A and B, if the resulting argument's premisses are true then so is its conclusion.

8.1.2 Previous Literature

The literature contains surprisingly little discussion of (what we call) **TLF**. The most relevant recent discussions we are aware of are by John Burgess, who writes 'The question which is the right validity logic has been answered at the sentential level. . . : it is the system known as S5. This result is essentially established already in Carnap (1946)' (Burgess 1999, 170), a claim repeated in Burgess (2003).

Carnap (1946) argues that the right propositional validity logic is S5. Carnap understands the sentential operator □ as 'It is logically valid that' and applies it to formal sentences of propositional logic. As Cresswell (2013) proves, on Carnap's understanding of propositional validity logic, the correct such logic is S5+. (See footnote 20 for details of S5+.)

Finally, Halldén (1962) argues that the propositional logic of logical truth contains S4's theorem set and speculates that it is identical to it. His understanding of 'It is logically true that' is proof-theoretic: to paraphrase him, a logical truth is a statement provable on the basis of a logical argument (from no non-logical premisses).

Although inspired by Burgess's discussion, unlike him we do not see the logic of logical form as a single entity. There is more than one such logic, the main subdivisions being first between different precisifications of truth by logical form, and second between logics that assume the corresponding operator is logical and those that do not. From Carnap we differ in that we consider natural-language readings of propositional formulas. Our interest is in an operator that applies to meaningful sentences rather than formal ones. And finally, unlike Halldén, our understanding of truth by logical form is not proof-theoretic.

Unlike previous writers, we are also careful to distinguish truth in virtue of logical form from logical truth. On a standard model-theoretic conception of the latter, the two notions do not exactly coincide. An example sentence on whose status they differ is A = 'There are least three things'.

[5] Assuming the conditional way of capturing validity.

Assuming A is true (e.g., because there are at least three planets) it must be true by logical form since the only sentence that shares A's form is A itself. But A is not a logical truth on the model-theoretic conception since its formalisation is false in some domains which, unlike the actual one, have fewer than three things in them.[6] Truth by logical form considers readings of a given form over the actual domain; model-theoretic logical truth consider readings of a given form over any domain (domains being furnished by a model theory, often ZFC).[7] Such examples notwithstanding, there is considerable extensional overlap between the two conceptions, so much so that most of what we say can be carried over to the logic of model-theoretic logical truth, something we aim to do in a sequel.[8]

8.1.3 Further Preliminaries

Naturally, TLF depends on what logic we take to be the 'one true logic'. For example, classical logicians think $p \vee \neg p$ is in TLF, whereas intuitionists disagree. Here, we adopt the mainstream classical view for simplicity and through space constraints. Fans of first-order logic, plural logic, second-order logic or any other classical logic, possibly supplemented with other operators (see below), can think of a sentence being true by logical form just when any reading of its formalisation in their preferred logic is true. Our assumptions below will hold for any of the usual candidates for this role.

The idea that there is such a thing as *the* logic of truth by logical form also rules out various forms of Logical Pluralism, as articulated for example in Beall and Restall (2006) or Shapiro (2014). If you see classical and intuitionistic logic as equally acceptable, you will take $p \vee \neg p$ to be in $\mathsf{TLF}_{\text{classical}}$ but not in $\mathsf{TLF}_{\text{intutionistic}}$. Pluralists may take the discussion to follow as applying to one of the several acceptable logics – *viz.* classical logic – and as drawing distinctions that will apply to any such.

Grant then that truth by logical form (and argumentative validity in virtue of form) should be cashed out in terms of a single logic. We let

[6] We are assuming here that logic includes at least first-order logic. A is first-order formalisable using only logical vocabulary as $\exists x \exists y \exists z(x \neq y \wedge x \neq z \wedge y \neq z)$.

[7] Essentially, our notion of truth by logical form coincides with Tarski's (1936) account of logical truth. The now-standard model-theoretic conception differs from Tarski's by allowing the domain to vary. There is some controversy over whether varying domains were *implicit* in Tarski's definition. See Paolo Mancosu (2010) for a good overview of the debate.

[8] For related discussion of logical consequence, see Griffiths and Paseau (2022).

$\mathcal{L}$ be a placeholder for this logic. One may still doubt that any sentence of natural language, or even any sentence of suitably cleaned-up natural language, has a unique formalisation into $\mathcal{L}$.[9] The very idea of truth by logical form, however, seems to assume that sentences have a logical form, that is, a unique formalisation into $\mathcal{L}$.[10] The discussion to follow may be easily amended to cover the case in which a sentence is *true-by-its-logical-forms*.[11] We leave this straightforward extension to the interested reader. So we take formalisation from natural language into $\mathcal{L}$ to be a function, and readings to be functions from $\mathcal{L}$ to natural-language sentences.

That $\mathcal{L}$ may be predicate logic, or even some higher-order logic, is entirely compatible with our interest in *propositional* logics of truth by logical form. We are interested in which MPL-formulas are true under the readings of interest. In all of these, $\Box$ is understood as 'It is true by logical form that', an expression analysed in terms of the logic $\mathcal{L}$.

8.2 Any TLF$^+$

TLF depends on whether 'It is true by logical form' is a logical constant. We call the assumption that it is, which applies throughout Sections 8.2–8.5 (without endorsement), the *constancy assumption*. In Section 8.6, we consider what follows on the contrary assumption, according to which 'It is true by logical form that' is *not* a logical constant. We call this resulting set of sentences TLF$^-$, and the corresponding set of sentences on the constancy assumption TLF$^+$. The superscripts indicate that the latter class is larger than the former.

Assume then (until Section 8.6) that 'It is true by logical form that' is a logical constant. On this assumption, the language of $\mathcal{L}$ contains $\Box$, whose interpretation is fixed. Without further specifying what the operator means, we may show that TLF$^+$ must be a propositional logic containing all propositional tautologies, all instances of the K and T axioms, and must be closed under *modus ponens*. (In fact, this also applies to the propositional logic of model-theoretic logical truth.) This section provides the

9 The cleaning-up will remove ambiguity, and perhaps other features such as vagueness.

10 Up to substitution equivalence; e.g., the first-order formalisations *Fa* or *Gb* of 'Albert is fair' are first-order substitution equivalents. Two formal sentences are substitution equivalents iff the second sentence can be obtained from the first by uniform substitution of some of its non-logical terms by other terms of the same respective categories, and vice versa.

11 By definition, *A* is true by its logical forms iff any of *A*'s formalisations admits only true readings. Conversely, *A* is not true-by-its-logical-forms if some reading of one of *A*'s formalisations is not true.

generic argument for this claim, so that it needn't be repeated for our two main disambiguations of 'It is true by logical form that' (on the constancy assumption) in Sections 8.4–8.5. We come back to Necessitation later.

(a) Suppose ϕ is a substitution instance of a propositional tautology, i.e., the result of uniformly replacing the sentence letters in a propositional tautology with MPL-formulas. Then the fact that the interpretations of Boolean connective symbols in ϕ are their corresponding truth-functional connectives guarantees that any reading of ϕ is true. Thus ϕ is in TLF^+.

(b) K-axioms: instances of $\Box(\phi \to \psi) \to (\Box\phi \to \Box\psi)$.
Suppose it is true by logical form that if A then B and that it is true by logical form that A, where A is a reading of ϕ and B of ψ. (Recall that A and B are variables ranging over natural-language sentences.) B follows from these two sentences by *modus ponens* for the interpreted material conditional. *Modus ponens*, however, preserves truth by logical form. Otherwise there would be a reading A^* of A's form and B^* of B's form such that A^* and 'If A^* then B^*' are both true but B^* is false. It must therefore be true by logical form that B. Thus any reading of an instance of $\Box(\phi \to \psi) \to (\Box\phi \to \Box\psi)$ with $\Box$ taken as 'It is true by logical form that' is true, i.e., any such instance is a member of TLF^+.

(c) T-axioms: instances of $\Box\phi \to \phi$.
Clearly, if it is true by logical form that A then A.[12]

(d) *Modus Ponens*: Suppose ϕ and $\phi \to \psi$ are in TLF^+; then so is ψ.
If B is a reading of ψ then for any reading A of ϕ, A and 'If A then B' are true; so B is true.

However exactly we understand 'It is true by logical form that', then, the set of MPL-formulas true under all intended readings is closed under *modus ponens* and contains all substitution instances of all propositional tautologies as well as all instances of the K and T schemata.

Finally, we observe that TLF^+ is closed under uniform substitution of sentence letters by MPL-formulas. The result of uniformly replacing sentence letters in an MPL-formula by other MPL-formulas we call a *substitution*; this is not to be confused with a *reading* of an MPL-formula which here means a natural-language construal of the formula,

[12] McGee (1991, ch. 2) rejects some instances of the T-axioms for logical necessity. However, his intuitive understanding of logical necessity differs from ours. We understand it as 'truth by logical form' whereas McGee's understanding is closer to analyticity. Further, he argues that some instances of the T-axioms fail to be *necessarily* true, not that they fail to be plain true.

as explained. Thus $\Box(p_1 \wedge p_2) \vee \Box r$ results from $\Box p \vee q$ by substitution of $(p_1 \wedge p_2)$ for p and $\Box r$ for q, whereas 'It's true by logical form that it's raining and it's cloudy, or it's true by logical form that it's sunny' is a reading of $\Box p \vee q$, in which p is read as 'It's raining and it's cloudy', q as 'it's true by logical form that it's sunny', $\Box$ as 'It's true by logical form that' and $\vee$ as 'or'. The reason TLF^+ is closed under uniform substitution is that a reading of the MPL-formula $\phi(\alpha_1/p_1, \cdots, \alpha_n/p_n)$ is also a reading of $\phi(p_1, \cdots, p_n)$. Thus if any natural-language reading of $\phi(p_1, \cdots, p_n)$ is true when $\Box$ is understood in this way (i.e., $\phi(p_1, \cdots, p_n) \in \mathsf{TLF}^+$), so is any reading of $\phi(\alpha_1/p_1, \cdots, \alpha_n/p_n)$.

8.3 Truth by Narrow and by Wide Form

To determine TLF^+ more precisely, we need to clarify what 'It is true by logical form that' means. One way to hear the claim that a natural-language sentence A is true by logical form is that any sentence of the *very same* logical form as A (in $\mathcal{L}$) is true. Another reading is that *any* sentence that interprets the logical form of A is true. Two examples illustrate this important difference. Consider first the statement 'It's raining' whose formalisation let us assume is the sentence letter p.[13] On the first view, to determine the status of 'It's raining' one needs to consider all readings of p as a sentence whose formalisation is also a sentence letter, that is, readings that are atomic sentences such as 'It's sunny', but not complex sentences such as 'It's cloudy or it's not cloudy'. On the second view, we must consider *all* readings of p, including those whose formalisations are not sentence letters, e.g., the reading of p as 'If John is tall then Jill is tall'. We call any sentence true by logical form in the former sense *narrowly true* by logical form, and in the latter sense *widely true* by logical form.

There are thus two notions of truth by logical form (on the constancy assumption):

(a) narrow-truth-by-logical-form, with corresponding set of MPL-formulas TLF^+_N;
(b) wide-truth-by-logical-form, with corresponding set of MPL-formulas TLF^+_W.

[13] This assumption is inessential; we may equally take the $\mathcal{L}$-formalisation of 'It's raining' to be, say, *Fa*.

The narrow and wide construals differ in the readings of formal sentences they countenance, with the narrow one countenancing *fewer* readings.

Let's write $\text{Form}_{\mathcal{L}}(A)$ for *A*'s $\mathcal{L}$-formalisation (up to substitution equivalence). For example, $\text{Form}_{\mathsf{MPL}}(A)$ is the **MPL**-formalisation of the meaningful sentence *A*, $\text{Form}_{\mathsf{FOL}}(A)$ is *A*'s first-order formalisation, and $\text{Form}_{\mathsf{FOL}\Box}(A)$ is *A*'s formalisation in the language of first-order logic augmented with $\Box$. Thus if $\mathcal{L}$ is first-order logic plus $\Box$ (recall the constancy assumption), i.e., **FOL**$\Box$, then $\text{Form}_{\mathsf{FOL}\Box}$(It's true by logical form that Barbara is great) would be $\Box Gb$; if $\mathcal{L}$ is second-order logic plus $\Box$, i.e., **SOL**$\Box$, then $\text{Form}_{\mathsf{SOL}\Box}$(There's something Charlie isn't) would be $\exists P \neg Pc$; and so on. Recall that in our discussion, *A*, *B* and *C* are variables ranging over meaningful sentences, and ϕ and ψ variables for (uninterpreted) **MPL**-formulas.

When *A* and *B* have the same form (relative to some logic, here $\mathcal{L}$), we say that they are *formal twins*. And when *B* is a reading of *A*'s form, say that *B* is *A*'s *formal relative*. Evidently, if *A* and *B* are formal twins they are also each other's formal relatives. But *A*'s formal relative need not be its formal twin. If, for example, *A* is 'It's sunny', then every sentence is *A*'s relative since *A*'s formalisation is the sentence letter p and a sentence letter can be read as anything; but 'It's rainy or it's snowing', though *A*'s formal relative, is not its formal twin, since its formalisation, $q \vee r$, differs from *A*'s formalisation. Unlike human relatives, note, the relation of being a formal relative is not symmetric, as the example demonstrates: *B* is *A*'s formal relative yet *A* isn't *B*'s formal relative. Clearly though, the relation of being a formal twin is symmetric.

Consider now the **MPL**-formula $\Box(p \vee q)$ and assume for example's sake that $\mathcal{L}$ is **MPL**, i.e., propositional logic supplemented with $\Box$. We consider all readings of $\Box(p \vee q)$ in which $\Box$ is interpreted as 'It is true by logical form that' and p, q are read as any (declarative) sentences. Let's focus on the reading of $\Box(p \vee q)$ as 'It is true by logical form that London is big or Paris is big'. Intuitively, this sentence is false, so that $\Box(p \vee q)$ should not appear in either TLF^+_N or TLF^+_W. Let's show this more formally and thereby illustrate the difference between narrow and wide truth by logical form.

The (up to substitution equivalence) formalisation of 'London is big or Paris is big' in **MPL** is the disjunction $r \vee s$, where r and s are sentence letters; so to determine whether 'London is big or Paris is big' is true by logical form, we must consider all relevant readings of $r \vee s$. On the narrow option, these readings construe r and s as sentences whose **MPL**-formalisation is

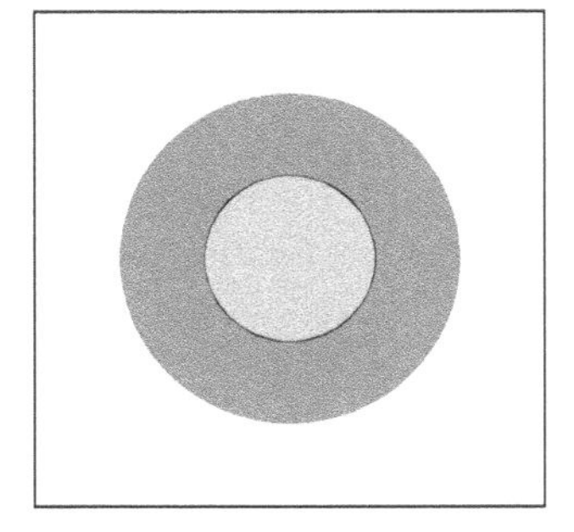

Light grey = narrow readings of $r \vee s$ (formal twins of 'London is big or Paris is big')

Dark grey = wide but not narrow readings of $r \vee s$ (formal relatives but not twins of 'London is big or Paris is big')

Dark or light grey = wide readings of $r \vee s$ (formal relatives of 'London is big or Paris is big')

Figure 8.1 Wide versus narrow readings

also a sentence letter, an example of a narrow reading of $r \vee s$ then being $A =$ 'Madrid is in Italy or Madrid is in France' (since 'Madrid is in Italy' and 'Madrid is in France' are MPL-formalised as sentence letters).[14] These readings are all and only the sentences that have exactly the same form as 'London is big or Paris is big', i.e., its formal twins. On the wide option, we consider all readings of $r \vee s$, i.e., all and only sentences whose main connective is a disjunction, since r and s can be read as anything, an example of one such reading being $B =$ 'It's raining and it's cloudy or it's snowing and it's cold'. These are, after all, all the sentences that result from interpreting the form of 'London is big or Paris is big' – the sentence's formal relatives. The difference may be displayed in a simple Venn diagram (Figure 8.1).

Now in this case, since A is false (because Madrid is neither in Italy nor France), we see that it is *neither* narrowly-true-by-logical-form *nor* widely-true-by-logical form that London is big or Paris is big, from which we deduce that $\Box(p \vee q)$ is in neither TLF^+_N nor TLF^+_W. Although there's agreement in this case, the distinction between the narrow and wide construals does make a difference, so that $\mathsf{TLF}^+_N \neq \mathsf{TLF}^+_W$, as we are about to see.

We hope readers now have a clear grasp of the distinction between wide and narrow truth by logical form, for it will be crucial in what follows. The plan for the rest of the chapter is that we discuss the narrow option in Section 8.4 and the wide one in Section 8.5. Given plausible assumptions about the logic $\mathcal{L}$, TLF^+_N is S5; in contrast, TLF^+_W is a superset of S4 but not a superset of S4M, S4.2 or S4.1.[15] As we shall see for example, on a wide

[14] Burgess (2003) calls narrow readings 'instantiations*' and distinguishes them from 'instantiations' which in our terminology are wide.

[15] 'Superset' means throughout 'proper or improper superset'.

construal, 'It's not true by logical form that it's raining' is true, whereas 'It's true by logical form that it's not true by logical form that it's raining' is false, so that $\neg\Box p \rightarrow \Box\neg\Box p$ is not in TLF^+_W. Following discussion of TLF^+_N and TLF^+_W in Sections 8.4 and 8.5, respectively, we suspend the constancy assumption in Section 8.6 and investigate TLF^-. We conclude in Section 8.7 with a word about which of TLF^+_N, TLF^+_W or TLF^- if any is *the* logic of truth by logical form.

Finally, we know from Section 8.2 that TLF^+ is a set of sentences containing all propositional tautologies and instances of the K and T schemata and closed under modus ponens and uniform substitution of sentence letters, so we need not repeat this part of the argument in Sections 8.4–8.5. We focus on how the logics TLF^+_N and TLF^+_W go beyond this set of principles.

8.4 The First Answer: TLF^+_N

We begin with the notion of narrow-truth-by-logical-form. The argument that TLF^+_N is S5's theorem set largely spells out the compressed presentation in section 2 of Burgess (1999). Our account will be much more explicit and complete. It also brings the assumption of narrowness out into the open, thereby pinpointing the difference between TLF^+_N and TLF^+_W.

- S4 axioms: instances of $\Box\phi \rightarrow \Box\Box\phi$.
  Assume that A, a reading of ϕ, is narrowly-true-by-logical-form.[16] Suppose that 'It is narrowly-true-by-logical-form that B' is a narrow reading of $\Box\text{Form}_{\mathcal{L}}(A)$. Clearly, B is a narrow reading of $\text{Form}_{\mathcal{L}}(A)$ and, precisely because readings are narrow, $\text{Form}_{\mathcal{L}}(B) = \text{Form}_{\mathcal{L}}(A)$. In other words, B and A are formal twins. Thus any (narrow) reading of $\text{Form}_{\mathcal{L}}(B)$ is a (narrow) reading of $\text{Form}_{\mathcal{L}}(A)$. As all narrow readings of $\text{Form}_{\mathcal{L}}(A)$ are true (since it is narrowly-true-by-logical-form that A, by assumption), it follows that all narrow readings of $\text{Form}_{\mathcal{L}}(B)$ are also true. This shows that it is narrowly-true-by-logical-form that B. As our argument works for any such B, it follows that all narrow readings of $\Box\text{Form}_{\mathcal{L}}(A)$ are true, so that it is narrowly-true-by-logical-form that it is narrowly-true-by-logical-form that A.
- S5 axioms: instances of $\neg\Box\phi \rightarrow \Box\neg\Box\phi$.
  Assume A, a reading of ϕ, is not narrowly-true-by-logical-form. Suppose that 'It is not narrowly-true-by-logical-form that B' is a narrow reading of $\text{Form}_{\mathcal{L}}$(It is not narrowly-true-by-logical-form that A) $=$

[16] Note that A is *any* reading of the instance of ϕ, not just a narrow one. See footnote 20 for more on this point.

$\neg\Box\text{Form}_{\mathcal{L}}(A)$. Clearly, B is a narrow reading of $\text{Form}_{\mathcal{L}}(A)$, and, precisely because readings are narrow, $\text{Form}_{\mathcal{L}}(B) = \text{Form}_{\mathcal{L}}(A)$, so that A and B are formal twins. Thus any (narrow) reading of $\text{Form}_{\mathcal{L}}(B)$ is also a (narrow) reading of $\text{Form}_{\mathcal{L}}(A)$ and vice versa. From the fact that it is not narrowly-true-by-logical-form that A, we know there is a false sentence C with $\text{Form}_{\mathcal{L}}(C) = \text{Form}_{\mathcal{L}}(A)$. So $\text{Form}_{\mathcal{L}}(B) = \text{Form}_{\mathcal{L}}(C)$, whence 'It is not narrowly-true-by-logical-form that B' is true, as B shares its $\mathcal{L}$-logical form with the false sentence C (i.e., B and C are formal twins). The argument applies to any such B, thereby demonstrating that all narrow readings of $\text{Form}_{\mathcal{L}}$(It is not narrowly-true-by-logical-form that A) are true.

- All instances of $\Box\phi$, where ϕ is a propositional tautology.
  Clearly, if A is a reading of a propositional tautology then 'It's true by logical form that A' is true.

A standard argument now shows that any set of MPL-formulas that contains all instances of the K, T, S4, S5 schemata, all propositional tautologies as well as necessitations thereof, and which is closed under modus ponens, contains all the theorems of S5.[17] We may thus conclude that Thm(S5) is a subset of TLF_N^+.

To finish off the discussion, a quick completeness argument may be given from the result of Scroggs (1951), who proved that every quasi-normal proper extension of S5 is the logic of a single finite frame.[18] (A quasi-normal logic extends K and is closed under modus ponens; it is normal if it is also closed under Necessitation.) If TLF_N^+ properly extended Thm(S5), it would contain the non-S5-validity

$$(*)\ \Box p_1 \vee \Box(p_1 \rightarrow p_2) \vee \cdots \Box((p_1 \wedge \cdots \wedge p_n) \rightarrow p_{n+1})$$

as an element, for some finite n. That is clearly not the case, as may be verified by interpreting the p_i from $i = 1$ to $n + 1$ as formally independent sentences. This we may do for any n, as there is an infinite number of atomic sentences that are mutually independent in this way. More precisely, there are infinitely many atomic sentences between whose Boolean combinations there are no non-trivial entailments.[19] (Note in particular that each sentence in a set of formally independent ones is neither true nor false by logical form.)

[17] See, e.g., Williamson (2013, 110).

[18] As noted in Burgess (1999, 177), though without the qualification that the extensions must be quasi-normal. The argument in this paragraph adapts one on p. 111 of Williamson (2013).

[19] An example of a trivial entailment is 'A_1 and not-A_1 entails B', for any B. In particular, for any n of these statements all 2^n truth-values of their Boolean combinations are formally possible.

It's plausible that English has infinitely many propositionally atomic sentences formally independent of one another. Assuming logicism is false, the sentences 'o is odd', 'The successor of o is odd', 'The successor of the successor of o is odd', etc. are a case in point. Non-mathematical examples include: 'John's children are tall', 'John's grandchildren are tall', 'John's great-grandchildren are tall', etc.

This completes the argument that TLF_N^+ is Thm(S5): TLF_N^+ is a superset of Thm(S5) but does not properly contain it.[20]

8.5 The Second Answer: TLF_W^+

Turn now to wide-truth-by-logical form. A sentence A is widely-true-by-logical-form when $\text{Form}_{\mathcal{L}}(A)$ is true under all readings in which $\Box$ is construed as 'It is widely-true-by-logical-form'; that is, when all of A's formal relatives are true. Unlike the case of narrow-truth-by-logical-form, progress on this question must depend on assumptions about $\mathcal{L}$, which helps explain why it has been overlooked in the literature. Our conclusion in this section will be that, under fairly uncontroversial assumptions about $\mathcal{L}$, TLF_W^+ is a superset of S4 but not of S4.1, S4.2 or S4M. As noted at the end of Section 8.2, TLF_W^+ is closed under uniform substitution of propositional letters by MPL-formulas.

8.5.1 S4 Axioms

Start with the case in which the language of $\mathcal{L}$ is that of MPL, for expository ease. That is to say, suppose that a sentence is true by logical

[20] TLF_N^+ is the set of MPL-formulas that are true however the sentence letters are read, on a subsequently *narrow* understanding of truth by logical form. The reading of MPL-formulas itself, however, is wide, in keeping with the motivation in Section 8.1. It is only the understanding of 'It is true by logical form that' in these MPL-formulas' readings that is narrow. A 'doubly narrow' construal would take ϕ to be in TLF^+ just when any of its *narrow* readings A is narrowly-true-by-logical-form, that is, when the readings of the sentence letters in ϕ are *atomic* natural-language sentences, i.e., sentences whose MPL-formalisation is a sentence letter. For example, a doubly narrow reading of $\Box p$ might be 'It's narrowly-true-by-logical-form that it's raining'; the reading 'It's narrowly-true-by-logical-form that it's sunny or it's not sunny' would be unacceptable, because 'It's sunny or it's not sunny' is not MPL-formalisable as a sentence letter. Although our motivation for investigating the logic of truth by logical form, set out in Section 8.1, rules out this approach, it is of formal interest to remark that the resulting set of MPL-formulas, which we may call TLF_{NN}^+, is the set S5+ (in Max Cresswell's terminology). This set contains $\neg\Box p$, for example, assuming no atomic sentence A is true by logical form, since 'It's not true by logical form that A' is true for any atomic A. The argument for this conclusion (not provided here) builds on that in Cresswell (2013, 53–7).

form just when its MPL-formalisation is a tautology, under an appropriate semantics for the language of MPL; just what this semantics might be will not matter, the only relevant assumption in the next few paragraphs being that $\mathcal{L}$'s language is the language of MPL. We come back to the more general case later.

The aim of our argument is to show that any instance of $\Box\phi \to \Box\Box\phi$ is in TLF^+_W; that is, any reading of such an instance in which $\Box$ is read as 'It is widely-true-by-logical-form that' (and the connectives are interpreted as usual) is true. Suppose then that A is a reading of the MPL-formula $\phi(p_1, \cdots, p_k)$ resulting from the reading of ϕ's sentence letters $p_1, \cdots, p_k$ as $A_1, \cdots, A_k$ respectively. Then

$$\mathrm{Form}_{\mathsf{MPL}}(A) = \phi(\alpha_1/p_1, \cdots, \alpha_k/p_k),$$

where each α_i is the MPL-formula formalising A_i; for notational simplicity, we may write $\phi(\alpha_1, \cdots, \alpha_n)$ for $\phi(\alpha_1/p_1, \cdots, \alpha_k/p_k)$.

Suppose then that it is widely-true-by-logical-form that A, i.e., that every reading of $\mathrm{Form}_{\mathcal{L}}(A)$ is true, not just readings B such that $\mathrm{Form}_{\mathcal{L}}(B) = \mathrm{Form}_{\mathcal{L}}(A)$. We must show that it is widely-true-by-logical-form that it is widely-true-by-logical-form that A, i.e., that any wide reading of $\mathrm{Form}_{\mathcal{L}}$(It is widely-true-by-logical-form that A) is true. Note that $\mathrm{Form}_{\mathcal{L}}$(It is widely-true-by-logical-form that A) $= \Box\mathrm{Form}_{\mathcal{L}}(A)$.

Suppose next that 'It is widely-true-by-logical-form that B' is a wide reading of $\Box\mathrm{Form}_{\mathcal{L}}(A)$. We must show that 'It is widely-true-by-logical-form that B' is true. Under the assumption that the language of $\mathcal{L}$ is that of MPL, write $\mathrm{Form}_{\mathcal{L}}(A)$ as $\phi(\alpha_1, \cdots, \alpha_k) = \phi(\alpha_1/p_1, \cdots, \alpha_k/p_k)$, where $p_1, \cdots, p_k$ are $\mathrm{Form}_{\mathcal{L}}(A) = \mathrm{Form}_{\mathsf{MPL}}(A)$'s sentence letters. Since B is a reading of $\mathrm{Form}_{\mathsf{MPL}}(A)$, its MPL-formalisation is $\phi(\alpha_1(q_1^1, \cdots, q_1^{n_1}), \cdots, \alpha_k(q_k^1, \cdots, q_k^{n_k}))$, where $\alpha_i(q_i^1, \cdots, q_i^{n_i})$ is the MPL-formalisation of sentence A_i that's a reading of p_i, and where $q_i^1, \cdots, q_i^{n_i}$ are the n_i sentence letters in α_i. Thus any reading of $\mathrm{Form}_{\mathsf{MPL}}(B)$ in which sentence B_i^j is a reading of the sentence letter q_i^j is a reading whose MPL-formalisation is

$$\phi(\alpha_1(\beta_1^1, \cdots, \beta_1^{n_1}), \cdots, \alpha_k(\beta_k^1, \cdots, \beta_k^{n_k}))$$

where $\beta_i^j = \mathrm{Form}_{\mathsf{MPL}}(B_i^j)$. As the notation crisply demonstrates, the highlighted formula can be obtained by substituting $\alpha_i(\beta_i^1, \cdots, \beta_i^{n_i})$ for p_i in $\phi(p_1, \cdots, p_k)$, for $i = 1$ to n. Now a reading of $\mathrm{Form}_{\mathsf{MPL}}(B)$, being a reading of $\mathrm{Form}_{\mathsf{MPL}}(A)$ where by assumption A is widely-true-by-logical-form, must be true. This shows that it is widely-true-by-logical-form that

B and hence that any reading of $\text{Form}_{\mathsf{MPL}}$(It is widely-true-by-logical-form that A) is true.

The crux of the argument just given was the following fact about **MPL**: a substitution instance of a substitution instance of a formula ϕ is a substitution instance of ϕ. Consequently, any reading C of the $\mathcal{L}$-formalisation of a reading B of the $\mathcal{L}$-formalisation of A is also a reading of A's $\mathcal{L}$-formalisation. The same argument can be rerun for any logic $\mathcal{L}$ with this property, which we may call the *transitivity of readings*.

The language of first-order logic has the property just noted, as does that of any other logic ever seriously touted as a candidate for $\mathcal{L}$. An example in the first-order case: if ϕ is $Fa \vee Gb$ then replacing F with $H \wedge I$ and b with $f(c)$ yields, after rearrangement to keep the formula well-formed, $(Ha \wedge Ia) \vee Gf(c)$. Now replacing H with $J \wedge K$ and G with $L \vee M$ yields, again following rearrangement $((Ja \wedge Ka) \wedge Ia) \vee (Lf(c) \vee Mf(c))$. This last formula can be obtained directly from $Fa \vee Gb$ by replacing F with $(J \wedge K) \wedge I$, G with $L \vee M$, and b with $f(c)$.

Since it is hard to imagine a candidate for $\mathcal{L}$ lacking transitivity-of-readings, it is a reasonable assumption that S4's theorem set is a subset of TLF^+_W. Naturally, we cannot unconditionally conclude this without knowing what the logic $\mathcal{L}$ is, a question we must remain neutral on here. We simply highlight the assumption about $\mathcal{L}$ invoked:

> Any reading C of $\text{Form}_{\mathcal{L}}(B)$ where B is a reading of $\text{Form}_{\mathcal{L}}(A)$ is also a reading of $\text{Form}_{\mathcal{L}}(A)$.

Using earlier terminology, this condition can be phrased more succinctly: if C is B's formal relative and B is A's formal relative then C is A's formal relative.

8.5.2 Necessitation

The Necessitation rule is: if ϕ is an element of the set in question then so is $\Box\phi$. If we assume that the language of $\mathcal{L}$ (properly or improperly) extends that of **MPL**, then TLF^+_W is closed under Necessitation. For suppose that ϕ is in TLF^+_W, so that every reading A of ϕ is true. Letting $\text{Form}_{\mathcal{L}}(A)$ be A's formalisation in $\mathcal{L}$, any reading B of $\text{Form}_{\mathcal{L}}(A)$ is a reading of the **MPL**-sentence ϕ. It follows that B is true, and hence that it's true by logical form that A. We may conclude that $\Box\phi$ is also in TLF^+_W.

The hypothesis about $\mathcal{L}$ just relied on is:

> Any reading B of $\text{Form}_{\mathcal{L}}(A)$ where A is a reading of the **MPL**-formula ϕ is also a reading of ϕ.

The hypothesis holds if $\mathcal{L}$ is an extension of **MPL**, which it will be on the constancy assumption (assuming logic contains propositional logic).

8.5.3 Not S5

So far, we have shown that Thm(S4) $\subseteq$ TLF^+_W, on the noted assumptions about $\mathcal{L}$. We now prove that Thm(S5), which corresponds to TLF^+_N and which has been touted as the logic of truth by logical form, is not TLF^+_W. We show that $\neg\Box p \rightarrow \Box\neg\Box p$, an instance of the distinctive S5 axiom scheme $\neg\Box\phi \rightarrow \Box\neg\Box\phi$, is not an element of TLF^+_W.

Consider $\neg\Box p \rightarrow \Box\neg\Box p$, in which we read p as, say, 'It's raining'. Clearly, it's not widely-true-by-logical-form that it's raining. But is it widely-true-by-logical-form that it's not widely-true-by-logical-form that it's raining? To determine the truth-by-logical-form status of the subsentence

> It's not widely-true-by-logical-form that it's raining,

assume to begin with that the language of $\mathcal{L}$ is that of **MPL**. The question turns on whether all wide readings of this sentence's **MPL**-formalisation, $\neg\Box p$, are true. Interpreting p as a tautology, e.g., 'It's sunny or it's not sunny', the resulting reading of $\neg\Box p$ is then *false*. That is to say, the sentence

> It's not widely-true-by-logical-form that it's sunny or it's not sunny,

a wide reading of $\neg\Box p$, is false. It follows that it's not widely-true-by-logical-form that it's not widely-true-by-logical-form that it's raining. This shows that $\neg\Box p \rightarrow \Box\neg\Box p$ is not an element of TLF^+_W, equating $\mathcal{L}$ with **MPL**.

The same argument shows that $\neg\Box p \rightarrow \Box\neg\Box p$ is not an element of TLF^+_W on the plausible assumption that the $\mathcal{L}$-formalisation of some non-formal truth (i.e., a sentence that is true but not by form) such as 'It's raining' or 'Tokyo is big' has some formal truth as a wide reading. Suppose $\mathcal{L}$ extends first-order logic and 'Tokyo is big' is $\mathcal{L}$-formalised as Fa. We may then interpret F as 'is orange or is not orange' and a as 'Donald Trump', resulting in the formal truth 'Donald Trump is orange or is not orange'. On that assumption, we may conclude that although it's not widely-true-by-logical-form that Tokyo is big, it's nevertheless not widely-true-by-logical-form that it's not widely-true-by-logical-form that Tokyo is big.[21]

A quicker way to reach the same conclusion is to notice that, on a plausible assumption about $\mathcal{L}$, TLF^+_W provides the rule of disjunction: if $\Box\phi_1 \vee \cdots \vee \Box\phi_n \in \mathsf{TLF}^+_W$ then $\phi_i \in \mathsf{TLF}^+_W$ for some i. The assumption about $\mathcal{L}$ is that there are meaningful sentences whose $\mathcal{L}$-formalisation can be (widely) read as any sentence whatsoever. If $\mathcal{L}$ includes propositional logic, for example, any sentence such as 'It's raining' which is formalisable as a single sentence letter in $\mathcal{L}$ will do, since sentence letters may be read as any sentence whatsoever. Given this assumption, suppose that

> It's widely-true-by-logical-form that A_1 or ... or It's widely-true-by-logical-form that A_n,

is true for any $A_1, \cdots, A_n$ that respectively interpret $\phi_1, \cdots, \phi_n$. Choose particular A_i (i running from 1 to n) such that $\mathrm{Form}_{\mathcal{L}}(A_i) = \phi_i$, that is to say, the sentence letters in each ϕ_i (for i from 1 to n) are read as sentences whose $\mathcal{L}$-formalisation is sentence letter. Assume without loss of generality that for this particular choice of A_1 *to* A_n, the first disjunct is true, i.e., that it's widely-true-by-logical-form that A_1. It follows that any B that's a reading of $\mathrm{Form}_{\mathcal{L}}(A_1) = \phi_1$ is true. By the choice of the A_i, the wide readings of $\mathrm{Form}_{\mathcal{L}}(A_1)$ include all the readings of ϕ_1. Conclusion: ϕ_1 is an element of $\mathsf{TLF}^W_{\mathcal{L}}$.

Observe that S5 does not provide the rule of disjunction, since for example $\neg\Box p \rightarrow \Box\neg\Box p$, equivalent to $\Box p \vee \Box\neg\Box p$, is an element of its theorem set though neither p nor $\neg\Box p$ is.

8.5.4 *Not S4M*

We saw that, on the assumptions noted in Sections 8.5.1–8.5.2, TLF^+_W contains S4's theorem set. Yet as we observed in Section 8.5.3, it does *not* contain S5's theorem set. A *McKinsey axiom* is any instance of the schema $\Box\Diamond\phi \rightarrow \Diamond\Box\phi$. S4 plus all instances of this axiom is known as S4M. Assuming that $\mathcal{L}$ contains first-order logic (plus $\Box$ on the constancy assumption), we show that TLF^+_W does *not* contain $\Box\Diamond p \rightarrow \Diamond\Box p$.

[21] At this point, we diverge from Burgess (1999), who believes that the characteristic S5 axiom is true when '$\Box$' is interpreted as 'it is true by logical form'. His argument (1999, 84–8) is essentially that spelt out in Section 8.4, which holds for the narrow but not the wide construal.

$\Box\Diamond p \rightarrow \Diamond\Box p$ is equivalent to $\neg\Box\Diamond p \vee \Diamond\Box p$ and thus to $\neg(\Box\Diamond p \wedge \Box\Diamond\neg p)$, which in any system extending K is equivalent to $\neg\Box(\Diamond p \wedge \Diamond\neg p)$. Hence it suffices to show that, for some sentence A, which is the reading of p, it is widely-true by logical form that: it's not widely-true by logical form that not-A and it's not-widely-true-by logical form that A (replacing not-not-A with A). In other words, for any reading B of this sentence's form, it's neither widely-true-by-logical-form that B nor widely-true-by-logical-form that not-B.

Suppose $\mathcal{L}$ contains first-order logic. Then there are some sentences A such that every wide reading of A's $\mathcal{L}$-formalisation is neither a logical truth nor a logical falsehood. Any sentence of the first-order form $a = f(a)$ will do, where a is a constant and f a monadic function symbol. An example of such a sentence is 'Al is Al's father'. The sentence's formal relatives all have $\alpha = \Phi(\alpha)$ as their form, where α is a (possibly complex) term and Φ is a (possibly complex) functional term. As is easy to check, $\alpha = \Phi(\alpha)$ has both true readings and false ones, since we can interpret Φ as a function that maps the referent of α to itself or alternatively as a function that maps α's referent to a distinct entity. Hence no reading of the formalisation of 'Al is Al's father' (i.e., of $a = f(a)$) nor any reading of its negation's formalisation is true by logical form. The assumption here is that $\mathcal{L}$ contains first-order logic, something virtually all logicians would go along with.

8.5.5 Neither S4.2 nor S4.1

In (a corrected version of)[22] the containment diagram on p. 367 of Hughes and Cresswell (1996) of normal modal systems, the three systems directly above S4 are S4M, S4.1 and S4.2. We have already seen that Thm(S4M) is not a subset of TLF^{+}_{W}. We now show that neither Thm(S4.2) nor Thm(S4.1) is either.[23]

The logic S4.2 adds to S4 all instances of the McKinsey axiom's converse

$$\Diamond\Box\phi \rightarrow \Box\Diamond\phi$$

If $\Diamond\Box p \rightarrow \Box\Diamond p$ were an element of TLF^{+}_{W}, so would $\Box\neg\Box p \vee \Box\Diamond p$ be (writing $\rightarrow$ in terms of $\vee$ and $\Diamond$ as $\neg\Box\neg$). On the assumption that TLF^{+}_{W} provides the rule of disjunction, we may then infer that either $\neg\Box p$ is in TLF^{+}_{W}, or that $\Diamond p$ is, each of which leads to a contradiction. To see this, interpret p in the sentence $\neg\Box p$ as a formal truth (i.e., a sentence true by

[22] There should be no arrow in the diagram between S4.2 and S4.1, since the former is not a supersystem of the latter. Thanks to Max Cresswell for drawing our attention to this error.

[23] We are grateful to Weng Kin San for the S4.1 argument.

logical form); and p in $\Diamond p$ as a formal falsehood. This shows that TLF^+_W is not a superset of S4.2. So we have a yet further way to establish that TLF^+_W is not Thm(S5), since S5 extends S4.2.

We could reach the same result without assuming that TLF^+_W provides the rule of disjunction. Read p in $\Diamond\Box p \rightarrow \Box\Diamond p$ as an atomic sentence such as 'It's raining'. Some reading of the formalisation of 'It's true by logical form that it's raining', i.e., $\Box p$, is true (simply interpret p as a formal truth). But not every reading of the formalisation of 'It's formally possible that it's raining', i.e., $\Diamond p$, is true (simply interpret p as a formal falsehood). So $\Diamond\Box p \rightarrow \Box\Diamond p$ is not in TLF^+_W, because if we read p as 'It's raining' and $\Box$ as 'It's true-by-wide-logical-form that', the resulting conditional has a true antecedent and a false consequent.

Turning next to S4.1, this logic adds to S4 all instances of:

$$\Box(\Box(\phi \rightarrow \Box\phi) \rightarrow \phi) \rightarrow (\Diamond\Box\phi \rightarrow \phi),$$

As shown in Goldblatt (1973), S4.1 in turn extends S4.01 which can be obtained by adding to S4 each instance of

$$\Diamond\Box\phi \rightarrow \Box(\Box\Diamond\phi \rightarrow \Diamond\Box\phi),$$

which (as in the last but one paragraph) is equivalent to $\Box\neg\Box\phi \vee \Box(\Box\Diamond\phi \rightarrow \Diamond\Box\phi)$.[24] On the assumption that TLF^+_W provides the rule of disjunction, from the hypothesis that $\Box\neg\Box p \vee \Box(\Box\Diamond p \rightarrow \Diamond\Box p)$ is an element of TLF^+_W, we deduce that one of $\neg\Box p$ or $\Box\Diamond p \rightarrow \Diamond\Box p$ also is. But the latter is an instance of the McKinsey axiom, which we showed earlier is not in TLF^+_W. And $\neg\Box p$ is clearly not in TLF^+_W either (interpret p as a formal truth). It follows that $\Diamond\Box p \rightarrow \Box(\Box\Diamond p \rightarrow \Diamond\Box p) \notin \mathsf{TLF}^+_W$, so TLF^+_W is not an extension of S4.01 nor, therefore, of S4.1.[25]

8.5.6 Upshot

Under the stated assumptions about $\mathcal{L}$, Thm(S4) is a subset of TLF^+_W; however, none of Thm(S4M), Thm(S4.2) or Thm(S4.1) is a subset of TLF^+_W. A corollary of particular interest is that TLF^+_W is not identical to Thm(S5).

Although we have not identified the set TLF^+_W precisely, we have narrowed down the possibilities to a fairly limited range. Given our assumptions, TLF^+_W is a normal modal logic at least as strong as S4 but

[24] This is the axiom Goldblatt (1973, 567) labels Γ2.

[25] In contrast to the case of S4.2, we are not aware of a proof that does not use the disjunction rule.

strictly weaker than S4M, S4.2 and S4.1 (and S4.01). In headline terms, TLF^+_W is either S4 or some very weak extension of S4.

8.6 The Third Answer: TLF^-

So far we granted the constancy assumption for the sake of argument and investigated TLF^+_N and TLF^+_W, which rested on it. We turn now to the third and final answer, on the assumption that 'It is true by logical form' is *not* a logical constant. In other words, when we consider whether ϕ is in TLF^-, we must interpret $\Box$ as usual to mean 'It is true by logical form that'; but *B* may now be *A*'s formal relative without containing any occurrences of 'It is true by logical form that', even if *A* does. For instance, 'It's probable that it's raining' is a formal relative of 'It's logically true that it's raining', as the former is a reading of the latter's formalisation, $\Box p$, in which $\Box$ is read as the sentential operator 'It's probable that'. More generally, replacing 'It is true by logical form that' in any (declarative) sentence *A* with any sentential operator results in a formal relative of *A*.

Clearly, every instance of a propositional tautology is a member of TLF^-, since any of its readings is true however $\Box$ is interpreted, and a fortiori when $\Box$ is read as 'It is true by logical form that'. Any instance of the K-axiom scheme is also an element of TLF^-: if 'If *A* then *B*' and *A* are both true by logical form then so is *B*; the argument here is exactly as in Section 8.2. Similarly, the set TLF^- is closed under modus ponens as well as any other valid propositional inference rule. It also contains any instance of the T-axiom, since if it is true by logical form that *A* then *A*.

Finally, TLF^- contains any formula of the form $\Box\phi$ where ϕ is a substitution instance of a propositional tautology, since readings of any such sentence are true by logical form.

TLF^-, however, is not closed under the rule of Necessitation, nor does it contain the S4-instance $\Box p \to \Box\Box p$, nor the S5-instance $\neg\Box p \to \Box\neg\Box p$. To see why Necessitation fails, observe that $\Box\Box(p \vee \neg p)$ is *not* in TLF^-. Consider the sentence

> (*A*) It is true by logical form that it is true by logical form that it's raining or it's not raining,

a reading of $\Box\Box(p \vee \neg p)$. For this sentence to be true, every formal relative of

> (*B*) It is true by logical form that it's raining or it's not raining

must be true. One of B's formal relatives – given that 'It is true by logical form' is no longer regarded as a logical constant but may be replaced by any sentential operator – is the sentence $G =$ 'Intuitionists believe that Goldbach's Conjecture is true or that Goldbach's Conjecture is not true'. As intuitionists have no grounds for asserting this instance of the Law of Excluded Middle, G is false. So B has a false formal relative, and thus A is false.

Similarly, the necessitated T-axiom-instance $\Box(\Box p \rightarrow p)$ is not in TLF^-. It is true, but not by logical form, that

> If it's true by logical form that it's raining then it's raining,

since for example the sentence 'If there ought to be no wars then there are no wars', which is also a reading of $\Box p \rightarrow p$ (as the constancy assumption has been jettisoned), is false, assuming its antecedent is true and its consequent false. Similar arguments show that the sentences $\Box p \rightarrow \Box\Box p$ and $\neg\Box p \rightarrow \Box\neg\Box p$ are not in TLF^-.

A subtlety is worth noting. TLF^- is the set of MPL-formulas true under any reading in which $\Box$ is interpreted as 'It is true by logical form that', on the assumption that this operator is not logical. To determine whether an MPL-sentence ϕ is an element of TLF^-, we check the truth-values of ϕ's readings when $\Box$ is thus read. But once ϕ has been so interpreted, as A say, then since (on this assumption) 'It is true by logical form that' is *not* a logical constant, the determination of A's truth-value must take into account readings of A's formalisation in which the sentential operator need no longer be exclusively construed as 'It is true by logical form that'. Readings of MPL-formulas in which $\Box$ is always understood as 'It is true by logical form that' should *not* be confused with readings of the resulting sentences' $\mathcal{L}$-formalisations, which need not construe $\Box$ in the same way.

The rest of Section 8.6 presents an algorithm for TLF^--membership. We describe the algorithm, illustrate it with two examples, and conclude the section with a brief justification of why it works.

8.6.1 *Definitions*

As is familiar, the *modal degree* of an MPL-formula is its highest number of embeddings of $\Box$.[26] An *ultimate occurrence* of $\Box$ in an MPL-formula is

[26] More formally, the modal degree of a sentence letter is 0, the modal degree of a Boolean combination of formulas is the maximum of these formulas' degree, and the modal degree of $\Box\phi$ is one greater than ϕ's modal degree.

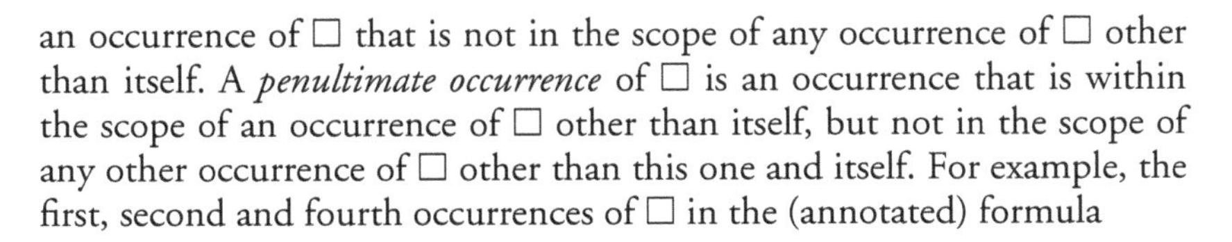

an occurrence of $\Box$ that is not in the scope of any occurrence of $\Box$ other than itself. A *penultimate occurrence* of $\Box$ is an occurrence that is within the scope of an occurrence of $\Box$ other than itself, but not in the scope of any other occurrence of $\Box$ other than this one and itself. For example, the first, second and fourth occurrences of $\Box$ in the (annotated) formula

$$\underset{u}{\Box} q_1 \vee \underset{u}{\Box}\underset{p}{\Box} q_2 \vee \underset{u}{\Box}\underset{p}{\Box}\Box\Box q_3$$

are ultimate occurrences (u), the third and fifth are penultimate occurrences (p), and the sixth and seventh are neither ultimate nor penultimate occurrences. We note that if ϕ has modal degree $N \geq 2$ then ϕ contains at least one penultimate occurrence of $\Box$; moreover, at least one such occurrence is the leading connective of a subformula of modal degree $N-1$.

8.6.2 *The Algorithm*

Given an **MPL**-sentence ϕ, define another **MPL**-sentence ϕ^* from it as follows. If the modal degree of ϕ is 0 or 1, set $\phi^* = \phi$. If ϕ's modal degree is $N \geq 2$, enumerate all of ϕ's subformulas of the form $\Box\psi$ in which the initial box in $\Box\psi$ is a penultimate occurrence of box in the formula ϕ. Suppose this enumeration is $\phi_1, \cdots, \phi_M$. N.B. Each ϕ_i, as i runs from 1 to M, corresponds to *all* occurrences of the same subformula $\Box\psi_i$ in which the initial box in $\Box\psi_i$ is a penultimate occurrence of box in ϕ. For example, both occurrences of $\Box p$ in $\Box(r \rightarrow \Box p) \wedge \Box(\Box r \rightarrow \Box p)$ correspond to ϕ_1 and the (single) occurrence of $\Box r$ in the second conjunct corresponds to ϕ_2. Let ϕ^* be the result of uniformly replacing each occurrence of $\Box\psi_i$ in ϕ by the sentence letter q_i, for $i = 1$ to M, where the q_i are distinct from one another and from any sentence letters appearing in ϕ. In the example just given, ϕ^* would be $\Box(r \rightarrow q_1) \wedge \Box(q_2 \rightarrow q_1)$. Clearly, ϕ^* is always of modal degree 1 or 0 (it is 0 iff ϕ's modal degree was 0). Now run one of the algorithms for Thm(S5)-membership on ϕ^*: if according to this algorithm, ϕ^* is an element of Thm(S5) then $\phi \in \mathsf{TLF}^-$; if not, $\phi \notin \mathsf{TLF}^-$.

8.6.3 *Examples*

For our first example, take ϕ to be

$$\neg p_1 \vee \Box p_2 \vee \Box(p_1 \rightarrow \Box(p_2 \rightarrow p_3)) \vee \Box(\Box(\neg p_1 \vee p_4) \leftrightarrow \Box(p_1 \rightarrow p_4))$$

ϕ consists of four disjuncts and its modal degree is 2. The subformulas of ϕ which start with a penultimate occurrence of $\Box$ are $\Box(p_2 \rightarrow p_3)$, $\Box(\neg p_1 \vee p_4)$ and $\Box(p_1 \rightarrow p_4)$. Replacing these with q_1, q_2 and q_3 in ϕ yields

$$\phi^* = \neg p_1 \vee \Box p_2 \vee \Box(p_1 \rightarrow q_1) \vee \Box(q_2 \leftrightarrow q_3)$$

Running an algorithm for Thm(S5)-membership shows that ϕ^* is not in Thm(S5), from which we may conclude that ϕ is not in TLF^-.

If instead ϕ were

$$\neg p_1 \vee \Box p_2 \vee \Box(p_1 \rightarrow \Box(p_2 \rightarrow p_3)) \vee \Box(\Box(p_1 \rightarrow p_4) \leftrightarrow \Box(p_1 \rightarrow p_4)) \vee \Box\Box\Box p_5,$$

ϕ^* would be

$$\neg p_1 \vee \Box p_2 \vee \Box(p_1 \rightarrow q_1) \vee \Box(q_2 \leftrightarrow q_2) \vee \Box q_3$$

Since $\Box(q_2 \leftrightarrow q_2)$ is an element of Thm(S5), the original formula in this second example is in TLF^-.

8.6.4 *Justification Sketch*

The justification for the algorithm is in three parts. We sketch each of the parts, leaving details to the interested reader.

1) First, replace the subformulas $\Box\psi_1, \cdots, \Box\psi_M$ uniformly by the new sentence letters $q_1, \cdots, q_M$, where the initial $\Box$ in $\Box\psi_1$ is a penultimate occurrence of $\Box$ in ϕ.[27] This is justified by the fact that all these formulas are in the scope of $\Box$, so that their readings are in the scope of 'It is true by logical form that'. When determining whether a sentence of the form 'It is true by logical form that *A*' is true, formalisations of further occurrences in *A* of 'It is true by logical form that' are reinterpretable as any sentential operator whatsoever. *This is simply a consequence of rejecting the constancy assumption.*

To illustrate the point, consider the formula $\Box(\Box p \leftrightarrow \Box\neg\neg p)$. The formula $\Box(\Box p \leftrightarrow \Box\neg\neg p)$ is in TLF^- iff any reading of $\Box p \leftrightarrow \Box\neg\neg p$ is true. Given that $\Box$ is not a logical constant, one such reading of $\Box$ is 'Alogical Alf believes that'. Now alogical Alf is a character whose belief

[27] When ϕ's modal degree is less than 2 this vacuous uniform replacement results in $\phi^* = \phi$ since ϕ contains no penultimate occurrences of $\Box$.

system violates the logical equivalence '*A* iff not-not-*A*': for some *A*, Alf believes that *A* but does not believe not-not-*A*. Since $\Box p \leftrightarrow \Box\neg\neg p$ is not true under the reading of $\Box$ as 'Alogical Alf believes that' and of *p* as *A*, it's *not* true by form that:

It's true by form that *A*
iff
It's true by form that not-not-*A*,

as this biconditional's form ($\Box p \leftrightarrow \Box\neg\neg p$) has false readings (once we reject the constancy assumption). Given the existence of such readings of $\Box$, $\Box(\Box p \leftrightarrow \Box\neg\neg p)$ is in TLF^- just when $\Box(q_1 \leftrightarrow q_2)$ is; and it is readily seen that the latter is not a member of TLF^-. More generally, one can come up with similar readings of $\Box$ in the formalisations of readings of MPL-formulas, so as to 'break' any logical equivalences. The constraint of uniformity nevertheless applies since any occurrence of $\Box\psi_1$ must be interpreted in the same way. For example, $\Box(\Box p \leftrightarrow \Box p)$ *is* in TLF^-, since whatever the reading of $\Box p$, a biconditional whose antecedent and consequent are identical is true by logical form.

2) The second part of the justification is to explain why, when ϕ has modal degree 0 or 1, the following equivalences obtain:

$$\phi \in \mathsf{TLF}^- \text{ iff } \phi \in \mathsf{TLF}^+_W \text{ iff } \phi \in \mathsf{TLF}^+_N$$

Assume in this paragraph and the next two that ϕ has modal degree 0 or 1. Suppose first that $\phi \notin \mathsf{TLF}^+_N = \text{Thm(S5)}$. As S5 has the finite model property, ϕ is false at a world w_1 in a model with $n+1$ worlds $w_1, \cdots, w_{n+1}$. Let $A_1, \cdots, A_n$ be n formally independent sentences of English with no occurrences of 'It is true by logical form that' such that A_1 is true and the others are false.[28] Associate A_i with w_i for $i = 1$ to n and not-(A_1 or ... or A_n) with w_{n+1}. For example if $n = 2$, A_1 might be 'China's flag is mainly red', A_2 'China's flag is mainly blue' and not-(A_1 or A_2) 'It is not the case that China's flag is either mainly red or that it is mainly blue'. If a sentence letter *p* is true in worlds $w_{k_1}, \cdots, w_{k_m}$ in the specified countermodel for ϕ, interpret *p* as the disjunction of those sentences among $A_1, \ldots, A_n$, not-(A_1 or ... or A_n) associated with the worlds $w_{k_1}, \cdots, w_{k_m}$; e.g., if *p* is true at w_2, w_3 and w_{n+1}, interpret *p* as 'A_2 or A_3 or not-(A_1 or ... or A_n)'. We also interpret any occurrences of $\Box$ in ϕ as 'It is true by logical form that'. An

[28] See the discussion in Section 8.4. Note that the truth of A_1 guarantees the falsehood of not-(A_1 or ... or A_n).

easy check shows that the resulting reading of ϕ is false. As an inductive argument on complexity demonstrates, the truth-value of ψ under this reading, where ψ is a subformula of ϕ (and thus has modal degree 0 or 1), is ψ's truth-value at w_1 in the countermodel for ϕ, so that the resulting reading of ϕ is false, as ϕ is false at w_1 in the countermodel. Since ϕ is of degree 0 or 1, the logical constancy or non-constancy of the operator 'It is true by logical form that' doesn't come into it; and since the A_i are formally independent, the issue of narrow vs wide construals does not arise either. Intuitively, we turned an assignment of truth-values to formulas of degree 0 or 1 under which ϕ is false, as specified by the truth-values at w_1 in the countermodel, into a false natural-language reading of ϕ. In sum, we've shown that if $\phi \notin \mathsf{TLF}^+_N = \text{Thm}(S5)$ then $\phi \notin \mathsf{TLF}^+_W$ and $\phi \notin \mathsf{TLF}^-$.

For the second half of the second step, suppose that $\phi \in \mathsf{TLF}^+_N$, where as explained earlier $\mathsf{TLF}^+_N = \text{Thm}(S5)$. Now if ϕ is of modal degree 0 or 1, $\phi \in \text{Thm}(S5)$ iff $\phi \in \text{Thm}(T)$.[29] Thus $\phi \in \text{Thm}(T)$. As we saw in Section 8.5, since $\text{Thm}(T) \subseteq \mathsf{TLF}^+_W$ (under the stated assumption about $\mathcal{L}$), it follows that $\phi \in \mathsf{TLF}^+_W$. It remains to show that if $\phi \in \mathsf{TLF}^+_N = \text{Thm}(S5)$ then $\phi \in \mathsf{TLF}^-$.

Turning to this argument, any **MPL**-formula of degree 0 or 1 is equivalent to a conjunction of disjunctive formulas of the form

$$\beta \vee \Box\gamma_1 \vee \cdots \vee \Box\gamma_n \vee \Diamond\delta,$$

where $\beta, \gamma_1, \cdots, \gamma_n, \delta$ are all propositional formulas. This standard result (see e.g., Hughes and Cresswell (1996, 104)) follows immediately from propositional logic's Conjunctive Normal Form Theorem and the fact that $\Diamond(\delta_1 \vee \delta_2)$ is equivalent to $\Diamond\delta_1 \vee \Diamond\delta_2$.[30] Moreover, if $\phi \in \text{Thm}(S5)$ then at least one of $\beta \vee \delta, \gamma_1 \vee \delta, \cdots \gamma_n \vee \delta$ is a propositional tautology,[31] for each of ϕ's conjuncts of this form. Now if $\beta \vee \delta$ is a propositional tautology, any interpretation of $\beta \vee \Diamond\delta$, and a fortiori of $\beta \vee \Box\gamma_1 \vee \cdots \vee \Box\gamma_n \vee \Diamond\delta$, is true, since from the fact that '*A* or *B*' interprets a propositional tautology it follows that '*A* or it's formally possible that *B*' is true, for any *A* and *B*.

[29] If ϕ is in Thm(T) then ϕ is in Thm(S5), whatever ϕ's modal degree. For the converse when ϕ is of modal degree 0 or 1, suppose ϕ is not true at a world w in some model based on a reflexive frame. Then ϕ is not true at the same world in the model whose worlds are the w-accessible worlds in the original countermodel for ϕ and whose accessibility relation is stipulated to be universal (keeping assignments of truth-values to sentence letters at worlds the same). N.B. This converse crucially relies on ϕ being of modal degree 0 or 1.

[30] The word 'equivalent' in the previous two sentences means: equivalent in any system which contains all instances of the K axiom schema $\Box(\phi \rightarrow \psi) \rightarrow (\Box\phi \rightarrow \Box\psi)$ for propositional ϕ and ψ, which contains all instances of ϕ and $\Box\phi$ for ϕ a propositional tautology, and which is also closed under modus ponens. TLF^- is such a system.

[31] Hughes and Cresswell (1996, 103–5).

Similarly, if $\gamma_i \vee \delta$ is a propositional tautology, any reading of $\Box\gamma_i \vee \Diamond\delta$ is true, since from the fact that it's true by logical form that (*A* or *B*), it follows that it's either true by logical form that *A* or that it's formally possible that *B* (i.e., *B* is not false by logical form). Thus $\phi \in \mathsf{TLF}^-$.

3) The final part of the justification involves determining whether ϕ^* is an element of Thm(S5). Since S5 has the finite model property, membership in this set is decidable. An efficient algorithm for membership of modal degree 0 or 1 formulas in Thm(S5) exists and its familiar details were alluded to in the previous paragraphs. These details may be found in Hughes and Cresswell (1996, 103–5).

8.7 Closing

The chapter's main aim was to show that TLF, the set of MPL-formulas true by logical form, depends on how we understand this notion. Based on the stated assumptions about $\mathcal{L}$, the three answers, given in Sections 8.4–8.6, were:

(a) Constancy assumption, narrow construal of truth by logical form: $\mathsf{TLF}_N^+ = \text{Thm(S5)}$.

(b) Constancy assumption, wide construal of truth by logical form: TLF_W^+ is a normal modal logic that is a superset of Thm(S4) but not of Thm(S4M) or Thm S4.2 or Thm(S4.1). A fortiori, $\mathsf{TLF}_W^+ \neq \text{Thm(S5)}$.

(c) Reject the constancy assumption: TLF^- is a weak non-normal modal logic; we provided an algorithm for TLF^--membership.

As mentioned earlier, many of our results can be carried over to the case of model-theoretic logical truth.

Whether the constancy assumption holds depends of course on what the logical constants are. This is a vexed issue in the philosophy of logic, on which progress has been made in the past few decades but which remains unsettled.

Whether the narrow or wide construal, on the constancy assumption, is preferable remains moot. A consideration in favour of the narrow construal is that it is natural to hear '*A* is true by logical form' as restricting attention to sentences whose form is the same as *A*'s form. *A* may be argued to be true by logical form if any sentence *B* with the *very same form – the exact same form* – as *A* is true. So only *A*'s formal twins need be considered. Admittedly, this construal is quite sensitive to the operator's linguistic expression.

And such intuitive judgements are heavily theory-laden, so it is perhaps unwise to rest too much on them.

A consideration in favour of the wide construal is that it would be odd to call A true by logical form if A has a false formal relative, as A's form would then lend itself to a false reading. Standard logical practice certainly allows wide readings of predicates, constants and sentences. For example, we take any natural-language readings of $Fa \vee \neg Fa$ to be true by form, not just readings in which a is interpreted as an atomic term and F as an atomic predicate, whatever exactly that would mean. If we call the demand that atomic sentence letters be read as atomic sentences *atomicity*, then in most discussions of logical form, atomicity is not assumed. When, for example, we claim that 'A and B' implies A where A and B are natural-language sentences, we of course intend this to include all atomic instances like 'It's raining and it's cloudy' implies 'It's raining'. But we also intend to include all non-atomic instances like

It's raining and it's cloudy, and it's Monday.

———————————

It's raining and it's cloudy.

Similarly, when we offer grammatical or semantic characterisations of propositional logic, atomicity is not assumed.[32]

Arguably, though, there is no need to choose between truth-by-wide-logical-form and truth-by-narrow-logical-form. They are both coherent notions of truth by logical form, which spell out different ways of understanding the informal notion (on the constancy assumption). At any rate, our aim in this paper has been less to gesture towards a particular set of **MPL**-formulas as *the* logic of truth by logical form than to explain why there are potentially several different such.[33]

[32] As noted in Burgess (2003), to make a slightly different point. There are rare cases when atomicity may be desirable. Nevertheless, the demand of atomicity is sufficiently rare that, when required, it is reasonable to demand that it be flagged. Smiley's (1982) discusses this and related issues in more detail.

[33] Thanks to students in our Hilary 2017 Logical Consequence seminar for comments, to an audience at the IHPST in Paris, to John Burgess, Weng Kin San and Tim Williamson, to the editors Gil Sagi and Jack Woods III both for comments and for including our piece in their volume, and finally special thanks to Max Cresswell for many helpful comments and encouragement.

CHAPTER 9

Reinterpreting Logic

Alexandra Zinke

According to the classical interpretational definition of logical truth, going back to Alfred Tarski, a sentence is logically true if and only if it is true under all interpretations of the non-logical terms. Obviously this definition rests on a distinction of logical from non-logical terms. Which sentences turn out to be logically true hinges on which terms are deemed logical and thereby exempted from reinterpretation. Tarski considers the *problem of the logical terms* the core problem of his definition (and the bulk of the literature has followed him in this[1]): 'Underlying our whole construction is the division of all terms of the language discussed into logical and extra-logical. This division is certainly not quite arbitrary. ... On the other hand, no objective grounds are known to me which permit us to draw a sharp boundary between the two groups' (Tarski 1936, 418–19).

While the problem of logical terms is indeed important, I think that it has eclipsed other, more fundamental difficulties. As will be illustrated in the next sections, the interpretational definition cannot demand truth under literally *all* interpretations of the non-logical terms, unless there be no logical truths whatsoever. But if, besides the restriction of reinterpretation to non-logical terms, we need additional restrictions on interpretations, we must revise the definition. The most non-committal interpretational definition of logical truth runs as follows:

> (LT) A sentence is logically true if and only if it is true under all *admissible* interpretations.

This formulation is neutral with respect to different possible restrictions. The central problem, then, is that of demarcating the admissible from

I am grateful to Gregor Betz, Wolfgang Freitag, Gila Sher, Wolfgang Spohn, and especially Gil Sagi and Jack Woods for very helpful comments.

[1] For surveys of the relevant literature, see, e.g., Gómez-Torrente (2002), Bonnay (2014), and MacFarlane (2017). But see also Sagi (2014), who challenges the centrality of logical terms for the notions of logical truth and formality.

the inadmissible interpretations. I call this the problem of admissible interpretations. The problem of logical terms is just a subproblem of this more general difficulty.

It has long been acknowledged in the literature that a viable interpretational definition of logical truth calls for restrictions beyond that concerning logical terms. However, the additional restrictions are usually only mentioned in passing, and the need to provide a rationale for them has often been overlooked. This would not be a problem if we could fill the lacuna easily. But we can't, at least not easily. I will argue that there is no straightforward justification for the common line of demarcation between admissible and inadmissible interpretations. As a consequence, the extension of 'logically true' rests on what appear to be unjustified choices.

Before we can unfold the problem of admissible interpretations, some clarifying remarks regarding the target of the discussion are necessary. In particular, we have to discuss whether the problem of admissible interpretations applies only to logical truth in natural language, or whether it also crops up for formal languages. Consider the traditional problem of logical terms. As formal languages have an explicitly defined syntax and semantics, they come with a fixed set of logical terms. In first-order predicate logic (PL), for instance, the logical terms are '$\forall$', '$\neg$', '$\vee$', and, possibly, '$=$'.[2] All other terms, i.e., individual and relation constants and individual variables, are deemed non-logical. This division of terms is also semantically encoded: an interpretation function for PL is defined only on the non-logical terms, i.e., the individual and relation terms, while the logical terms get their meaning via the recursive truth-definitions. Thus, if we consider a sentence under all interpretations of PL, we automatically only consider reinterpretations of the non-logical terms. More generally, an interpretation function of a formal language is defined only on the non-logical terms, and each formal language comes with a predefined set of logical terms. Because of the former, no explicit restriction confining reinterpretation to non-logical terms seems necessary, and because of the latter, the impact of the (non-explicit) restriction is already determined: in PL, for instance, interpretations that reinterpret, e.g., the '$\vee$' as expressing the operation of conjunction, are inadmissible.

However, recourse to formal languages yields only apparent relief from the demarcation problem. Different formal languages come with different

[2] For reasons of simplicity, here and in what follows we assume that the alphabet of PL only contains these logical terms as primitives, and not, e.g., '$\exists$', '$\wedge$', '$\rightarrow$', '$\leftrightarrow$'.

sets of logical terms, as shown by existing formal languages, and we could invent new formal languages with yet alternative selections of logical terms. Thus we still have a choice between different sets of logical terms within the realm of formal languages, though it is somewhat indirect: we choose the logical terms by choosing the formal language. Moreover, not only do we have to make our choices, e.g., decide which terms are the logical ones and decide not to reinterpret the logical terms, but we also have to substantiate them in order to justify the delineation between logical and non-logical truths. For formal languages, the problem of logical terms consists in justifying why we prefer one formal system with an associated set of logical terms over an alternative formal system with a different set of logical terms.

The same holds true for the problem of admissible interpretations generally. Many restrictions on reinterpretation are already implicitly encoded in our standard formal languages. For example, we will introduce a restriction that confines the admissible interpretations to the grammatically faithful ones. The interpretation function of, say, PL automatically interprets the terms according to those grammatical categories that PL itself establishes, namely the categories of individual constants and relation constants (of different arities). Thus, for PL and similar formal languages, no explicit grammatical restrictions seem necessary. But, again, this does not resolve our quandary. PL provides one classification of terms and thus encodes some grammatical restriction on interpretations. However, different formal languages may come with different grammatical categories, and thus encode different grammatical restrictions.[3] The original problem of choosing and justifying a particular grammatical restriction turns into the problem of justifying the choice of the formal language.

Recourse to formal languages does not help to resolve the problem of admissible interpretations. In unfolding the problem, I will therefore concentrate on natural language – or the philosopher's toy fragment of natural language. This also serves another aim. I am a staunch supporter of the view that, as philosophers, we are ultimately interested in logical truth in natural language, and logical validity of natural-language arguments.[4]

[3] See Freitag and Zinke (2012) for a general formal system that allows for arbitrary grammatical restrictions.

[4] Even Tarski, who so prominently confined his research to formal languages, remarks that the formal languages or 'formalized languages' he is interested in can be understood as 'fragments of natural language': 'I should like to emphasize that, when using the term "formalized languages", . . . I do not have in mind anything essentially opposed to natural languages. On the contrary, the only formalized languages that seem to be of real interest are those which are fragments of natural languages . . . or those which can at least be adequately translated into natural languages' (Tarski 1960, 68).

Thus I will deliberately use natural-language examples of logical truths, such as 'Mary is a mathematician or Mary is not a mathematician', and I will talk of the reinterpretation of say, 'Mary' as referring to Harry, or the reinterpretation of 'is a mathematician' as expressing the property of being a musician. Despite the focus on natural-language examples, however, I will also comment on the roles of the proposed restrictions on interpretations in formal languages.

Here is the outline of the paper. In the first section, I will make explicit the grammatical restriction on interpretations and discuss its role for the distinction between logical truth and structural truth. The second section addresses the demand for uniform reinterpretation and argues that there is a natural generalisation of this constraint that will cast into doubt the demarcation of analytically true sentences from logically true ones. The third and final section shows that the plausibility of the interpretational definition rests on certain non-trivial metaphysical background assumptions. If restrictions motivated by these background assumptions are imposed, even some alleged paradigm cases of merely metaphysically necessary sentences turn out to be logically true. Without a proper justification of the traditional delineation of admissible interpretations, the classical boundary between logical and non-logical truths crumbles.

Let me add a note of caution. While I will suggest specific answers to the problem of admissible interpretations as we proceed, the main purpose of the paper is not to defend any particular answers, but to unfold the challenge, to illustrate its scope, and to stress its relevance. If my specific answers should turn out to be wrong, the problem of admissible interpretations remains unsolved until more satisfactory answers are provided.

9.1 Grammatical Restrictions

According to (LT), a sentence is logically true if and only if it is true under all admissible interpretations. Let me now discuss several dimensions of the problem of admissibility, starting with the grammatical dimension.

A language, be it natural or formal, usually contains terms of different grammatical categories. For example, PL contains individual and relation constants. The grammatical category of a term determines its syntactic properties (given the syntax rules of the language). Grammatical categories typically correspond to semantic categories: terms of different grammatical categories are also of different semantic categories, i.e., refer to different types of entities, and vice versa. For example, in PL, terms belonging to the grammatical category of individual terms refer to elements of the domain, while terms belonging to the grammatical category of relation terms refer to sets of (n-tuples of) elements of the domain. I will assume such a 'harmony' between grammatical/syntactical categories and semantic categories here.[5]

Thus the grammatical categories of the terms play a crucial role in the division of interpretations into admissible and inadmissible ones. When we claim that a sentence is true under all interpretations, we not only ignore interpretations that reinterpret the logical terms, but also interpretations that do not respect the grammatical categories of the terms. For example, we ignore interpretations that assign to an individual term the semantic value of a predicate term or vice versa. If we did not disregard such interpretations, any sentence would under some live interpretation fail to express a meaningful proposition, and – given that meaningless sentences have no truth values – no sentence would be true under all admissible interpretations.[6] There always has been, and always must be, some *grammatical restriction* on interpretations. The crucial question is now, which grammatical restrictions should be in place and for what reason.

In its most common form, the grammatical restriction says that terms must be reinterpreted within the confines of their grammatical category. It thus rests on a classification of terms into grammatical categories. The chosen classification has severe implications for the class of logical truths. If we choose a fine-grained classification of terms, many interpretations will be deemed inadmissible and more sentences will turn out to be logically true, whereas if we choose a more liberal classification, fewer sentences will qualify as logically true. It is therefore all the more important to take a closer look at the grammatical restriction.

To begin with, let us consider the role of the grammatical restriction in formal languages. The first thing to note is that in common formal languages categorical differences are typographically encoded. In PL, for example, individual terms are usually lower-case letters taken from the beginning of the alphabet, such as a or c, while predicate and relation terms are capital letters like F or G.[7] Moreover, the grammatical restriction is already implicit in the definition of an interpretation function of PL and related formal languages: interpretation functions of PL map individual

[5] For more on this, see, e.g., Jackson (2017) and Trueman (2018).

[6] See, however, Magidor (2009), who argues against the meaninglessness of such type confusions.

[7] Also Tarski remarks that in those formal languages he considers, the category of a term can be read off its typographic form. See Tarski (1935, 217, fn. 1).

terms on elements of the domain and relation terms on sets of (tuples of) elements of the domain. Just as the interpretation function of PL is defined in such a way that it respects the distinction between the logical and the non-logical terms, so it is also automatically faithful to those grammatical categories PL itself establishes. If we only apply the interpretational definition to formulas of PL or related formal languages, no explicit grammatical restriction is called for.

However, PL is just an arbitrary formal language. Though a typographic encoding of categorical differences may have pragmatic advantages, formal languages that do not display the grammatical category of a term typographically surely seem possible. More importantly, we could opt for a formal language with different or more fine-grained grammatical categories. While particular grammatical restrictions are encoded in individual formal systems, and thus no explicit grammatical restriction is necessary, we still need a rationale for our choice of a logical system. We find ourselves in the same dialectical situation as with respect to the problem of logical terms: recourse to formal languages does not liberate us from addressing the demarcation problem. The interpretationalist has to provide not only a demarcation of the logical from the non-logical terms, but also a justified classification of terms into grammatical categories.

I will here not attempt to defend a particular solution to the demarcation problem, i.e., I will not propose and justify a specific categorisation of terms into grammatical categories. Rather, I want to illustrate the significance of our choice by reference to so-called structural truths: there seem to be reasonable grounds for categorising the terms in such a way that certain paradigm cases of merely structural truths turn out to be logically true. Let me explain.

A sentence is *structurally true* if and only if its truth 'depends merely upon the kind of semantic elements out of which a sentence is constructed, and its manner of construction' (Evans 1985, 60–1). Paradigm examples of structural truths are sentences such as the following: 'If Mary quietly left school, Mary left school' and 'If Snowball is a black cat, Snowball is a cat.' In response to these examples it is often argued that they are not logically true, as there are reinterpretations which make them false: reinterpreting 'quietly' as having the usual semantic value of 'allegedly' in the first example, and reinterpreting 'black' as having the semantic value of 'fake' in the second, can make the respective sentences false. Thus structural truths are not automatically logical truths.

This reasoning rests, however, on the assumptions that 'quietly' and 'allegedly', and 'black' and 'fake', respectively, belong to the same grammatical category. Yet, perhaps it is too crude to classify all adverbs into one and the same category. In particular, it is still an open question whether manner adverbs like, e.g., 'quietly', 'quickly', and 'slowly' belong to the same grammatical category as 'allegedly'-type adjectives. They do not seem to satisfy substitutability *salva congruitate*, i.e., substitution of the one for the other does not always preserve grammaticality: substitution of 'allegedly' by 'slowly' turns the well-formed sentences 'Mary allegedly knows Snowball' and 'Mary is allegedly allergic to cats' into ones that are not well-formed. Turning from syntax to semantics, it is unclear whether manner adverbs (always) function as verb-phrase operators, i.e., whether the meanings of manner adverbs determine functions from verb-phrase meanings to verb-phrase meanings. Alternatively, modification by a manner adverb can be understood as an attributive use of an event-level adjective. As this is implausible for 'allegedly'-type adverbs, manner adverbs would then have a different kind of semantic value than 'allegedly'-type adverbs have. They are not verb-phrase operators but predicates of events. This counts in favour of fine-graining the category of adverbs such that manner adverbs and 'allegedly'-type adverbs are not grouped together.[8]

If we accept the view that 'slowly' and 'allegedly' do not belong to the same grammatical category, then the sentence 'If Mary quietly left school, Mary left school', a putative paradigm example of a merely structurally true sentence, may well be true under all admissible interpretations and hence a logical truth: if 'quietly' may only be reinterpreted as having the semantic value of some other manner adverb, such as 'slowly' or 'loudly', then there seems to be no reinterpretation under which the sentence is false. Similarly, if we treat attributive adjectives like 'black' and 'fake'-type adjectives as belonging to different categories, also the second cited paradigm case of a structural truth may turn out to be a logical truth. Which interpretations are admissible, and thus which sentences turn out to be logically true, depends on the chosen classification of terms into grammatical categories.

9.2 Semantic Restrictions

Grammatical restrictions on interpretations are necessary, but they are not sufficient. We also need *identity restrictions* because interpretations must be 'uniform': all occurrences of a term must be interpreted alike. Otherwise,

[8] See especially Jackson (2017) for an overview of the differences in the syntactic and semantic behaviour of these adverbs.

no (or only very few) sentences will turn out to be logically true.[9] Take an example. 'If Tarski is married, then Tarski is married' generally comes out as a logical truth only if the two occurrences of 'Tarski' are always interpreted uniformly.

Of course, the requirement of uniform reinterpretation is not new and indeed part of our common logical practice. However, the identity restriction has usually received a very superficial treatment at best. This is quite unfortunate, as the identity restriction is not as innocent as it might seem at first glance. It crucially involves the notion of term identity: which interpretations are deemed inadmissible depends on the conditions under which two term occurrences count as being occurrences of the same term. The interpretationalist therefore has to provide identity conditions for terms.[10]

To begin with, consider the role of the identity restriction and the conditions for term identity in formal languages. As the division between logical and non-logical terms and the classification into grammatical categories is already implemented in the standard formal languages, so is the identity restriction: the interpretation function is defined on the set of term types. It maps term types, not term occurrences, on semantic values. Hence all occurrences of a given term are automatically interpreted alike. Furthermore, in familiar formal languages term identity is typographically displayed. Two occurrences of the same term type are of the same typographic type, and vice versa. Observe also that in most formal languages term occurrences belonging to the same typographic type have the same semantic value, too. Nevertheless, one could easily invent languages in which this is not the case and which are still plausibly regarded as formal languages.[11] They may be hard to handle and hence not very practical. But they are not ruled out as such.

That typographically identical term occurrences have the same semantic value seems to be a contingent feature of formal languages at best. We surely cannot rely on it to be present in natural language: natural

[9] I here assume a reasonable choice of logical terms and a sensible categorisation of terms into grammatical categories. Also there might be some sentences, e.g., 'a is self-identical', that could come out as true under all interpretations even if we do not insist on uniform reinterpretation. (See Gajewski (2009) for more on logical truth without uniform reinterpretation.)

[10] Of course, the proponent of the interpretational definition does not have to spell out the conditions of word identity in all generality, but can concentrate on the particular notion of term identity figuring in the identity restriction: we are looking for that notion of term identity that is relevant for determining the logical properties of a sentence. This notion need not coincide with the (or a) common or pre-theoretic notion of word identity.

[11] For an example, see (Freitag 2009, 308–9).

language contains homonyms. Thus, in natural language, sameness of typographic type alone cannot determine sameness of term type. Otherwise all homonyms would be of the same term type and far too many sentences would turn out to be logically true. Let us look at an example discussed by Peter Strawson:

> [C]onsider two typographically identical occurrences of the sentence "He is sick." In one occurrence the sentence might be used to attribute a property of mind to one person, in the other occurrence to attribute a condition of body to a different person. (Nor is this fact altered by replacing the pronoun "he" by a proper name, say, "John".) If now, keeping in mind two such uses for this expression, we frame the sentence "If he is sick, then he is sick", we obtain something which may be used to make statements some of which may be true and others false. (Strawson 1957, 16)

The sentence 'He is sick' allows for different readings. That the pronoun 'he' can refer to different persons is a matter of the vagaries of indexical reference which I do not want to discuss here. However, according to Strawson, the sentence remains doubly ambiguous even if we substitute 'John' for 'he'. If we nevertheless take the two occurrences of 'sick' and of 'he' (or 'John') in the sentence 'If he is sick, then he is sick' to be occurrences of the same term, then they may only be interpreted alike. Thus the sentence 'If he is sick, then he is sick' (or 'If John is sick, then John is sick') would be declared logically true, even if used to say something that is only contingently true.

Strawson's example illustrates that typographically identical natural-language expressions might have different semantic values. His argumentative aim, however, is more ambitious. His paper is a critical reply to Quine's 'Two Dogmas of Empiricism', in which Quine famously argues that the notion of an analytic sentence rests on the notion of synonymy: an analytic sentence is one that can be transformed into a logical truth by substituting synonyms for synonyms. Yet, Strawson argues, Quine would have to extend his critique to logical truths, as his definition of logical truth also rests on the notion of synonymy. As a consequence, Strawson says, 'Quine's characterization of logical truth can be made coherent, and made to do its job, only by implicit use of notions belonging to the group which he wishes to discredit' (1957, 126).

Strawson argues, by reference to the above example, that sameness of typographic type is not sufficient for that notion of a term type that underlies the definition of logical truth. If Strawson's critique generalises to all non-semantic criteria, e.g., syntactic or phonetic ones, it follows that the relevant identity conditions of natural-language expressions cannot be given independently of semantic notions. Strawson's positive suggestion is that two occurrences of one term need not only be of the same typographic type, but must also have the same Fregean sense (and thereby also the same reference) (Strawson 1957, 24).

I agree with Strawson that term identity cannot be spelled out without recourse to semantic notions. However, *contra* Strawson, I want to tentatively propose that the semantics does *all* the work in determining the relevant term type: two occurrences are occurrences of the same term iff they have the same meaning. I want to reject typographic identity even as a merely necessary criterion for term identity, because it encounters some serious difficulties: it is obvious that we cannot literally demand *sameness* of typographic appearance for sameness of term type. Even the two occurrences of 'Mary' in 'Mary is a magician or Mary is not a magician' probably do not have the same physical appearance in every detail. More than typographic *similarity* cannot be achieved. However, similarity is a vague notion, and we should not base our definition of logical truth on vague concepts. More importantly, the typographic criterion of term identity gets us involved in questions that seem to have nothing to do with logic at all. When we discuss the logical properties of a sentence, we do not want to talk about such things as font size or colour. These aspects seem to be wholly irrelevant for the logical properties of a sentence.[12] As this critique seems to generalise to further non-semantic criteria, I tentatively suggest that the relevant notion of a term type must be given in purely semantic terms.

A natural choice is to claim that two term occurrences are occurrences of the same term if and only if they have the same meaning. As a result of this semantic criterion for term identity, the sentence 'If John is sick, then John is sick' is logically true if and only if both occurrences of 'John' and both occurrences of 'sick' are identical in meaning. This seems to be the desired outcome. However, the purely semantic criterion of term identity has also more revolutionary – if not more contestable – consequences. Which exactly these are depends on the particular notion of meaning involved. If the meaning of, say, a proper name is given by its

[12] As has been pointed out to me by Gil Sagi, philosophers such as Wittgenstein or Carnap might disagree. Wittgenstein (1922, §6.113) writes: 'It is the characteristic mark of logical propositions that one can perceive in the symbol alone that they are true; and this fact contains in itself the whole philosophy of logic.' Carnap (1937, 186) concurs: 'It is certainly possible to recognize from its form alone that a sentence is analytic; but only if the syntactical rules of the language are given.' I think, however, that Wittgenstein and Carnap are not referring to typographic form here, but to logical form. But if the logical form of a sentence is given, so are the relevant term identities.

referent, then sentences like 'If Hesperus is a planet, then Phosphorus is a planet' will come out as logically true. If, however, the meaning of a proper name is something akin to a Fregean sense, then 'Hesperus' and 'Phosphorus' may be interpreted independently and the sentence does not come out as logically true. I cannot here dive into the intricacies of the semantics of proper names, and even less into natural-language semantics in general. However, as we definitely do not want to identify predicate terms that happen to have the same extension, their meaning must here not be understood purely extensional. Beyond this claim I conveniently remain silent about the meaning of 'meaning', and thus about the exact consequences of the semantic criterion for term identity. For the present purposes it suffices to recognise that the identity restriction on interpretations rests on the notion of term identity, which is, I claim, a semantic notion. It is or involves thus a semantic restriction on interpretations since it refers to the semantic relations between term occurrences.

I will now tentatively propose a further semantic restriction that also exploits weaker semantic relations between terms. Consider the following paradigm case of an analytic sentence: 'Mothers are female.' This sentence is not declared a logical truth by the interpretational definition, if we allow independent reinterpretation of the terms involved. This result has been welcomed by most philosophers, who take the sentence to be a paradigm case of a merely analytic truth. But some disagree. Stephen Read, for example, claims that in letting the interpretation of terms like 'mother' and 'female' vary independently we 'overlook the fact that their interpretation is linked – their interpretation is not independent'. (1994, 250). He sees the need for a theory that respects these semantic relations: 'What is lost in interpretational semantics is the analytical linkage between expressions. . . . [W]e would need a theory which took account of these connections' (1994, 253). I will now provide the grounds for such a theory.

An analytic sentence is famously characterised by Kant as a sentence in which 'the predicate *B* belongs to the subject *A* as something that is (covertly) contained in the concept *A*' (Kant 1781, A6–7). The sentence 'Mothers are female' is accordingly analytically true because the predicate 'female' is contained in the subject-term 'mother'. Of course, the term 'female' is not contained in 'mother' as 'other' is contained in it: 'female' is not a syntactical part of 'mother'. As Kant says, in an analytic sentence the predicate is 'covertly' contained in the subject term. I understand the relevant notion of containment as semantic containment.

No doubt the Kantian containment analysis of analyticity has its drawbacks. It at best provides an analysis of a subclass of analytic sentences, as analyticity is obviously not always grounded in containment,[13] and we lack a precise analysis of the notion of containment itself. Nevertheless, a certain intuitive grasp of semantic containment can safely be presupposed: it is a platitude that the predicate 'female' is semantically contained in the predicate 'mother'. And it seems to be just another platitude that the containment relation and the identity relation are closely connected: semantic identity is but mutual semantic containment.[14] Terms are identical (in the relevant sense) if and only if they semantically contain each other.[15]

Just as we demand that reinterpretations respect the identity relation between the terms, so we might want them to respect their containment relations. More precisely, we might want to impose the following restriction on interpretations: if term A semantically contains term B under the actual interpretation, then only those interpretations are admissible under which A still contains B.[16] Such a restriction secures that interpretations respect the analytical linkage between the terms (or at least some aspects of the analytical linkage).

If term identity is fleshed out in semantic terms, the containment restriction is a straightforward generalisation of the identity restriction. The containment restriction then entails the identity restriction. However, if we opt for the containment restriction, paradigm cases of allegedly merely analytical truths will come out as logically true. I tentatively endorse this consequence. Others might take it as a *reductio* of the containment restriction (or the semantic criterion of term identity). But rejecting the containment restriction puts you in a bind. If the containment restriction is but a natural generalisation of the identity restriction, abandoning the former either means abandoning all restrictions based on semantic relations, and hence also such restrictions pertaining to two different occurrences of the same term, or it calls for an explanation why some semantic

[13] Kant was aware of this and intended his analysis of an analytic judgement in terms of containment to be restricted to affirmative judgements. See Kant (1781, A154/B193).

[14] Interestingly, already Kant must have seen a close connection between the identity relation and the relation of semantic containment, as he also gives the following characterisation of analyticity: 'Analytic judgments (affirmative ones) are . . . those in which the connection of the predicate is thought through identity, but those in which this connection is thought without identity are to be called synthetic judgments' (Kant 1781, A7/B10).

[15] Or, devised for term occurrences: Two term occurrences are occurrences of identical terms if and only if their respective term types semantically contain each other.

[16] Devised for occurrences, this reads as follows: If A is an occurrence of term A^*, and B is an occurrence of term B^*, and A^* semantically contains B^*, then only those interpretations are admissible under which A's term type still semantically contains B's term type.

restrictions are to be accepted (e.g., the identity restriction) while others are to be rejected (e.g., the containment restriction). Either there is a proper rationale for the traditional choice of imposing only the identity restriction, or the distinction between analytical truths and logical truths is unwarranted, indeed arbitrary.[17]

9.3 Metaphysical Restrictions

Let's address a different dimension of our quest: metaphysics. Logic is often portrayed as being free of metaphysics and in this sense completely neutral. At a first glance, the interpretational definition of logical truth seems to satisfy this demand. It reduces logical truth to truth *simpliciter*, i.e., truth in the actual world. The interpretational definition avoids recourse to modal notions – such as 'truth in all possible worlds' – by reducing logical truth to actual truth (under all interpretations).[18] The purpose of this section is to show, however, that the interpretational definition is imbued with metaphysics. More precisely, I will first argue that the plausibility of the interpretational definition rests on certain non-trivial metaphysical assumptions, and then tentatively propose a further restriction on admissible interpretations which is motivated by these background assumptions. The new restriction would attack the classical division of sentences into merely *metaphysically* necessary and *logically* necessary ones: if the restriction were imposed, many allegedly merely metaphysically necessary sentences would be declared logically true by the interpretational definition.[19]

To uncover the metaphysical assumptions of the interpretational definition, consider some contingently true atomic sentence, e.g., the sentence 'Tarski is married.'[20] Any definition of logical truth that declares

[17] Lycan (1989) also argues against a principled distinction between logical truth and structural or analytic truth and suggests that logicalness is a matter of degree. However, he does so for reasons different from the ones discussed here.

[18] Tarski's original definition of logical consequence (Tarski (1983b)) must surely be read interpretationally (for a defence, see Etchemendy (1990)). After all, Tarski aimed at a mathematically precise definition of the concept of logical consequence that didn't involve any modal notions. More contemporary approaches to logical consequence that are not purely interpretational, but identify logical truth with, e.g., truth under all interpretations in all worlds, are not the target of the objection developed in this section.

[19] Compare Hanson (1997), who also rejects the view that logical and metaphysical necessity are two different kinds of necessity. However, he assumes a modalised interpretational definition of logical truth, which of course changes the whole picture.

[20] I here disregard aspects of tense: 'X is married' is true if and only if X has at some time been married.

this sentence to be logically true must be discarded as obviously inadequate – at least if the definition still aims at capturing something even remotely resembling our pretheoretic notion of logical truth. The interpretational definition is thus prima facie plausible only if there is an admissible interpretation under which this sentence is false.

At a first glance, there seem to be many such interpretations. For example, interpret 'Tarski' as referring to Kant. Under this interpretation the sentence says that Kant is married. As far as we know, this is false. However, we might be mistaken. It might actually be the case that Kant was secretly wedded, unknown to anyone but the couple and without a noticeable trace for posterity. But if Kant was actually married, the proposed reinterpretation does not make the sentence false. Of course, the possible failure of a single reinterpretation does not constitute any problem yet, as there seem to be further, indeed many, reinterpretations that yield false sentences. Let 'is married' express the property of being a musician. Then the sentence says that Tarski is a musician, which is again false. Or is it? Maybe some secret society or an evil demon makes us all believe that Tarski spent his time on logic, mathematics, and philosophy, while actually he was sitting in some bar, playing the guitar . . .

Let me generalise. According to the interpretational definition the sentence 'Tarski is married' fails to be a logical truth if and only if there is some admissible interpretation under which the sentence is false. Yet might it not be the case that, contrary to our expectations, every state of affairs the sentence describes under some admissible interpretation does in fact obtain? Suppose this to be the case. Then the interpretational definition of logical truth declares our sample sentence logically true. This would be an untenable consequence.

I sense that, at this stage, the attentive reader will get slightly impatient. Isn't there a trivial way to show that some reinterpretation exists which renders the sentence false? Assuming an extensional semantics, we simply map 'is married' onto a set that does not contain the respective semantic value of 'Tarski'. To make it simple, we can let the interpretation function map 'is married' onto the empty set. So, one might conclude, there are no worries for the interpretational definition of logical truth.

Sure, from a purely formal perspective this is correct. The existence of a model mapping 'is married' onto the empty set is guaranteed. However, this has nothing to do with my objection to the interpretational definition. According to the interpretational approach, each model stands for an interpretation of the language. The interpretationalist must say which informal interpretation is encoded by the model that maps 'is married' onto the empty set; he must provide an interpretation under which 'is married' stands for a property that as a matter of actual fact is not instantiated. If all properties are actually instantiated, however, this is impossible.

Nevertheless, there seems to be a different, equally trivial way to argue for the existence of an interpretation that makes the sentence 'Tarski is married' false and that is not confronted with the above problem. We could reinterpret 'Tarski' as referring to, say, the apple on my desk, and consider one reinterpretation of 'is married' as expressing the property of being green and another where 'is married' stands for the property of being red. As the apple on my desk cannot be both green and red (all over at the same time), the sentence is false under at least one of the proposed interpretations. Thus the original sentence is shown not to be logically true, which is the desired result. There seems to be no serious problem.[21] Before I address, and indeed (in some sense) embrace, this position, however, it is important to spell out the alleged challenge in all its generality. Only then can we see what it takes to answer it.

Firstly, the problem does not only apply to our sample sentence and further sentences of subject–predicate form, but generalises to atomic sentences of all 'forms': For illustration, assume again that Kant indeed died as a bachelor. In this case, the sentence 'Tarski is married' is not a logical truth, as it is false if we interpret 'Tarski' as referring to Kant (assuming the intended interpretation of 'is married'). However, this type of contingency does not guarantee that there is an admissible reinterpretation that makes, say, the atomic sentence 'Mary is married to Harry' false. Under some reasonable categorisation of terms into grammatical categories, we cannot interpret this sentence as expressing the atomic state of affairs that Kant is married. To illustrate this pint, let us divide the terms into individual and relation terms, and let atomic states of affairs take the form of atomic Russellian propositions each consisting of an n-place relation and n individuals. If we impose the grammatical restriction of PL, individual terms must get individuals assigned as semantic values and n-ary relation terms must stand for n-ary relations. As a consequence, sentences can only describe propositions of the corresponding form. It then holds that for every n for which there is a sentence of the form '$R^n a_1, \cdots a_n$,' the

[21] Maybe this is already unnecessarily complicated. Could we not leave the interpretation of 'Tarski' as it is, but interpret 'married' as expressing the property of being unmarried? Then, necessarily, the sentence is true under the original interpretation if and only if the reinterpreted sentence is false. However, one could possibly object to this reply that there is no genuine property of being unmarried, but only the absence of the property of being married, and thus that 'married' cannot be interpreted as standing for the property of being unmarried. As I am not sure what to make of this objection, I avoid such subtle discussions by choosing a less contentious example.

interpretationalist has to assume that there is a relation R and that there are n individuals $o_1, \cdots o_n$, such that the proposition R, $o_1, \cdots o_n$ does not obtain.[22] To prevent that an atomic sentence of a given grammatical form is declared logically true, there must be a state of affairs of the corresponding form that does not obtain. If also the interpretational definition of logical *falsehood* should be prevented from massive overgeneration, we must additionally assume that for every relevant form, there is an *obtaining* state of affairs.

Secondly, even if there happen to be obtaining and non-obtaining states of affairs of every relevant form, the proponent of the interpretational definition is not yet home free. Atomic sentences would then not be logically true or logically false, which is the desired result. But this outcome could still be a matter of mere luck. As long as it is *possible* that all relevant states of affairs of a given form are obtaining or that all are non-obtaining, the plausibility of the interpretational definitions rests on contingencies.

Modal talk is tricky here, so let us be a bit more careful: A sentence is logically true if and only if there is no interpretation under which it is false, i.e., if and only if there is no interpretation under which it expresses a state of affairs that is not obtaining. Let w_α be the actual world. A given sentence is not logically true iff there is an interpretation under which the sentence expresses a state of affairs that is not obtaining in w_α. Assume that there is such an interpretation. If, however, there is a possible world w^* in which, say, all atomic states of affairs are obtaining, then the definition of logical truth yields the right extension by luck only: if we were inhabitants of w^* (i.e., if w^*, rather than w_α, were the actual world), we would have to evaluate the sentence not in w_α, but in w^*, and the sentence 'Tarski is married' would turn out as logically true. If one wants to defend the view that the plausibility of the interpretational definition does not depend on contingencies, then one does not only have to assume that there are actually enough obtaining and non-obtaining states of affairs, but one also has to show that it cannot be otherwise, i.e., that this is *necessary*. To sum up: given that the interpretationalist wants to show that his definitions work by necessity, he has to assume that, necessarily, there is an obtaining and a non-obtaining atomic state of affairs of every relevant form.

To illustrate that this is no innocent assumption, let me show that there are metaphysical positions that are incompatible with it. For example, assume a Tractarian ontology of atomic states of affairs. According to

[22] The interpretationalist has to presuppose something even stronger, as the form of the proposition that can be expressed by a sentence is not determined by the grammatical categories of the terms alone, but also by the identity relations between the terms.

the view of Logical Atomism as proposed by the early Wittgenstein, each atomic state of affairs can be obtaining or non-obtaining, and there are no atomic states of affairs that are necessarily obtaining or necessarily non-obtaining. Furthermore, atomic states of affairs are said to be independent of one another: for any atomic state of affairs p and any set P of atomic states of affairs with $p \notin P$, it holds that the obtaining of the elements of P entails neither the obtaining nor the non-obtaining of p.[23] Thus any subclass of the class of all atomic states of affairs may obtain. Such a metaphysics does not exclude extreme cases. It allows for the possibility that none of the atomic states of affairs obtains as well as for the possibility that they all obtain. As argued above, both possibilities generate unsurmountable problems for the interpretationalist.

Now that the challenge is clear, let me finally address the obvious reply to it. There is one interpretation under which the sentence 'Tarski is married' expresses the proposition that the apple on my desk is green, and another interpretation under which the sentence expresses the proposition that the apple is red. As the apple cannot be both green and red (all over at the same time), the sentence must be false under at least one of the mentioned interpretations. Thereby it is shown not to be logically true. I think that this reply is on the right track. By necessity, there will always be an interpretation under which an atomic sentence is false: we must only choose two interpretations of the sentence such that the state of affairs expressed under the first interpretation is incompatible with the state of affairs expressed under the second. However, we now see that the 'obvious reply' to our problem must rely on the fact that there are 'incompatible states of affairs', i.e., that there are states of affairs such that the obtaining of the one entails the non-obtaining of the other. The assumption that there are dependent atomic states of affairs is, however, no innocent claim of logic, but pure metaphysics. The interpretational definition rests on genuine metaphysical assumptions.[24]

Although it is not my aim here to provide a full-fledged justification of the interpretationalist's background assumptions, let me hint at a possible strategy for doing so. (For reasons of brevity, I will here solely concentrate on the implicit assumptions underlying the interpretational definition of logical truth and suppress the discussion of logical falsity.) To justify the

[23] See Wittgenstein (1922), §2.062 and §5.135. Wittgenstein gives up on this independence principle in Wittgenstein (1929).

[24] I am not the first who holds that the interpretationalist has to make certain extra-logical assumptions. Most famously, Etchemendy (1990, 110) claims that 'Tarski's account remains dependent on completely non-logical facts'. I compare Etchemendy's critique with mine in Zinke (2018).

dependence of atomic states of affairs, the interpretationalist could draw on William E. Johnson's famous theory of determinables and determinates (Johnson 1921, ch. 11). According to this view, the world unfolds in a determinate–determinable structure, which roughly means that some properties stand in the 'determination' relation to other properties. For example, the property of being green and the property of being red each determine the property of being coloured, and are thus determinates of the determinable property of being coloured, whereas the property of being round and the property of being triangular are determinates of the determinable property of having a shape. There is a number of open questions concerning the notions of determinables and determinates. The concepts would have to be further justified and developed. For now, however, let us suppose that the theory can be spelled out properly, and consider one of the core principles of the theory. It states that a determinable property can only be determined by a single maximally determinate property: an object cannot be green and red (all over and at the same time).[25] Thus, according to the determinable–determinate theory, it holds necessarily that if an individual instantiates a certain property F that is a maximal determinate of a determinable property H, then this individual does not instantiate a distinct property G that is also a maximal determinate of H. Therefore the state of affairs that a is F and the state of affairs that a is G cannot both obtain.

If we subscribe to the determinable–determinate theory, we can explain why some atomic states of affairs exclude one another, i.e., why they cannot obtain simultaneously. Supposing that this generalises to relations of all relevant arities, we have established the assumption that there must be some non-obtaining state of affairs of every relevant form. There seems hence to be a plausible way of justifying the implicit assumptions underlying the interpretational definition of logical truth. However, the plausibility of the background assumptions does not change their metaphysical nature. The interpretational view of logic, often praised for its independence from metaphysics, relies on genuine metaphysical assumptions about the structure of the world.

The proponent of the interpretational definition must make this background assumptions explicit. I want to suggest even more, namely that the interpretationalist adopts a restriction on interpretations that is motivated by her metaphysical background assumptions. If, for instance,

[25] I here assume that being green and being red are maximally determinate properties, ignoring the fact that there are different shades of green and red.

one entertains the determinable–determinate theory in order to prevent the massive overgeneration of the interpretational definition, one could also restrict the admissible interpretations to those that respect this structure. If this suggestion is heeded, then an interpretation may only reinterpret predicates expressing maximally determinates of one determinable as still expressing maximally determinates of some, possibly different, determinable. Like the semantic restrictions discussed in Section 9.2, this restriction constrains the possible simultaneous reinterpretation of terms. Consider some examples. The suggested restriction excludes a reinterpretation of 'If the apple is green, the apple is not red' as saying that if the apple is green, it is not tasty: whereas being green and being red are determinates of the same determinable, the properties of being green and of being tasty are not. The restriction allows, however, a reinterpretation of the sentence such that it says, e.g., that if the apple is green, it is not brown, or if the apple is big, it is not small, or even as saying that if Tarski is in New York, he is not in a bar in Warsaw.[26] As a consequence of this restriction the sentence 'If the apple is green, the apple is not red' turns out to be true under all admissible interpretations and thus logically true.[27] However, many would deem this sentence a paradigm case of a merely metaphysically necessary sentence.

The proposed restriction demands that the determinable–determinate structure must be preserved under reinterpretation. If predicates stand for determinates of the same determinable under the actual interpretation, then only interpretations under which they still do so are admissible. Of course the interpretationalist is not obliged to impose this restriction. However, once it is acknowledged that the interpretational definition is imbued with metaphysics, there seem to be at least no theoretical grounds not to enforce the metaphysical restriction. The metaphysical dependencies must be assumed by the interpretationalist anyway. If the restriction is imposed, however, many allegedly merely necessary truths are turned into logical truths.

Conclusion

The interpretational definition of logical truth is based on a prior demarcation of the admissible from the inadmissible interpretations. The

[26] Sagi (2014) discusses a very similar constraint. However, her justification for the constraint differs from the one proposed here.

[27] Here I presuppose that different terms may not be interpreted in the same way. The sentence 'if the apple is green, the apple is not red' may not be reinterpreted as saying that if the apple is red, it is not red.

usual demarcation – which normally remains implicit – seems to lack a proper rationale. As a consequence, we also lack a rationale for classifying some truths as logical and others as merely structural, analytic, or metaphysically necessary truths. The central task for the proponent of the interpretational definition is to provide a solution to the problem of admissible interpretations. If no justified demarcation is put forward, the choice of admissible interpretations seems arbitrary, and so does the cleavage between logical and non-logical truth.

PART III

Logic and Natural Language

CHAPTER 10

Models, Model Theory, and Modeling

Michael Glanzberg

For some time now, I have been exploring the relations between model theory, logic, and the semantics of natural language. Some might think these are all nearly the same thing. A view might hold that model theory is the fundamental way we understand logic, and that it is equally the basic tool in semantics, and that indeed doing semantics is just the same as doing logic. We can do it for formal languages, or natural ones. This is a fairly strong and contentious set of claims, but forms a natural and I think appealing view. (The latter claims about natural language were at least held by Montague (1970).) The number of books and papers whose titles have phrases like 'model-theoretic semantics' or 'logic and language' might also make one think this view has some currency. Regardless, in earlier papers (Glanzberg 2014, 2015) I have urged caution in these matters. I have argued for a real but limited role for model theory in semantics of natural language, and I have likewise argued for limited connections between natural language and logic proper.

In this paper, I shall return to the relations between logic and semantics of natural language. My main goal is to advance a proposal about what that relation is. Logic as used in the study of natural language – an empirical discipline – functions much like specific kinds of scientific models. Particularly, I shall suggest, logics can function like analogical models. More provocatively, I shall also suggest they can function like model organisms often do in the biological sciences, providing a kind of controlled environment for observations.

My focus here will be on a wide family of logics that are based on model theory, so in the end, these claims apply equally to model theory itself. I do not think this makes model theory unique. Virtually all the tools of contemporary logic have proved fruitful in studying natural language: model theory, proof theory, recursion theory, set theory, intensional and extensional logics, classical and non-classical ones,... But, model theory does offer a particularly important set of applications, so I shall focus on it.

At the same time, model theory offers a particular way of understanding what is basic about logic. Along the way towards arguing for my thesis about models in science, I shall also try to clarify the role of model theory in logic. At least, I shall suggest, it can play distinct roles in each domain. It can offer something like scientific models when it comes to empirical applications, while at the same time furthering conceptual analysis of a basic notion of logic.

The plan of the paper is as follows. In Section 10.1, I shall begin with logic itself. I shall argue that model theory, even some of the complex mathematics we find in modern model theory, can help us understand some basic issues about logic. It can even support a broad conceptual analysis of the nature of logic. In Section 10.2, I shall turn to applications to natural language; and briefly, to other aspects of cognition. I shall argue there that model theory provides tools that function like scientific models for the study of empirical phenomena. Finally, in Section 10.3, I shall ask what sorts of scientific models logic and model theory provide. I shall argue that they provide two sorts: they provide both analogical models, and models that function much the way model organisms function in the life sciences.

Before proceeding, let me mention some unfortunate terminological problems. Clearly, the term 'model' is being used multiple ways. These uses are well established, and so it is pointless to try to introduce new terminology. Sometimes I shall have to just let context disambiguate. But for the most part, when I say 'model' I mean what model theory – the branch of logic – has in mind. I shall try to say 'scientific model' or 'model in science' to distinguish such models from model theory. Also, for the most part, I shall discuss a number of logics, and be quite liberal in what I count as a logic. But at some points, I shall ask about a fundamental philosophical issue of what really counts as logic, and which, or how many, of the logics mathematics provides are really logic. I shall write 'LOGIC' in capital letters to mark this philosophically fundamental notion (assuming that indeed it is a coherent and well-defined one). Neither of these notational fixes is elegant, but they will suffice.

10.1 Logic and Model Theory

Before getting to empirical applications, I shall consider some ways of thinking about logic, and perhaps LOGIC.

What is logic? There are many potential answers to this question. One will be especially important to empirical applications, but there are others. Taking inspiration from work of Cook (2002), we can consider several options:

The instrumentalist view of logic: Logic is just mathematics. Like any mathematical machinery, you can use it however you like. It can be used in an instrumentalist fashion, to roughly track some phenomena, perhaps in language or reasoning, but offering no real explanations of the phenomena in question. Any implications from the mathematical machinery can be viewed as convenient fictions.

Logic as description: Logic describe what is really going on with the truth conditions, consequence relations, etc. of various discourses.

Logic as conceptual analysis: There are fundamental facts about consequence and other logical properties. These facts are revealed by conceptual analysis, and logic provides that analysis. Pluralists might think there are many such sets of facts, singularists will hold there is only one.

Logic as modeling: Logic can provide models (in the sense of scientific models) that help us understand various phenomena, especially in language and reasoning. It can be a fruitful way to represent and study these phenomena. But it is one tool among many, and there are many ways phenomena can be modeled, with different benefits and limits.[1]

As Cook notes, these positions are extreme, and it is not clear if anyone has held them in the forms stated. But even if this range of options is something of a caricature, it offers a useful structure within which to ask very general questions about logic.

If we assume there is such a thing as LOGIC – a philosophically fundamental notion – then it seems to go most naturally with the Logic as conceptual analysis option, though perhaps some empiricists might opt for the Logic as description option. I shall argue that in the end, the Logic as modeling option is the best one for thinking about empirical applications of logic.

With these options in mind, I shall turn to the question of the relation of model theory to logic. Model theory as it is done these days includes

[1] Cook builds on work of Shapiro (1998), who in turn notes a suggestion of Hodes (1984). I have modified Cook's presentation in several ways, but kept to the spirit of his taxonomy. I added the conceptual analysis option. My statement of the instrumentalist view is somewhat broader than Cook's. He has in mind a fairly specific set of applications to vagueness (which is the main focus of his paper), while I have a wider range of applications in mind. One might worry that not any old piece of mathematics is *logic*. I agree, but given the range of different mathematical techniques that do seem to count, I am avoiding that demarcation issue here.

a lot of advanced mathematics. It is usually done as a branch of abstract mathematics, typically within set theory, but with many applications, to algebra, geometry, and so on. Especially if we are thinking of LOGIC, we might worry that such mathematics and LOGIC have little to do with each other. The point is made vividly by Sacks (1972, 1):[2]

> Part of the blame belongs to B. Dreben who once asked with characteristic sweetness: "Does model theory have anything to do with logic?" It is true that model theory bears a disheartening resemblance to set theory, a fascinating branch of mathematics with little to say about fundamental logical questions...

In contrast, many philosophical logicians find model theory to be essential to logic, or at least to LOGIC. The point is made vividly by a wonderful quote from Routley and Meyer (1973, 199) via Restall (2000):

> Yea, every year or so Anderson and Belnap turned out a new logic, and they did call it E, or R, or $E_{\overline{I}}$ or $P - W$, and they beheld each such logic, and they were called relevant. And these logics were looked upon with favor by many, for they captureth the intuitions, but by many they were scorned, in that they hadeth no semantics.

(That is, until Urquhart (1972) provided one.)

I shall assume throughout that giving a semantics for a logic is an exercise in model theory. Hence, the two views could not be more different. On the one hand, we question of whether model theory has anything to do with logic; on the other, we insist that without a model theory a logic is somehow not good enough.

If we are instrumentalists about logic, of course, there is no issue here. Any mathematics is fine. Model theory is, so is the proof theory that framed many relevance logics, and so on. If we think of logic as description, we might worry more, about whether various pieces of mathematics help with such descriptions. But the question is most vivid for the Logic as conceptual analysis view. There, the questions of whether one must have a model theory as part of one's conceptual analysis, and if so, what role such mathematics could play in any conceptual analysis, seem urgent.

I shall propose a way of bridging these two opposing views, that will allow us to see sophisticated mathematics as playing a role in conceptual analysis, at least a background role. This will show that instrumentalist and conceptual analysis views of logic can work together.

[2] This is from the introduction to Sacks's book *Saturated Model Theory*, which covers a great deal of the mathematics of model theory as it was done at the time.

The key observation is one that Sacks (1972) already made. The quotation above continues:[3]

> But the resemblance is more of manners than of ideas, because the central notions of model theory are absolute, and absoluteness, unlike cardinality, is a logical concept. That is why model theory does not founder on that rock of undecidability, the generalized continuum hypothesis, and why the Łos conjecture is decidable... Łos conjectured and Morley proved that if a countable theory is κ-categorical for some uncountable κ, then it is κ-categorical every uncountable κ. The property "T is κ-categorical for every uncountable κ" is of course an absolute property of T. (Sacks 1972, 1–2)

Model theory has properties that do lend themselves to the study of fundamental logical notions.

Sacks takes an abstract, and somewhat technical, view of this issue, but the point is quite general. Many model-theoretic notions, starting with model-theoretically defined consequence relations, definability properties, many model-theoretic properties of theories, and so on, are relevant to how we understand fundamental aspects of logic, and perhaps even LOGIC. And indeed, the way we can find absolute properties in model theory is a good clue to this fact.

To fix ideas, let us start with a familiar way of thinking about logic that supports a conceptual analysis view. We take the most fundamental aspect of logic to be logical consequence, and take the model-theoretic approach to logical consequence pioneered by Tarski (1935, 1966/86) to be the core of a conceptual analysis of logic.

This is sometimes called the 'semantic conception of logical consequence', and it embodies a long-standing tradition of thinking about the nature of LOGIC (the philosophically fundamental notion), and is well viewed as a kind of conceptual analysis of LOGIC. I'll briefly review some of the key ideas of this view.

LOGIC, according to this view, describes something fundamental: valid arguments, which are a fundamental constraint on good reasoning. The tradition from Tarski (1936) places two constraints on LOGIC: *necessity* and *formality* (cf. Beall and Restall 2009; Etchemendy 1990; Sher 1991):

- Necessity: If S is a consequence of a set X of sentences, then the truth of the members of X necessitates the truth of S.
- Formality: Logical consequence holds 'in virtue of the forms' of sentences.

[3] This passage is also discussed by Kennedy (2015).

Note, that even before spelling these out further, they are enough to distinguish logical consequence from, e.g., inductive support.

Articulating these requirements more fully can be done via the notion of a 'case' from Beall and Restall (2006). Providing a logic can be done by providing a range of cases of some sort. The range must be sufficient to capture necessity: preserving truth in all cases must suffice to establish necessity. Cases also need to support formality, by allowing fixed treatment for logical constants. Validity of arguments is characterized by what Beall and Restall (2006) call the *generalized Tarski Thesis*:

> GENERALIZED TARSKI THESIS (GTT): An argument is valid$_x$ if and only if, in every case$_x$ in which the premises are true, so is the conclusion.

Beall and Restall are logical pluralists, and so they subscript valid$_x$ and case$_x$ to allow many different instances. But regardless, what we have here is a fairly common conceptual analysis of logical consequence. This can be seen as a conceptual analysis of LOGIC. It is one of many contenders, but will serve our purposes here well as an illustration of a conceptual analysis.

Model theory enters to tell us what cases are. In the tradition of classical logic, cases are just models in the usual sense from model theory. Beall and Restall also consider possible worlds, situations, and so on, to give a wider range of logics. But model theory has the mathematical wherewithal to capture any of these.

So far, it looks like model theory only relates to the conceptual analysis of logic, or to LOGIC, in a very modest way. It tells us what cases are, but that barely scratches the surface of the rich mathematics of model theory.

But this is to underestimate the complexity of the mathematics of these 'cases', which is of course model theory. Even if we fix the standard classical notion of a model, there is a huge range of different ways we can use them as cases, and they produce different logics. This is the family of *model-theoretic logics* as explored, e.g., by Barwise and Feferman (1985). Examples include:

- Classical first-order logic.
- Second-order logics of various strengths.
- Logics of generalized quantifiers $\mathcal{L}(Q)$ for various families of quantifiers Q.
- Smaller infinitary logics like $\mathcal{L}_{\infty,\omega}$ (with arbitrary conjunctions and disjunctions but finitely many quantifiers in any sentence), and its fragments.
- $\mathcal{L}_{\kappa,\lambda}$ (with $< \kappa$-sized conjunctions and disjunctions, and $< \lambda$-many quantifiers in any sentence).

We can study the properties of these in depth, including forms of compactness, interpolation, etc. We can sometimes, as with Lindström's theorem (Lindström 1969), characterize these logics in substantial ways. As Sacks notes, we find reasonable degrees of absoluteness in many of these cases (though not all!). We can, of course, find even more options if we depart from classical models, including many intensional, relevance, and other logics.

We use substantial mathematical resources – substantial parts of the mathematical subject of model theory – to formulate and understand these options. So sophisticated mathematics can play a role in our conceptual analysis. It can help us understand a wide range of options for what formally fits with our conceptual analysis. Of course, there are a number of further questions for our conceptual analysis. Which of these are LOGIC, if there is such a thing? Can one or many be LOGIC? Are there important properties that help us decide? I suspect there are. Compactness properties are a good example. Though I am not sure that every aspect of model theory plays a role here (would the applications to geometry?), we do see that substantial mathematics interacts fruitfully with conceptual analysis. Model theory does have much to do with logic understood as conceptual analysis, and maybe even has something to do with LOGIC.

We see here ways that two approaches to logic, the Logic as instrumental and the Logic as conceptual analysis approaches, can fruitfully interact. Perhaps much model theory in mathematics is done simply as pure mathematics, with little more than the instrumentalist view in mind (what else would mathematicians do than mathematics?). But mathematics can be applied in many ways, and it turns out that model theory has interesting, and I think rich, applications to Logic as conceptual analysis.

The application can be seen as follows. The mathematics of model theory give us many candidate logics. At least to some extent, they meet the conditions conceptual analysis provides. So, they seem to be candidates for being LOGIC, and we then have to explore, sometimes with real mathematical sophistication, which candidates are really the right ones.[4]

[4] Not every exercise in the mathematics of building models seems to offer the kinds of conditions our conceptual analysis requires. I am not sure that some models of Lambek calculi or Linear logics do. But I shall not argue that here.

Let us call the kinds of logics that we generate this way model-theoretic logics.[5] Model-theoretic logics have a home in pure mathematics, which perhaps may suggest an instrumental view, but they can be part of a conceptual analysis view as well.

With that, let us ask the next question: what does model theory have to do with natural language or other empirical phenomena?

10.2 Relations to Language (and Cognition)

So far we have focused on foundational or conceptual matter: what is logic, or even LOGIC, and what role does model theory play in it. I argued that we can view model theory as having instrumental roots, but that it supports conceptual analysis views of logic as well. Indeed, these come together in an interesting way. But I did not consider either Logic as description or Logic as modeling approaches. Logic as modeling does not seem to have any place when we address only such conceptual questions. There is no empirical phenomenon to model. But I shall argue, it is the best option when we turn to empirical phenomena. I have already argued in effect that Logic as description is a mistake in other work, that I shall review here. My main thesis for this section is that when we come to empirical applications, we should rely on Logic as modeling exclusively.[6]

Let me begin with the Logic as description view. It presupposes that natural languages, or discourses formed in them, have a logic to begin with. This is what in earlier work (Glanzberg 2015, 75) I called the *Logic in natural language thesis*:

> A natural language, as a structure with a syntax and a semantics, thereby has a logical consequence relation.

I take the Logic as description view to be stronger than the Logic in natural language thesis. The latter says we can find a logic in natural language. I take the former to hold that it is THE logic, or perhaps LOGIC.

The Logic in natural language thesis is in effect endorsed byMontague (1970, 222):

[5] This includes the sorts of logics called model-theoretic by Barwise and Feferman (1985), but is wider, as it also includes intensional logics and many others.

[6] I suspect most researchers working in empirical areas will not be surprised by such a conclusion. Scientific models are common tools for them, and being told that logic is one may not be a surprise. But among philosophical logicians, the Logic as modeling idea from Cook (2002) is indeed surprising and controversial.

> There is in my opinion no important theoretical difference between natural languages and the artificial languages of logicians; indeed, I consider it possible to comprehend the syntax and semantics of both kinds of languages within a single natural and mathematically precise theory.

I am not sure if Montague was assuming a broader descriptivist, instrumentalist, or conceptual analysis view of his 'single natural and mathematically precise theory', but descriptivists should also endorse Montague's claim.

One of the main claims in my (Glanzberg 2015) was that the Logic in natural language thesis is false. If so, this rules out descriptive approaches to the application of logic to natural language. I argued there that what we find in natural language is not really logical consequence. Descriptively, natural language does not hand us a logic. If this is correct, Logic as modeling is our only option for these applications. But moreover, I shall argue here that the way of thinking about the relation of logic to language I outlined in that paper really is an instance of logic as modeling.

I argued for the claim that the Logic in natural language thesis is false at (perhaps excessive) length already, and I shall not try to repeat the full arguments here. But to get to the new points I wish to make, it will be helpful to make reference to the specific arguments from my earlier work. I presupposed that the conditions of formality and necessity discussed above are constraints on what counts as logic (and so perhaps engaged in some conceptual analysis). With that, I gave three distinct arguments.

First was the argument from absolute semantics. As noted by Davidson (1967) and Lepore (1983), semantics for natural language must be absolute, in that to give correct and non-vacuous truth conditions for sentences we must fix the correct reference and satisfaction conditions for their constituents. This leaves no use for a space of models. We only need real-world reference and satisfaction. For instance, recall that an atomic sentence *Fa* will be assigned arbitrary extensions for *F* and *a* across models. That does not give any substantial truth conditions. If you want to give the truth conditions of *Sam is happy* you need the real-world reference of *Sam* and the real-world extension of *happy*. Hence, what natural language gives us does not satisfy necessity, and we find no genuine consequence relation.

Second was the argument from lexical entailments. I argued lexical entailments will not count as logical consequences (as they either fail necessity or formality).

Finally, I offered the argument from logical constants. I argued that there is no linguistically distinguishing marks of logical constants.

All this made me claim that we do not find any genuine consequence relations in natural language, and so, we cannot take any plausible consequence relation to be the one of natural language. Nor can we take the semantics of natural language to be the semantics that gives any such consequence relation. The Logic in natural language thesis must be false, and if we assume some minimal constraints on what counts as logic, so must any descriptive approach.

This was never to say that we cannot find useful clues to logics within our languages. Historically, we have. But we need to do more than just describe the semantics of our languages to find them. In Glanzberg (2015) I argued that we need to go through a substantial three-fold process to get from a language to a logic. It includes abstraction from absolute semantics to get an appropriate domain of models or cases. In many situations, we do this in a specific way. Absolute semantics provides specific meanings to non-logical expressions. Abstraction allows these to vary, and as has been much discussed, we also usually allow the domain of quantification to vary. We also have to identify the logical constants. Natural language fails to do this, but we must. To achieve formality, we will need to hold the meanings of the logical constants fixed in abstraction. Abstraction and identification work together, and doing both typically yields a common post-Tarskian understanding of a model-theoretic consequence relation.

I argued that this is not sufficient. Another step, of *idealization*, is needed. Even after we have performed abstraction, we are still going to be stuck with some idiosyncratic and quirky features of natural language grammar, that we will not want to contaminate our logic. Rather, we want our formal languages to display uniform grammatical properties in important logical categories. So, we must idealize these quirks away. Idealization like this is familiar from modeling in science.

What can result from this process of abstraction, identification, and idealization? The familiar history suggests that classical logic seems to be one likely outcome. First, we identify logical constants to be *or, and, if, not, every, some*. Then, we use set theory to freely vary extensions of non-logical terms and predicates. This is easy and natural. But we also do substantial idealization. For instance, we modestly idealize the meaning of *or*, and more substantially the meaning of *then* to get $\vee$ and $\longrightarrow$. We make structural idealizations as well. For instance, the standard first-order quantifiers depart from the syntax of natural language, and the scope behavior of quantifiers in natural language differs from what we have in logic. Historically, these idealizations were driven by applications to mathematics, as well as reasons of simplicity and uniformity.[7]

We might find more than classical first-order logic. For instance, plural constructions might lead you to second-order quantifiers (Boolos 1975; Linnebo 2003; Rayo 2002; Uzquiano 2003). The structure of quantified noun phrases might lead you to $\mathcal{L}(Q_{(1,1)})$, the logic of (conservative) binary quantifiers (Barwise and Cooper 1981; Keenan and Stavi 1986). Modals, conditionals, and tenses can get you to intensional logics. We might also be led to depart from classical logic in some ways. Presupposition can get you to K_3 or even LP (e.g., Beall and van Fraassen 2003). Many model-theoretic logics are potential results of the process. Perhaps there are good reasons to rule some of these examples in or out, but we have prima facie reasons to consider all of them, if not more. This does not mean any logic will arise by this process directly. It is hard to see infinitary logics resulting from the process. Rather, they seem to arise as generalizations once we have a logic in hand.

Now, we can return to logic as modeling. We can observe that any instance of this process describes a modeling exercise, and the results are best understood as certain kinds of scientific models. As I mentioned, idealization is a familiar aspect of modeling. Abstraction is a form of generalizing, which is also an aspect of modeling. Identification is too. You identify things you are modeling and things you are not, in any instance of modeling. Without those steps, we do not get a recognizable logic, so getting logic in the way I described should best be thought of as an exercise in a certain kind of modeling.

Though I have focused on language, it is worth noting briefly that we might say similar things about the role of logic in human reasoning. A good example is the PSYCOP model from Rips (1994). This is a model, based on a specific (classical) natural deduction system, with a generous rule set. It is a memory-based model, as it models reasoning as constructing proofs in working memory. It is part of a larger model, with goal and control systems. In this case, not surprisingly, Rips extensively tested his model against subjects' judgements of validity for arguments, judgements of validity of proof steps under timed conditions, and tested failures of reliability for conditionals and negation, and so on. Logics can provide scientific models in many forms, applying to language and other aspects of cognition.

So, I claim, when logic is applied to language and cognition, we should adopt the Logic as modeling approach. We find logics for natural language

[7] For a general historical overview of the emergence of first-order logic, see Ewald (2019).

by a process of scientific modeling. But if so, what kinds of models do we get, and what do they tell us about the phenomena? And, returning to one of our main themes, what is the role of model theory in this kind of scientific modeling?

One role model theory plays in Logic as modeling is just the same one it played when we considered conceptual analysis. Model theory can provide us a broad and rich range of model-theoretic logics. These are candidate models, and understanding them will help us with our modeling exercise. Indeed, the process I described above is one tailored to produce model-theoretic logics, and the mathematics of model theory can help us do that, and understand the results. So, we can see model theory as providing the mathematical resources that give us certain kinds of scientific model, and allow us to understand how they work. Those kinds of models are just the kinds we get by the process I described above.

The question then becomes, what value do these sorts of models have? Why model that way? One answer is not surprising. These models help us understand the behavior of expressions close to logical constants, of course.

Here is one example among many. Generalized quantifier logics, with their mathematically rich model theory, offer insights into the properties of natural language quantificational determiners: expressions like *most* (in languages that have them[8]).

Here is a logical property expressible in a logic with generalized quantifiers. We look at Q_M, the extension of a generalized quantifier in a model with domain M. This will be essentially a set of sets. We can then define such properties as

(1) EXT for type (1) quantifiers: For any $A \subseteq M \subseteq M'$, we have $Q_M(A) \longleftrightarrow Q_{M'}(A)$.

EXT is a strong form of domain restriction. It is not something we would find if we just specified the absolute truth conditions for sentences, as it requires varying domains for each quantifier. We have to engage in the kind of scientific modeling I just described to find this property. And yet, its analog for natural language determiners seems to hold. They are highly domain-restricted, in much the way this property describes. A classic theorem tells us that EXT plus permutation invariance implies isomorphism invariance (EXT + PERM implies ISOM). Natural language quantifiers

[8] It is uncertain whether all languages do. See Bach et al. (1995) and Keenan and Paperno (2012).

also show strong permutation invariance. So, we have learned something about our quantifiers, using our logic, i.e., our scientific model.[9]

We can keep adding to this list of examples. The entailment properties of quantifiers has proved useful in understanding the surprising behavior of expressions like *any*: so called negative polarity items.[10] Looking to intensional systems, we can say the same about modals, conditionals, tenses, and so on. Generally, we can fruitfully study the properties of logically rich expressions using the tools of model theory and model-theoretic logic as scientific models, in much the way I described above. I noted above that absolute semantics show us that building these models is not the immediate job of specifying the truth conditions of sentences and their parts, and so is not the most immediate task for semantics. But it can be useful nonetheless.

This particular sort of scientific model has its limits. That is not a surprise. Any model has limits. Let me mention one. I am skeptical of how much model-theoretic logics will help with lexical entailments, which is an important and data-rich area of semantics.[11] We can indeed, as Carnap (1952) and Montague (1973) both noted, sometimes capture lexical entailments as constraints on spaces of models. These appear as meaning postulates, like the familiar:

(2) $\forall x(Bachelor(x) \longleftrightarrow [Human(x) \wedge Male(x) \wedge Adult(x) \wedge Unmarried])$

The result is a step to partially interpreted languages, but their model theory is also a rich subject (cf. Barwise 1975).

Since work of Zimmermann (1999), many linguists have been cautious about meaning postulates, and tend to prefer to rewrite these rules as lexical decomposition rules. More importantly, many cases of lexical entailment do not indicate this kind of restriction on a space of models.

A good example is the dative alternation and its puzzles.

(3) a. Anne gave Beth the car. (Double Object (DO))
    b. Anne gave the car to Beth. (Prepositional Object (PO))

These are near synonyms. But not quite:

(4) a. Anne sent a package to London.
    b. # Anne sent London a package.

[9] See Peters and Westerståhl (2006) for more details and references. I have discussed this kind of use of model theory more in my (Glanzberg 2014).

[10] See Giannakidou (2011) for an overview.

[11] For an interesting discussion of what meanings can be captured with model-theoretic tools, see Sagi (2018).

We also see different entailments:

(5) a. Mary taught John linguistics.
ENTAILS
John learned linguistics.
b. Mary taught linguistics to John.
DOES NOT ENTAIL
John learned linguistics.

We have a complicates set of patterns here!

How should we understand them? This remains controversial. But let me illustrate briefly with one idea. We should look for a rich lexical decomposition that tracks some conceptual difference that this pattern displays. One approach posits a polysemy between the DO and PO cases (Krifka 1999; Pinker 1989):[12]

(6) a. DO: NP_1 CAUSES NP_2 TO HAVE NP_3
b. PO: NP_1 CAUSES NP_2 TO GO TO NP_3

These frames show different sorts of meaning: a meaning involving causing to have, versus a meaning involving causing to go to. The two different meanings are realized in English with different syntax.

One virtue of this proposal is that it helps explain why we get *# London a package* above. This is a DO form, and so requires HAVE. London does not have a package. It also explains the entailment we noted by assuming HAVE typically includes mastery. 'Successful transfer' is required for HAVE.

So, perhaps this is a good explanation. As I said, it remains controversial. But it is a plausible exercise in the kind of scientific modeling we might do for natural language. We have a complex phenomenon, and we build an abstract model that helps explain what is happening.

But this is not an exercise in logic, or scientific model-building with model-theoretic logics. Rather, it is an analysis of the components of a verb frame, which tells us how a class of verbs behave. Perhaps we could do some logic to understand the components like CAUSE or DO, or with locations, GO. But that would not give us the insight this theory offers. It offers a different kind of analysis, not one from logic-like modeling.

So, even when looking at entailments in language, logic as modeling may not be the most useful approach. Like all scientific models, this sort has its limits. Model theory and model-theoretic logics offer interesting

[12] For a critical discussion both of the data and the proposal, see Rappaport Hovav and Levin (2008).

abstract scientific models of some entailment patterns. These have proved useful for studying some 'logical', mainly functional, expressions. But they do not seem of much help for difficult problems like the lexical entailments I illustrated above.

But this raises a further question. When we can use models to model, what sorts of models do they provide? And how do they really help us to learn about phenomena?

10.3 Learning from Models

I have argued that when applied to empirical phenomena like natural language and human reasoning, we can usefully see logics, especially model-theoretic logics, as offering scientific models. But there are many sorts of models in science, and we do different things with them and learn from them differently. I shall conclude by suggesting that logics can work as models of at least two sorts. They can function as what are called analogical models, but also in ways similar to model organisms.

The way I described generating a logic starting with a natural language puts weight on idealization, along with identification and abstraction. This suggests that perhaps logics function as what are sometimes called idealized models in science.[13] The standard example is the frictionless plane, which idealizes away friction. Notably, we have surfaces of varying amounts of friction, and this idealized model simply pushes that to an idealized limit. As I recall from my college physics class, we use this model to help us work out basic laws of mechanics.

To some extent, logics can behave like this. A logic of generalized quantifiers idealizes away various features of natural language, including syntax and scope restrictions on quantifiers, to provide a general picture of what scope and quantifier meanings look like. We use this to understand the phenomena in question.[14]

But this is not quite right. After all, we do not stare at logics of generalized quantifiers and figure out the laws of scope in natural language. We cannot, as these logics have idealized away what we would need to do so.

A better picture comes from what are often called analogical models in science. The billiard ball model of gases is a standard example. These

[13] I am not a philosopher of science, and am very much working as an amateur when it comes to scientific models. As amateurs sometimes do, I am trying to just follow the experts, in this case with the help of the *Stanford Encyclopedia of Philosophy* entry on models in science by Frigg and Hartmann (2009).

[14] For an extensive overview of scope phenomena in natural language, see Szabolcsi (2012).

models analogically exploit similarities between systems. So, a system of colliding billiard balls is analogically related to a gas. These models also help us to find basic, sometimes idealized laws. But they are not simply limits of what is real. Gases are not made of billiard balls, even in a limit.

I think we rely on such analogical roles when we use logic to model natural language. Take quantifier scope, a messy phenomenon in natural language. We understand it best after we abstract away to a logic with quantifiers. We then look back and say that scope in natural language must be something like we see there. Not exactly, but by analogy.

My suspicion is that many applications of model theory to natural language and cognition are like this. Model-theoretic logics give us idealized and often analogical models of various aspects of language, or perhaps cognition. We can use them to try to formulate more accurate hypotheses about how language really works, and can learn from general comparisons or analogies. Note, for instance, that the idea that possible worlds are like models is already a clear kind of analogical modeling.

Analogical models are often useful to get inquiry going. But they have limits, of course. One is that they do not, typically, have parameters we can manipulate. Many models are systems of equations, or computer simulations, that have such manipulable parameters. (Basically, constants that can be adjusted to affect the behavior the model.) This is important, as it allows us to test models against data, and refine them by adjusting parameters. We can then use the models to generate new predictions, which are again tested against data. This cycle allows the building of more refined models, which help us improve our understanding of some phenomenon and make better predictions.

No doubt in some cases, logics can be used like this. A clear example is the PSYCOP model of Rips (1994), which was built to be just such a model. Perhaps there are others. We might think of the accessibility relation of a Kripke model as a kind of manipulable parameter, or perhaps the strength of a logic of generalized quantifiers. Perhaps this is right in some cases, but in many, I find the comparison strained. It is not clear that we simply adjust the accessibility relation as a parameter, with a set range of values, in response to data. We certainly cannot do that with our logics of generalized quantifiers. At the same time, I think there is a better way to see how these sorts of tools function as models beyond their analogical roles.

The range of what counts as models is very large, and I shall leave it to the philosophers of science to decide whether it forms a homogenous class or kind. But among what are called models are model organisms, often worms or mice. I shall suggest that logics often function more like model organisms than other sorts of models.

Model organisms are often carefully designed, by breeding or genetic engineering. For instance, if you want to study a human disease, you might design a model organism to simulate relevant aspects of the human. You can then observe the phenomenon in a controlled environment, typically by infecting your carefully designed mouse and seeing what happens.[15]

Animal models can be changed in the face of data, but in many cases, they do not have straightforward manipulable parameters. Certainly, when they are designed, allowing fairly straightforward changes is a valuable feature. But often when an animal model fails, it is a significant task to build an improved one

I think in many applications, logics function surprisingly like animal models. Take the case of a generalized quantifiers again. A logic of generalized quantifiers in part, but only in part, reflects the way quantifiers work in natural language. The same is true of animal models. We can use the logic as a controlled environment to study the behavior of quantifiers. Drop a quantifier into it, and we can see a set of entailment patterns, and a number of other logical properties, like monotonicity properties, definability properties, EXT mentioned above, and so on. We can watch these behaviors, and see how that compares to quantifier behavior in natural language. We can generate predictions from the model. Sometimes, we can change the model itself. We might restrict ourselves to conservative quantifiers, for instance. We can sometimes even find a manipulable parameter. Restricting ourselves to finite models can provide interesting results, especially about discourse processing. But then, the size of a domain can be manipulated.[16]

In practice, I suggest, we use logics to study natural language both like analogical models, and like model organisms. We can use lots of other tools, and we do. I have focused on model-theoretic logics, but proof theory has many applications. So do non-logical methods. Decision theory comes to mind. So do theories of concepts and categorization. We have lots of tools, but model-theoretic logics are among then, and often used as model-theoretic logics to provide scientific models in these two senses.

[15] This is a common technique among biomedical researchers. For an overview, see for instance Fox et al. (2007).

[16] See, for instance, results from Keenan and Stavi (1986).

I have illustrated specific roles for model theory in both foundational approaches to logic and to the study of natural language. In both, the rich mathematics of model theory can be important. I doubt these are the only roles for model theory in applications, but I think they are central ones. Model theory provides us a rich stock of model-theoretic logics. In foundational studies, these can be important options for conceptual analysis, and raise a number of foundational questions about the nature and scope of logic. Perhaps this might help us to understand the nature of LOGIC. When we turn to applications to empirical matters like language and cognition, these logics stand in the relation of models in science to phenomena. Specifically, I suggest, we use them like analogical models, and sometimes much like model organisms are used. Such models have valuable, but I think limited applications, as all good models do. But generally, when applied to language and cognition, we should think of logic and model theory as much more like models in science.[17]

[17] A preliminary version of this paper was presented at a workshop at the Munich Center for Mathematical Philosophy, February 2018. Thanks to all the participants there, but especially to Roy Cook, Gil Sagi, and Jack Woods, for helpful comments. Thanks also to Gil and Jack for extraordinary patience while I finished this paper, and for comments on an earlier draft. I learned about model organisms in science from the wonderful Ching-I Chen.

CHAPTER 11

On Being Trivial: Grammar vs. Logic

Gennaro Chierchia

The importance of the function/content dichotomy manifests itself in the recurrent and arguably successful attempts at explaining aspects of grammar in terms of logic. The claim emerging more and more forcefully in this connection is that many cases of linguistic deviance owe their status not to the violation of some syntactic well-formedness condition but to the fact the relevant structures are logically determined (i.e., logically true or logically false), and hence in some sense 'trivial'. Since classical tautologies or contradictions (of the form *Is John smart? Well, he is and he isn't*) are logically determined but are not perceived as ungrammatical, one immediately faces the issue of how to tease apart (possibly, as a matter of principle) trivialities rooted in grammar and perceived as ungrammatical (which I will call G-trivialities) from classical tautologies and contradictions (L-determined sentences). In what follows, I will go over an example, rather compelling in my view, of a class of phenomena best explained in terms of G-triviality. I will then present and discuss the way of conceptualizing the special status of G-trivialities vis-à-vis standard tautologies put forth by J. Gajewski (2002). I will point out a problem with Gajewski's approach and sketch a solution that embodies a somewhat different view of the relationship between grammar and logic.

As is well known, Negative Polarity Items (NPIs) like *ever* or *any* are restricted in their distribution to (roughly) 'Downward Entailing' (DE) contexts, i.e., contexts that license 'subset inferences' such as those in (1a) as opposed to (1b):

(1) a. I didn't eat pizza → I didn't eat pizza with anchovies *Subset inference* (DE)
b. I ate pizza with anchovies → I ate pizza *Superset inference* (UE)

Thanks to the participants of the 2018 Münich Workshop and to an anonymous referee for helpful comments. I am also grateful to Guillermo Del Pinal, Danny Fox, and Jon Gajewski for extensive discussions of these matters over time.

A representative set of contrasts involving NPIs is given in (2) vs. (3):

(2) a. There isn't any pizza left
 b. I doubt that there is any pizza left
 c. If there is any pizza left, we won't go hungry

(3) a. * There is any pizza left
 b. * I believe that there is any pizza left.
 c. * If you are hungry, there is any pizza left

One and the same string of words, namely *there is any pizza left* is deviant in a non-DE environment (e.g., under *believe* or in the consequent of a conditional) and becomes perfect in a DE one (e.g., embedded under negation, or under *doubt* or in the antecedent of a conditional). Notice that dropping the item *any* in the sentences in (3) renders them grammatical, which confirms that the occurrence of *any* is the culprit for their degraded status in (3). The deviance of the sentences in (3) is quite severe, comparable to an agreement mismatch or a basic word order violation of the form *boy the walked in*. In spite of this, a thesis that has consistently gained credibility is that there is nothing wrong with the syntax of the sentences in (3). The problem is wholly semantic: These sentences are unrescuably trivial. Let me flesh out this prima facie implausible claim, starting with a seemingly unrelated example of linguistic deviance.

Consider the dialogue in (4):

(4) Speaker A: How many of the twenty papers you have to grade do you think you will have graded in two hours?
Speaker B:
a. Possibly, even ten b. Possibly not even one c. * Possibly, even one

You will agree that while (4a) and (4b) are natural answers to Speaker A's question, (4c) is distinctly deviant. Why? Use of *even* generally conveys that the proposition *even* applies to (i.e., the prejacent) is regarded as the least likely among some relevant set of alternatives:

(5) a. I understood even Chomsky's paper
 b. Alternatives under consideration: Given some set A of contextually salient individuals, ALT = I understood a's paper: a ∈ A
 c. Presupposition of Sentence (a):
 Understanding Chomsky's paper $<_{LIKELY}$ Understanding a's paper (for any a ∈ A)

The way in which the alternatives are typically individuated is determined by the context (e.g., by some question under discussion), which in turn is often coded into the focal structure of sentences. In the context of the question in (4), the answer in (4a) will evoke a scale of the form:

(6) < (in two hours) I will have graded ten or more papers ,
I will have graded nine or more papers,
. . . ,
I will have graded one or more papers> ↓

The arrow in (6) indicates the entailment pattern holding among the alternatives, where if X is stronger than Y, X is true in fewer worlds and hence necessarily less likely than Y (or at most as likely as Y). So, sentence (4a) states that I will grade probably ten assignments and this is the least likely and best scenario option among the alternatives that stand a chance at being true, and this yields a felicitous response. Consider next (4b). The presence of negation in (4b) reverses the scale in (6):

(7) < (two hours from now) I won't have graded a (single) paper,
I won't have graded two or more papers,
. . . ,
I won't have graded ten or more papers > ↓

In this reversed scale, grading no paper becomes the strongest, and hence least likely member of the relevant alternative set, which again makes use of *even* appropriate. At this point, the reason why (4c) sounds weird becomes apparent. With respect to the scale in (6), naturally associated with positive sentences, the sentence *I graded one paper* is the weakest member and hence it cannot be the least likely. Claiming the contrary results in a contradiction (for a proposition cannot be less likely than its entailments). So, it looks like behind the immediacy of our reaction to (4c) there is a rapid computation that leads to the following conclusion: sentence (4c) triggers a contradictory (i.e., trivial) presupposition. This seemingly obscure corner of the grammar of *even* constitutes an illustration of what I mean by G-triviality.

This example of G-triviality is directly relevant to NPIs. Suppose that *any* means something like *even* + *one*, and associates with a scale analogous to (6). This would immediately explain why the sentences in (2) are fine. More specifically, replacing *any* in (2) with something like *even* + *one (single)* is grammatical and yields virtual synonyms of the original sentences:

(8) a. There isn't even one (single piece of) pizza left
b. I doubt there is even one (single piece of) pizza left
c. If there is even one (single piece of) pizza left, we won't go hungry.

The *any* = *even* + *one* hypothesis also explains why the sentences in (3) are ungrammatical: They are all contradictory, just like (4c) is. For example, the logical form of (3a), repeated here as (9a), would be something like (9b):

(9) a. There is any pizza left
b. $\texttt{even}_{\texttt{ALT}}\ [\exists_{\texttt{x}}[\texttt{one(x)} \wedge \texttt{pizza(x)} \wedge \texttt{left(x)}]]$
$\approx$ there is even ONE piece of pizza left
c. ALT = $\{[\exists_x[n(x) \wedge pizza(x) \wedge left(x)] : n \in N\}$
where the entailment in (c) goes from n to n-m, for any n and m.

It's as if (9a) was interpreted as *there is even ONE (single piece of) pizza left* in reply to a how many-question: (9a) triggers a contradictory presupposition. This account extends to all DE environments and provides an arguably elegant and simple explanation for the distribution of NPIs, that descends directly from their (hypothesized) semantics.

There is one striking fact that seems to support this hypothesis. In many languages NPIs are explicitly formed by composing focus-sensitive additive particles that mean roughly *even* with some item that expresses a low quantity like the first numeral 'one'. Here is a representative sample:

(10) Hindi: ek bhii — one even/also
Tagalog: anu-ma-ng — wh-even-CASE
Italian: neanche uno — negative agr + also one

(11) A Hindi example (from Lahiri 1998):
i. * ek bhii aadmii aayaa
one even man came — 'any man came'
ii. ek bhii aadmii nahiiN aayaa
one even man not came — 'no man came'

Historical change provides further evidence in favor of this view. *Any* comes from Old English *ænig* (lit. "one-y," "one-like"), which in turn is derived from ProtoGermanic **ainagas*. The Proto IndoEuropean source for this class of words is *oinos, the word for *one*. Now, *any*'s German cousin *einig* remained a plain vanilla indefinite without a negative polarity use. The Italian counterpart of *any/einig*, namely *alcuno/alcuni*, has a split behavior: the plural is a regular indefinite, while the singular is an NPI. It is not implausible to conjecture that when expressions of minimal amount start being used as NPIs, it is through the association with an adverbial particle like *even*, association which sometimes goes unexpressed.

All in all, it is clear that the combination *even* + *one* is a widespread source for NPI behavior, possibly even in English. Which directly lead us to the following (admittedly rhetorical) question: if the reason for why *even* + *one* is restricted to DE environments is not analogous to the reason why (4c) is deviant, then what is it?[1]

And here comes the issue of interest to our present concerns. Our account for NPI-violations relies on the fact that they turn out to be, under a plausible semantics, contradictions/trivialities. Which immediately raises the question of why aren't *all* trivialities ungrammatical. Why is it the case that (12a.ii) is odd in reply to (12a.i), while (12b.ii) is a perfectly natural answer to (12b.i)?

(12) a. i.How many papers can you grade by 5? ii. * Oh, even one
b. i. Is John smart? ii. Well, he is and he isn't

Why, moreover, the contradiction in (12a) requires analysis to unveil its contradictory nature, while the one in (12b) is readily accessible to introspection? As it turns out, these questions have principled answers, as we will see.

11.1 'Modulated' Logical Forms: A Restrictive Contextualism

What we have seen so far is that there are reasons to believe that NPIs in non-DE contexts are to be ruled out not because of some syntactic violation, but because they are contradictions (and hence useless in communicating). However, there are plenty of contradictions that sound fine and are in fact used in concrete communicative situations. So, how can the idea that NPIs in non-DE contexts are just contradictions be right? Gajewski (2002) argues that the solution is rooted in the distinction between function and content words and we are now going to review his proposal. Let us consider the key sentences in (13a) vs. (14a) and their respective syntactic structures.

(13) a. [Is John smart?] He is ~~smart~~ and he isn't ~~smart~~
b. Key: T = Tense; T',TP = Tense Phrases; SC = Small Clause;

1 The Landscape of Polarity Sensitive Items is much richer than the one sketched in the text. But I think that dealing with it in more detail still requires dealing in depth with the problem of how logicality affects grammar. Compare with, e.g., Chierchia (2013) and references therein for a more thorough investigation of the relevant issues.

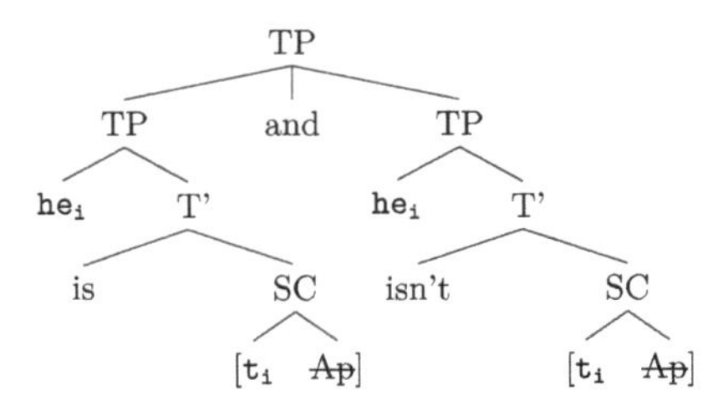

crossed out constituents go unpronounced but are used for interpretive purposes.

(14) a. There is any pizza left

b.

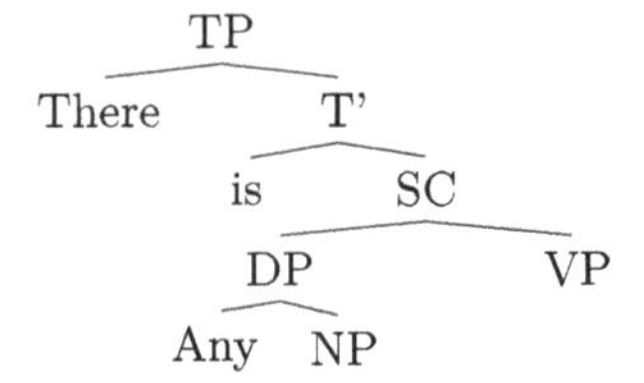

The trees in (13b) and (14b) constitute the rough syntactic analysis of the two key example sentences, along relatively uncontroversial lines, easy to translate into other popular approaches to these constructions. They may be viewed as the functional skeleta of sentences (13a) and (14a) respectively. I.e., they constitute the kind of structure that grammar would generate out of the functional elements alone. The pronoun he_i is treated as a variable, whose value is assumed to be contextually set, and variables qualify as logical elements. Within the framework of Distributed Morphology (see, e.g., Halle and Marantz 1993) sentences are composed by assembling their functional structure first; content words are inserted at a later stage so as to take into account the contribution of functional structure. This design is meant to make sense of the fact that the final shape of content words is sensitive to the functional structure in which they are inserted: think for example of the common place observation that PAST + V sequence spells out in English as V-ed if the chosen verb is regular like *loved* or *walked*, but the same grammatical information spells out differently if the verb is irregular, like *went* or *hit*. Within Distributed Morphology, which adopts this 'late lexical insertion' strategy, structures roughly like (13c)–(14c) correspond to an actual phase of the derivation of the relevant sentences.

With the notion of functional skeleton in place, Gajewski's approach to G-triviality is based on the following central generalization. A sentence like

(14a) turns out to be contradictory regardless of the choice of content words one inserts in the corresponding structure (14b), i.e., regardless of which N and which V one eventually selects from the lexicon. A sentence like (13a), on the other hand, comes out as contradictory only if one chooses the *same* adjective from the lexicon for the two instantiations of the category AP. This is the source of the distinction between ungrammatical vs. grammatical contradictions: G-trivialities are contradictory for *any* choice of content words.

Let us put this insight in slightly different terms. Consider the interpretations of (13a)/(14a) 'minus their content words', i.e., replacing the latter with variables of the appropriate type:

(15) a. $\mathtt{P}(\mathtt{x}_{\mathtt{i}}) \wedge \neg \mathtt{P}'(\mathtt{x}_{\mathtt{i}})$
b. $\mathtt{even}_{ALT}(\exists_{x}[\mathtt{one}(\mathtt{x}) \wedge \mathtt{P}(\mathtt{x}) \wedge \mathtt{P}'(\mathtt{x})])$
where $ALT = \{\exists_{x}[\mathtt{one}(\mathtt{x}) \wedge \mathtt{P}(\mathtt{x}) \wedge \mathtt{P}'(\mathtt{x})] : \mathtt{n} \in \mathtt{N}\}$

Formula (15b) is contradictory no matter how the variables P and P' are interpreted, while (15a) is contradictory only if P and P' are mapped onto the *same* property. Gajewski's original algorithm is a 'syntactic replacement' version of this very same idea: he proposes replacing in the relevant Logical Forms (i.e., the complete syntactic trees for (13a) and (14a)) all occurrences of content words with distinct variables of the same type. If the result is contradictory, the sentence is deemed as ungrammatical.

This account, besides being algorithmic, has an arguably natural functional basis in the pragmatics of communication. What suggests itself is that the reason why sentences like (13a) are perceived as grammatical is because it is possible, and indeed natural, to reinterpret the two occurrences of the (unpronounced) adjective *smart* in slightly different ways (e.g., as 'John is clever at his job, but he is not savvy in the way he manages people', or the like). But no such strategy can be of any help with (14a): no matter how we reinterpret the N or the V, contradictoriness persists.

Gajewski's proposal provides us with a principled way of distinguishing between grammatical and ungrammatical trivialities. Appeal to this distinction is by no means limited to the case of NPIs. Other phenomena that have been argued to require an account in terms of G-triviality include the distribution of *for-* vs. in- *X time* adverbials (Dowty 1979), the definiteness effect in there-sentences (Barwise and Cooper 1981), exceptive constructions (von Fintel 1993), the distribution of quantifiers in comparative constructions (Gajewski 2008), weak island violations (Abrusán 2014), and more. This is just a pointer to some of the relevant literature and sounds like a laundry list. The important fact is that resorting to ungrammatical contradictions to explain properties of grammar is a widespread practice that is proving to be more and more fruitful in figuring out how natural language works.

The distinction between function- vs. content-words is somewhat vague and obviously in need of further clarification, but it plays a crucial role both in traditional as well as in cutting-edge linguistic theories. We will come back to the function/content distinction in later parts of this paper. For now, what is important is that we need to reconsider the widespread stance that syntax determines well-formedness and semantics determines how well-formed sentences are interpreted. On the view I see myself forced to adopt here, grammatical sentences are constituted by well-formed structures *that are non-G-trivial*, and determining the set of G-trivial sentences involves resorting to an empirically determined cast of (interpreted) function (/logical?) words, as per Gajewski's algorithm.

Del Pinal (2019) argues for an interesting modification of Gajewski's proposal. His proposed modification is meant to directly reflect the functionalistic/pragmatic interpretation of Gajewski's proposal. Del Pinal's idea can be illustrated by way of example, representing the interpretations of our two key illustrative sentences (13) and (14) as follows:

(16) a. $\mathtt{g(smart)(x_i)} \wedge \neg \mathtt{g'(smart)(x_i)}$
b. $\mathtt{even_{ALT}}(\exists_x[\mathtt{one(x)} \wedge \mathtt{g(pizza)(x)} \wedge \mathtt{g'(left)(x)}])$
where ALT = $\{\exists_x[\mathtt{n(x)} \wedge \mathtt{g(pizza)(x)} \wedge \mathtt{g'(left)(x)}] : \mathtt{n} \in \mathtt{N}\}$

The assumption here is that the interpretation of content words can be modulated through the (optional) insertion of functions g, g',... which are contextually determined and map any semantic object into something of the same logical type. The default interpretation of these modulating functions (whenever present) is simply the identity map. However, when the default interpretation leads to a contradiction, as, say, with sentences like *John is and isn't smart*, the offending item, in this case a property, typically gets modulated on the basis of the intentions, communicative goals, etc. of the illocutionary agents, for example as in (17):

(17) John is g(~~smart~~) and isn't g'(smart)
g(smart) = clever at his job
g'(smart) = savvy in managing people

The strike through indicates that the second instance of the AP *smart* is elided under identity with the first, but present for interpretive purposes.

One might try to explain the acceptability of the prima facie contradictory sentences of this sort in different manners, e.g., through the resetting of some contextual parameter implicit in adjectives like *smart*, that require a 'comparison class' (smart with respect to what?). However, this alternative account doesn't extend in any obvious way to examples like *How is the weather? Well, it rains and it doesn't rain* ($\approx$ it rains on and off). Moreover, resetting grammatically determined variables (such as comparison classes) is generally banned in VP-ellipsis environments such as those in (17). The present proposal doesn't suffer from these drawbacks and is thus superior to alternatives relying solely on grammatically determined parameters.

We call logical forms/interpretations such as those in (16) 'modulated logical forms'.[2] We may regard Del Pinal's proposal as variant of Gajewski's that embeds the latter within a contextualist stance according to which the standard interpretation of content words is context dependent. On Del Pinal's modification, G-trivial sentences are those that are true/false for any value of the modulating functions, while classical tautologies/contradictions are those that are true/false when all modulating functions are interpreted as identity maps. Obviously, G-trivial sentences are a proper subset of the classical tautologies/contradictions. This approach requires constraints on modulation. Mapping some content word W into something g(W) of the same type is not enough. The mappings appealed to must be in some sense 'natural', address the communicative intentions of the illocutionary agents in pragmatically sensible ways, e.g., by resolving appropriately the questions under discussion (*how smart is John?*). Nobody has a fully worked out theory of what makes a modulation 'natural', beyond appealing to context, questions under discussion, and the like.[3] Still, with these limits acknowledged, Del Pinal's proposal provides a useful characterization of G-triviality and is more general than Gajewsky's, since resetting of basic word meanings happens extensively.[4] Here are two cases. One can be illustrated by a famous example, due to G. Nunberg:

[2] Del Pinal uses the terms 'rescaling' and 'rescaled logical forms' in this connection. I prefer the term 'modulation'. This terminological choice foreshadows a generalization of Del Pinal's approach in two ways, which will be developed in Section 11.4 below. First Del Pinal limits rescaling to predicates (or types that 'end in' the type $< e, t >$), while I generalize it also to individuals. Second, I think that the present proposal fits with and accommodates also a treatment of de re belief.

[3] Del Pinal suggests that modulation may be subsective, i.e., map a property into some subproperty. I believe this constraint to be too restrictive in light of examples like (18) and others considered below.

[4] Notice that Del Pinal's proposal does not make contradictions inexpressible in English, a worry expressed by an anonymous referee. It all depends on the intended interpretation of remodulation. In the following quote, from Aristotle's *Metaphysics*, for example, a contradiction is clearly intended and communicatively effective:

(18) a. [A waiter to a fellow waiter:] The ham sandwich wants his bill
b. wants(his bill)(ιx[g(hamsandwich)(x)])
where g(ham sandwich) = *person that ordered the ham sandwich*

The logical form in (18b) illustrates how Nunberg's example might be handled on an approach based on modulation. A second class of cases is exemplified by sentences like (19a), due to B. Partee:

(19) a. Tommy believes that clouds are alive
b. $\forall_w[BEL_{TOMMY,w0}(w) \rightarrow \forall_x[g(cloud)(w)(x) \rightarrow alive(w)(x)]]$
Where for any individual u and world w, $BEL_{u,w}$ is the (characteristic function of the) set of worlds compatible with u's beliefs in w.

A sentence like (19a) expresses a (typically *de re*) belief about instances of the cloud-kind, namely that they have life. Given how life is understood in our linguistic community, the belief attributed to Tommy constitutes a metaphysical impossibility that fails in every possible world (much like Hesperus cannot be different from Phosphorus, given how names work, etc.). Hence the semantics for (19a), say (19b), would condemn Tommy's belief state to incoherence (under the default interpretation of g as identity). Modulation offers a way out. The g-function in (19b) may map clouds into some cloud-like living creature, for example.

Del Pinal's proposal, besides providing a conceptual embedding of Gajewski's approach within an independently plausible form of contextualism, also has, perhaps, a further technical advantage. Consider yet again our toy NPI violation, repeated here.

(20) a. * There is any pizza left
b. $even_A LT(\exists_x[one(x) \wedge g(pizza)(x) \wedge g'(left)(x)])$
c. $ALT = \{\exists_x[n(x) \wedge g(pizza)(x) \wedge g'(left)(x)] : n \in N\}$

It is crucial that the interpretation of the non-logical words (*pizza, left*) be kept constant in the assertion (19b) and across all of the alternatives in (19c). On a substitutional approach like Gajewski's, where logical skeleta are obtained by replacing each occurrence of the non-logical words with distinct variables, some work is required to ensure that the replacement

(a) It is impossible that the same thing can at the same time both belong and not belong to the same object and in the same respect, and all other specifications that might be made, let them be added to meet local objections (Aristotle, *Metaphysics*, 1005b 19–23)

In (a), the word *belong* has to be remodulated via the identity map and hence the sentence in boldface expresses a genuine logical contradiction, as that is the choice that makes pragmatic sense.

is uniform across the alternatives. It is not hard to imagine a definition of logical skeleta that would NOT have such a property, thereby yielding wrong predictions. On Del Pinal's approach this issue doesn't arise.

On the basis of these considerations, I conclude that Del Pinal's proposal constitutes a friendly and useful amendment to Gajewski's original approach. But before attempting some general reflections on what this take on G-triviality tells about the relation between logic and grammar, we need to address a problem that both Gajewski's and Del Pinal's approach leave open.

11.2 The Problem of Bound Variables

Consider sentences of the following form:

(21) a. John is never himself
b. Yesterday, John managed to be more eloquent than himself

The first relevant observation is that these sentences are perfectly grammatical and communicatively useful. The second noticeable point is that, taken literally, they are contradictory. Third, these examples are beyond repair on the modulation approach adopted here, for the following reasons. Reflexive pronouns, comparative morphemes (*more*), and negation (*never*) are prototypical functional items. Thus the modulated structure of (21a,b) is going to be roughly as follows:

(22) a. g(John) is never himself
b. g(John) g(managed) to be more g(eloquent) than himself

The (pseudo) formulae in (22) are contradictory for any choice of g. Hence, sentences (21a,b) should be ungrammatical according to the characterization of G-triviality we are adopting. But they are not; they clearly do not have the same status as *there are any cookies left*.[5] Our proposal, as it stands, seems therefore to rule out too much.

The source of the problem seems to lie in the fact that reflexives are interpreted as variables bound to some suitable antecedent in their local syntactic environment; and bound variables are functional/logical items, if anything is. Hence they should not be targeted by replacement or modulation. This is why the Gajewski/Del Pinal approach appears to fail in

[5] This version of the bound variables problem was pointed out to me by Richard Larson, at a talk I gave at SUNY Stonybrook in 2015. Gajewski (2002, Section 4.2) is clearly aware of it. Also Del Pinal (2019) discusses it explicitly (cf. fn. 9 below).

its job of sifting ungrammatical vs. grammatical trivialities in the cases at hand. How can we modify such an approach, so as to retain its main merits and its principled character? In the present section I address this problem.

The syntax and semantics of reflexives and comparatives is a complex matter. In what follows, I will base my proposal on reflexives, sketching as much of their grammar as needed, in an as uncontroversial manner as possible. I will then indicate how my proposal extends to comparatives.

Let us say that reflexives are governed by Principle A of Chomsky's (1981) binding theory according to which they must be bound to an antecedent in their local syntactic environment, which for our purposes can simply be the smallest sentence containing the reflexives. Principle A is a syntactic axiom with semantic consequences. Binding is achieved, let us assume for the argument's sake, by assigning scope to the antecedent (via Quantifier Raising – QR – or the equivalent), which creates an abstract that binds the reflexives as illustrated in what follows.

(23) a. John is (not) $\mathtt{himself}_i$
b. $\mathtt{John}_i$ [$\mathtt{t}_i$ is (not) $\mathtt{himself}_i$]
$\rightarrow$ John λx_i [x_i is (not) $\mathtt{himself}_i$]
where $\mathtt{t}_i$ is the trace left behind by (string vacuous) raising of the subject.

Following a widespread practice, we analyze the index on *John* as an abstractor that creates the derived (reflexive) predicate in (23b).[6] Principle A ensures that the index on the subject in (23b) be the same as the anaphoric index on the reflexive. We assume that the copula winds up being interpreted here as identity.

We must briefly consider a further option at this juncture. One way of addressing our problem and making (23a) non-G-trivial might consist of treating the copula as a content item and modulate it by mapping identity into some other relation that doesn't yield a contradiction. I think this move is implausible. First, the copula is, syntactically speaking, a prototypical functional item. In many languages, copular sentences like (23a) do not exploit any overt item like the verb *to be* but are assembled by mere concatenation. This is true even of some English predicative 'small clause' constructions such as:

[6] See, e.g., Heim and Kratzer (1998).

(24) a. I consider [$_{SC}$ John a good player/his own worst enemy/finally himself again]
b. I regard [$_{SC}$ John as my best friend/his old self again]

Sentences like (24) seem to yield manifestations of the same problem as (23), in ways that does not rely on an overt copular verb. Moreover, in just about any language the item used in copular sentences, when attested, typically doubles up, just as in English, as a mere expression of tense and aspect (as in *John was in the bathtub/a good friend*). This behavior is symptomatic of functional elements. In fact, the most detailed attempt at analyzing the semantic side of copular construction namely Partee (1986), analyzes copular constructions as involving a (restricted) set of 'logical' type-shifting devices. Second, semantic criteria such as identity under domain permutations put identity among the logical constants.[7] Third, treating identity as a content item would not help with the case of comparatives, where the identity relation as such is not involved. We might as well look for a solution that covers also reflexives in comparatives, as the diagnosis of the source of the problem (namely, the presence of bound variables) seems to be the same.

If tinkering with the copula is of no help, the only other way to go is to modulate the bound variable itself, e.g., as follows:

(25) a. John is (not) $\mathtt{himself}_i$
b. John $\lambda x_i[x_i$ is (not) $g(\mathtt{himself}_i)]$
c. John is not the person he usually is/the way he usually is
d. $\lambda x_i \neg[x_i = \iota x[\mathtt{x\,behaves\,(in\,w)\,most\,similarly\,to\,how\,x_i\,usually\,behaves}]]$

In (25c) I exemplify typical ways of understanding sentences like (25a). The modulation of variables has to have an intensional character, which spelled out in a full-fledged compositional system would yield, for example, something like (25d).[8] Notice that the outcome of modulation of the reflexive

[7] See MacFarlane (2017) and references therein – esp. McGee (1996) and Sher (2003). Compare with also the discussion in Section 11.4 below.

[8] The semantic metalanguage I have in mind is Gallin's (1975) TY2 with overt world variables. I am assuming that predicates carry a world variables (e.g., *that is red* = *red(that)(w)* abbreviated as $red_w(that)$); modulation of individual variables maps individuals (of type e) into individual concepts of type $< s, e >$, e.g., in the case at hand, it picks 'the individual that behaves most similarly to how John usually behaves'. Such a concept winds up being applied in the end to the actual world. In other words, the proposition associated with (24) is something like:

(i) $\lambda w \neg[j = \iota x$ behaves (in w) most similarly to how j usually behaves$]$

in (24d) is a contingent property, which in turn ensures that sentence (25a) won't come out as G-trivial.[9]

While this modification perhaps yields an empirically adequate solution to the problem of bound variables, it seems to give up on the functional/logical vs. content distinction that we have been relying on so far, and appears to be less principled than the original. Not only content/non-logical items need to be modulated. Variables need to be modulated as well:

(26) New definition of G-triviality.
  a. Modulation: optionally insert a modulation function g on any content word or variable ($\approx$ bound pronoun or trace).
  b. A sentence is G-trivial iff it comes out as true/false on any modulation.
  c. A sentence is L-trivial iff it comes out as true/false for the default value of modulations (as identity).

Under this new definition, sentences like those in (21) come out, correctly, as *non*-G-trivial. I think that, in spite of appearances, this proposal in fact retains the original inspiration and principled character of Gajewski's and Del Pinal's. Variables constitute stand-ins for content expressions. If you think in model-theoretic terms and consider a canonical intensional model <U,W, F>, with U a set of individuals, W a set of worlds, and F an interpretation function, the items that can be modulated are the values of F and of the assignments to variables. Logical constants, on the other hand, have values that remain constant across models and cannot be modulated. Obviously, I am not giving here a characterization of the logical/non-logical divide. I am simply adopting a standard semantic practice and pointing out how my definition of modulated logical forms falls with respect to it. Here is the guiding principle of our proposal:

(27) The referential points of a logical form (/LF tree), namely the non-logical constants and variables, may be modulated.

While making this fully explicit may require some work, I trust that the chief idea is clear enough for our present purposes.

[9] Del Pinal (2019) proposes to address the problem with reflexives by modulating the property obtained by abstracting over the reflexive pronoun roughly along the following lines:

(i) John $g(\lambda x[x = x])$

This tantamounts to treating identity as a content item, which I have argued against, compounded with the complication that the property in (i) is a singleton property, that in every world maps each individual into the property of being identical to itself. It is unclear how modulation of such property can yield the right results.

The solution just sketched does extend to comparative constructions. The rough logical form of a sentence like (21b), repeated in simplified form in (28a), is as in (28b):

(28) a. John was more eloquent than himself
b. $\mathtt{John_i}\ \lambda \mathtt{t_i}[\mathtt{t_i}$ was MORE(eloquent) than $\mathtt{himself_i}$]
$= \lambda \mathtt{x_i}[\mathtt{MORE(eloquent)}(\mathtt{x_i})(\mathtt{x_i})](\mathtt{j})$
where for any u, MORE(eloquent)(u) is the property of being more eloquent than u defined as follows: u' has the property of being more eloquent than u iff there is some degree d such that u' is at least d-eloquent and u is not.

The analysis sketched in (28) relies on a degree semantics for comparatives, such the one explored in Kennedy (2007) and much related work. According to it, adjectives correspond to relations between individuals and degrees: John is d-tall iff John's height is at least d. The comparative morpheme, thus, says something about the respective maximal degrees to which two individuals have a certain gradable property. The logical form in (28b) can be modulated just like other sentences involving reflexives:

(29) a. $\mathtt{John_i}$ [$\mathtt{t_i}$ was MORE(eloquent) than g($\mathtt{himself_i}$)]
$= \lambda \mathtt{x_i}$ [MORE(eloquent) (g($\mathtt{x_i}$))($\mathtt{x_i}$)] (j)
b. John is (today) more eloquent than the degree to which he usually is
c. John is more eloquent than the individual whose eloquence is most similar to John's usual one.

The sentences in (29b–c) are possible informal renderings of the effect of modulating the reflexive pronoun in (29a).[10]

The treatment of variables I am proposing bears a non-accidental connection, I think, to the issue of de re (and de se) belief. I have already hinted at this in connection with example (19) above, *Tommy believes that*

[10] *Than*-complements are sometimes clausal. For example, (i) is best analyzed as (ii)

(i) John is taller than Bill is
(ii) John is taller than Bill is tall

If all *than*-complements were clausal, modulation of variables would not be necessary, for one might achieve the intended results by modulating the two occurrences of the adjective *tall* in (ii). However, it is not clear whether this is right for sentences involving reflexives, for the source is ungrammatical:

(iii) * John is taller than himself is ~~tall~~

Moreover, there are languages that have been argued to lack clausal comparatives such as Fijan (cf. Pearson 2010, and for a general overview of variation in comparison constructions, Beck 2011).

clouds are alive. Let me outline the connection more fully here through a simple example. Consider:

(30) John believes that his brother is not his brother.

a. John believes that his actual brother is in fact an impostor trying to steal John's inheritance.

b. John is at the dentist. While sitting on his dentist's operating chair, he spots a man acting as an aid to the main doctor. He forms the belief that that person is the new assistant to his dentist, without recognizing that he is in fact John's own brother. While knowing that his brother is a dentist too, John doesn't think that the person assisting his dentist is his brother.

On its non-contradictory interpretation, sentence (30) may be used to report a de re belief of John's towards his actual brother, compatible with (and appropriate to) a variety of scenarios, such as for example those in (30a–b). The issue of de re belief is of course intricate. One important tradition[11] addresses the problem by appealing to concepts through which the relevant *res* is accessed by the attitude holder. A belief is de re about an individual *u* whenever *u* reliably induces a concept about *u* in the belief holder *a*, which identifies *u* for *a* in *a*'s belief state. Such concepts for example (30) might be, say, *the man that wants to share John's inheritance* for context (30a) and *the man John is seeing* for (30b). Charlow and Sharvit (2014) have proposed an implementation of this kind of approach to de re in which logical forms for de re beliefs employ 'concept generators' that are inserted in the syntactic spot of the res and drive pragmatically the propositional content of the belief. In the case of (30a), for example, we might go for a logical form like (31a), with the g-function spelled out as in (31b–c):

(31) a. $\forall_w[\mathtt{BEL}_{j,w0}(w) \rightarrow \neg\mathtt{brother}_w(g(\iota x.\mathtt{brother}_{w0}(j)(x))(w)]$ where j = John and $brother_{w0}(j)(x)$ = x is brother of j in w_0.

b. Let u be John's brother in the actual world. Then: $g(u) = \lambda w.\iota x$[x wants to share j's inheritance in w]

c. $\forall_w[\mathtt{BEL}_{John,w0}(w) \rightarrow \neg\mathtt{brother}(\iota x$[x is the person who wants to share j' inheritance)(w)]

= John believes of the person who wants to share John's inheritance (namely his actual brother) that he is not his brother.

As in the case of variables, the definite description *his brother* (evaluated in the actual world) is modulated via a concept that mediates between John,

[11] See in particular Quine (1956), Kaplan (1968), Cresswell and von Stechow (1982).

the attitude holder, and the res his belief is about. The use of modulation for individual expressions proposed here can thus be viewed as an extension of Charlow and Sharvit's proposal for the semantics of de re belief in general.

In sum, the present proposal is that logical forms (which drive the compositional interpretation of sentences) can and sometimes must be modulated by the insertion of 'replacement functions' in their referential points. Referential points of an LF are the content words ('non-logical') and variables (whose values range on the denotations of content words). Modulation is necessary for a variety of reasons, most prominently to make sense of our belief-states and to resolve contradictions in a communicatively effective way, explaining why sometimes contradictions can be useful communication tools. There are, however, sentences that cannot be rescued in this way. Their LFs turn out to be contradictory for any modulation. These sentences are useless and can be regarded as on par with syntactically ill-formed sentences. The outcome is an arguably general and principled proposal in which a characterization of G-triviality stems from the independent need of reinterpreting certain sentences in context.

11.3 Grammar vs. Logic

Our approach to modulation is rooted in the distinction between function and content words, where function words subsume logical words. In the present section, I go over some issues in the characterization of this dichotomy.

As is well known, functional items do not have a clear cut, absolute characterization, but there a number of syntactic and semantic criteria that are reliably relevant to the distinction *functional vs. content*. Starting at the syntactic end of things, here is a (partial) list of functional categories and morphemes:

(32) Typical Functional categories and subcategories
  a. Determiners, Quantifiers, Classifiers, Complementizers, Coordinations, Negation, Comparative/Superlative markers, Tense and Aspect markers ($\approx$ 'Auxiliaries'), Modals, Focus and Topic markers, Discourse Particles, ...
  b. Gender, Number, Case, (in)definiteness markers, Verb/Noun-class markers, pronouns, wh-elements, ...

Those in (31a) are typical functional categories and subcategories; those in (31b) involve 'features' specific to certain items (like pronouns) or active in agreement patterns. To see what the items in (31) have in common, consider the main basic orders in the languages of the world:

(33) a. S O V b. S V O c. V S O
[where S = subject, O = Object, V = verb)

Typical functional elements such as those in (32) tend occur at the edges of the main clausal constituents in (33), relative to the basic word order a language chooses, where they may be realized as bound morphemes or as autonomous words ('free' morphemes). For example, the expression of PAST-ness occurs at the periphery of the VP and can be realized as a bound morpheme *lov-ed* (through, say, incorporation of the V into the PAST morpheme) or as an independent morpheme as in *has walked* (with semantic differences between the two options). To illustrate further, coordinators (*and, or*) typically connect clauses and hence they tend to occur at the edge of clausal structures rather than in the middle of them. Similarly, Discourse Particles (e.g., German *doch*, Greek $\mu\varepsilon\nu$) have to do with signaling discourse junctures related to topicality, backgrounding, etc. and are often placed at or near major constituent boundaries.[12] The way functional categories are conceptualized within current generative approaches is as a series of heads at the edge of NPs or VPs, forming the so-called 'functional spines' or 'extended projections' of the latter. I provide an example in (34):

(34)

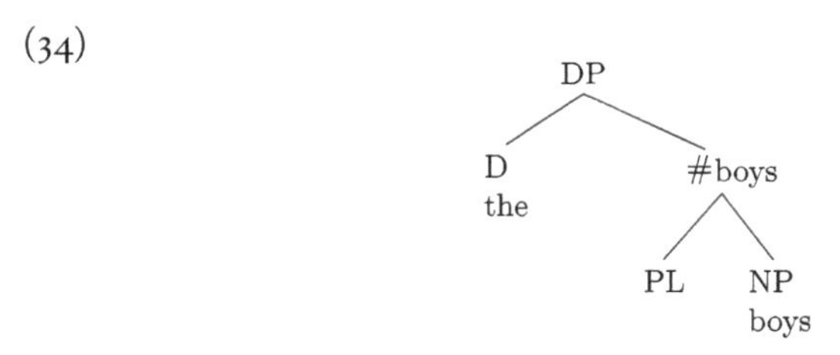

The structure in (34) represents the definite plural DP *the boys*, where #P ('number phrase') is the layer driving information about number. DP and #P are (part of) the functional spine or extended projection of NP.

Functional layers are drawn from a hypothesized universal inventory and are subject to (limited) parametric variations. Variation has to do with which members of the universal inventory are exploited by a given

[12] This rough characterization needs to take into account the fact that constituents can be moved from their base position.

language, matters of word order, and whether a language admits phonologically unrealized items of specific sorts. The special character of functional items determines a further series of associated properties, schematically:

(35) a. Frequency: Highest frequency in any language is associated with functionality. The most frequent fifty lemmas in English include no more than three or four content words (with *say* in the lead, at the nineteenth position).[13]
b. Historical change: content words change constantly without affecting language identity (whence the characterization 'open class'); changes in the functional layers involve grammatical change, which may affect the identity of a language (whence the label 'closed class').
c. Selective impairments: function words are often selectively impaired in a variety of language pathologies, like agrammatism or 'non-fluent' aphasia. (See, e.g., Caramazza and Hillis (1989), Friedman and Grodzinsky (1997) among many others.)

The properties in (35) are fairly self-explanatory, and perhaps unsurprising given the nature of function words. Notice that they are 'one-way' generalizations, i.e., conditionals, not biconditionals: if x is high frequency x has a high probability of being a function word. But there are of course relatively low frequency function words (e.g., *shall, ought*).

The syntactic characterization of functional items just reviewed relies primarily on the 'slots' they occupy within the clause. This main trait unavoidably comes with a cluster of semantic properties. Building on von Fintel (1995), I will briefly discuss here four such properties, namely:

(36) a. Having high types.
b. Being 'inference based'
c. Being subject to crosslinguistically widespread, sometimes universal constraints.
d. Being permutation-invariant.

Starting with (36a), nouns and verbs typically express first order properties and relations that subdivide domains of discourse into classes and relate individuals, events, etc. to one another. In type-theoretic terms, this is conceptualized by positing a basic type of individuals e, and relations over individuals of type <e, t>, or <e, <e,t>>, etc. Functional expressions

[13] Compare with www.wordfrequency.info.

find their natural conceptualization at higher types. For example, determiners can be viewed as associated with higher order relations of type «e, t>, «e,t>,t» between sets or classes. Similarly, one can think of propositions as carving an abstract space of possibilities (say, a set of worlds) into subregions and propositional connectives can be represented as higher order functions on sets of worlds. The property in (36b) is easy to grasp but hard to define. The basic idea is that while the meaning of every kind of expression is ultimately rooted in its entailments, presuppositions, and implicatures, content words are also causally linked to fairly tangible and localized regularities in our environment (the meaning of 'cats' is causally linked to cats, that of 'run' to running events, etc.). In contrast with this, the meanings of *every*, *or*, *only*, or *even* are way more abstract and only characterizable in terms of the inference patterns they give rise to. Such patterns are moreover subject to possibly universal structural constraints. For example, *only* and *even* are always 'alternative sensitive': they require identifying a class of alternatives with respect to which their prejacent is evaluated; determiners are conservative,[14] etc.

Turning next to permutation invariance, i.e., the idea the logical word remain constant across one-one mappings of the domain onto itself,[15] there is little doubt that it is a powerful criterion that identifies a natural semantic class. Items with this property systematically fall within the functional segment of the lexicon, to an extent that simply can't be accidental. Expressions with an arguably logical meaning that behave like content words are exceedingly few. They include verbs like *deny*, or *exist*, nouns like *majority*, adjectives like *mere* or *former*. But note that these words are all morphologically derived (e.g., major + ity) and they typically undergo a drift that gives them some non-logical content (e.g., *exists* is not just 'being the value of a bound variable', but drifts into something like 'having physical existence'). So the claim that permutation-invariant functions are expressed within the functional layer of syntax is, I think, born out.

Are function words limited to expressing permutation-invariant items? I'd say no. The clearest case is perhaps that of gender features (and more generally class agreement markers).[16] Grammatical gender systems can be quite complex; they typically code some anthropologically salient trait and extend it, often arbitrarily, in order to partition or classify the domain

[14] A determiner D is conservative iff for any A and B, $D(A)(B) = D(A)(A \cap B)$. See Barwise and Cooper (1981).

[15] See, e.g., McGee (1996), Sher (2003), and, for an overview, MacFarlane (2017).

[16] Current terminology uses the term 'φ-features' for elements of this sort.

of individuals. Since at least Cooper (1983),[17] the semantic side of feature information is treated presuppositionally:

(37) a. i. $||\text{fem}|| = \lambda x_e$: female($x_e$). x_e $||\text{male}|| = \lambda x_e$: male($x_e$). x_e
ii. $||\text{ragazz-a}|| = \lambda x_e$: fem($x_e$). young adult($x_e$)
iii. $||\text{ragazz-o}|| = \lambda x_e$: male($x_e$). young adult($x_e$)
b. $\forall_x$ [female(x) $\rightarrow \neg$male(x)]

The functions in (37ai) are restricted identity maps, defined only for female or male individuals, respectively; in (37a.ii) you see how such functions can be used to restrict the denotation of the words for *girl* vs. *boy* in a language with grammatical gender. The predicate in (37a.ii), for example, is defined only for female individuals; whenever defined, it is true of young adults and false of non-young adults. Use of features of this sort induces disjointness constraints such as (37b), which are among the most common across languages. This seems to require an extension of what counts as 'logical' to constraints that define 'subcategories' of various content words. Sagi (2014) provides an interesting general way of extending the notion of logicality by using constraints of this sort to relativize permutations to subcategories of content words.

So permutation invariance seems to have a core and a periphery. In very rough terms, the core is constituted by the items characterized by some strict definition of permutation invariance (say, bijections among domains of equal cardinality). The periphery takes into account more specific structural constraints on natural semantic categories like modals, tense, mass vs. count, etc. all the way to fairly idiosyncratic feature-based constraints. Jointly, core and periphery determine what might be viewed as a universal natural logic, specific to *Homo sapiens*.

Summing up, the syntax of function words systematically differs from that of content words. These syntactic differences correlate with semantic ones. Permutation-invariant items in the strict sense are systematically treated as functional by syntax and are obvious candidates for Universal Grammar membership. The remnant of the functional vocabulary seems to be constituted by a broader class of mostly inference based operations, relations, etc. subject to cross-linguistically stable structural constraints.

Our starting point has been that certain forms of linguistic deviance appear to be best made sense of in terms of logical truth or falsehood rather

[17] For a more recent version of the presuppositional treatment of grammatical features, compare with, e.g., Sudo (2012) and references therein.

than in terms of well-formedness. Assume some background logical framework, say the typed lambda calculus, and enrich it to a theory NatLog by some set of axioms/structural constraints on modals, event structures, countable vs. uncountable entities, etc. Imagine next using NatLog to specify the semantics of a natural language, say English, in the usual sense of a systematic, compositional mapping from the structures constructed by the grammar of English into formulae or statements of NatLog. Some English sentences will be logically false/true relative to their Logical Forms interpreted in NatLog. And a subset of the NatLog-logically false/true sentences will be perceived by speakers of English as 'not in the language', on a par with syntactically deviant structures. We have called such sentences 'G(rammatically)-trivial'. Our problem was to determine: (i) which sentences within the L(ogically)-determined ones are G-trivial, and (ii) why. We have proposed a modification of Gajewski's and Del Pinal's proposal that addresses these issues.

Be that as it may, the search for the components of the functional/logical lexicon and the ways in which they may vary, while still daunting, is definitely no longer mere speculation, but a well-defined and exciting research program, which delivers constantly new results. And the discovery that forms of 'ungrammaticality' are in fact due to logical inference (rather than to syntactic ill-formedness) is actually game-changing, and shows in very tangible and fruitful ways how interconnected grammar and logic are.

CHAPTER 12

Grammaticality and Meaning Shift

Márta Abrusán, Nicholas Asher and Tim Van de Cruys

12.1 Introduction

Acceptable sentences are all alike; every unacceptable sentence is unacceptable in its own way. Observe the following examples:

(1) a. *There is every fly in my soup. (Barwise and Cooper 1981)
b. *Some students but John passed the exam. (von Fintel 1993)
c. *Mary is taller than no student is. (Gajewski 2008)
d. *There are any cookies left. (Chierchia 2013)
e. *How fast didn't you drive? (Fox and Hackl 2007)
f. *How tall do you regret that you are? (Abrusán 2007, 2014)

For each of the above sentences it has been argued that their unacceptability follows once we recognize that their semantics expresses something logically false or true. Yet, if logical triviality leads to ungrammaticality, then what distinguishes the examples above from the examples in (2), which are acceptable or at worst semantically anomalous?

(2) a. Every woman is a woman.
b. This table is red and not red.

An answer to this question was proposed in Gajewski (2002). According to Gajewski, there is a formally definable subset of trivial sentences, namely L-trivial sentences, whose members are systematically ungrammatical. L-triviality is calculated in the following way: replace non-uniformly all the non-logical vocabulary by a fresh variable in the logical form of a sentence S. If the resulting representation is trivial, then S is L-trivial.

We are grateful to the editors of this volume for the invitation and to an anonymous reviewer as well as the audience at Sinn und Bedeutung 22 (Berlin/Potsdam, September 2017) for helpful comments. The research reported here was supported by a Marie Curie FP7 Career Integration Grant, Grant Agreement Number PCIG13-GA-2013-618550, a grant overseen by the French National Research Agency ANR (ANR-14-CE24-0014), and by ERC grant number 269427.

This proposal has been quite influential in the literature (cf. Fox and Hackl 2007; Chierchia 2013; Abrusán 2007, 2014, among others). However, it has also been observed that the theory needs to accommodate a number of non-trivial restrictions if it is to be applied to explain data as in (1) (cf. Abrusán 2014). In particular, on closer examination of these types of data, the calculation of L-triviality needs to be supplemented with a number of ad-hoc assumptions in order to restrict the replacement procedure in one way or another. Another problem we think is the radical modularity assumption that follows from Gajewski's proposal. Grammar, viewed as a logical deductive system, is encoded in functional/logical vocabulary and is blind to the content of lexical words. This suggests that grammar is insulated not only from conceptual systems, general world knowledge, but also from most of the information encoded in lexical items. One consequence of this view is that we need to be able to distinguish logical from non-logical vocabulary on principled grounds and thus the logical and conceptual aspects of meaning need to map into two distinct types of vocabulary: the logical vocabulary and the lexical vocabulary. This consequence, as it was noted by Gajewski himself, is non-trivial. A second consequence, emphasized in Del Pinal (2019), is that a deductive system that operates on logical skeletons is a rather exotic system for which most classical formulas and rules of inference are invalid. As Del Pinal (2019) argued, this is problematic for some of the accounts that are based on the idea of L-triviality. Finally, the idea that the judgments of ungrammaticality in (1) are due to problems of *evaluations* of logical form seems conceptually wrong; it's not that these sentences are false in all models consistent with the meanings of logical words or trivial; they are word salad. They aren't even evaluable.

So although we believe a semantic explanation of the judgments in (1) is the right way to go, we don't believe that this has to do with isolating a particular kind of logical form or a class of models that such logical forms would yield. It has rather to do with semantic constraints that must be met, in order for a logical form to be constructed. More particularly, we suggest that the examples in (1) reflect a compositionality problem; in this sense they are similar to well-known examples of semantic anomaly, e.g., (3):

(3) a. #Tigers are Zermelo-Frankel sets.
 b. #Colorless green ideas sleep furiously.

Following Asher (2011), we propose that a semantic anomaly is the result of a type presupposition that cannot be satisfied.

This suggestion immediately gives rise to two questions: Firstly, not all type conflicts lead to unacceptability. What is the difference between resolvable type conflicts and unresolvable ones? Second, there is an intuitive difference between classic examples of semantic anomaly such as the ones in (3) and the seemingly ungrammatical examples in (1). What explains this difference?

The process of integrating a lexical meaning into an interpretive context, can also shift a lexical meaning or introduce new meaning components as a result of the integration, depending on the type presuppositions of predicates and their arguments. Coercions are an example of a construction where type conflicts can be resolved, because the predicate licenses the addition of material needed to accommodate the type presuppositions of the argument. To a limited degree, we can also resolve type conflicts in the case of semantic anomaly. To do so, we need to apply meaning shift to relax the type presuppositions in question. Depending on the semantic distance between the actual and the target types, type shift allows us to make sense of examples of semantic anomaly to a greater or smaller degree.

(4) a. #Tigers are human.
b. #Tigers are teabags.
c. #Tigers are Zermelo-Frankel sets.

The higher the type that is involved in the type clash, the more difficult it is to shift the meaning of predicates to resolve type conflict. With very high types, meaning shifts will not be able to help. Both the classic examples of semantic anomaly in (3) and the examples in (1) involve high-level types.

In order to get a better understanding of the nature of meaning shift involved in semantic composition, as well as the restrictions on meaning shift alluded to above, we turn to a computational method called distributional semantics (DS). This computational method can find low-level types and capture corresponding meaning shifts. It does not, however, predict shifts with high-level types because either (a) the types in question denote context-invariant logical meaning that is simply invisible for distributional methods or (b) the conflicting type-clash is attached to a type so high in the type hierarchy that the type has no neighbors that share the same syntactic/semantic dependencies and so there is nowhere in the space for its meaning to shift. The fist case corresponds to examples in (1), the second case to classic examples of semantic anomaly such as (3).

Another aspect that might underlie the difference between semantic anomaly and (1) is the locality of the type conflict. In the case of classic examples of semantic anomaly, type conflict (presupposition failure) arises at the level of predicate-argument composition. If shifting were to occur, it would also happen at this level, where the nature of the type conflict is clear and lowest common types are easy to calculate. In the case of examples in (1), the type conflict arises at a more global level with more linguistic elements (and types, as a consequence) that are involved. Calculating lowest common types is harder, and it is more likely there are simply none.

Our proposal assumes that logical and lexical aspects of meaning do not map neatly to two different types of words (see also Abrusán et al. 2018). There is no purely logical vocabulary, and purely lexical/conceptual vocabulary. Instead, this distinction cross-cuts word boundaries; all words have both aspects of meaning. In particular, the meanings of lexical as well as logical words have both logical aspects (their model-theoretic meaning) and lexical/conceptual aspects. We cannot neatly separate grammar and conceptual knowledge because they are packaged together within lexical entries. However, we can distinguish conceptual content that is contextually invariant from shiftable conceptual content. Conceptual content that supports logically valid inferences for first order definable quantifiers (whose conceptual content in the form of proof rules can determine all logical consequences of such quantifiers) should always be contextually invariant, since a particular context should not render logically valid inferences incorrect. We would never expect meaning components that lead to logically valid inferences to shift in context. Other types of conceptual content of logical words, however, can be shiftable; that is, context might affect the conceptual content and the extension of a predicate or its argument as well.

12.2 Logical Forms, Skeletons, Triviality and Grammaticality

In the analytic tradition, logical form is assumed to be a translation of the meaning of a sentence of natural language into a more ideal language that allows to calculate inferences of the original sentence more perspicuously than the natural language sentence itself. Under this conception, logical relations of the original sentence can be formally explained if they can be deduced from a formal principle that applies to the logical form of the sentence involved. In the tradition of Chomskyan generative grammar, logical form is a syntactic/semantic representation that registers the effect of certain operations that do not have a phonological effect, e.g., quantifier raising. On this conception, logical form is a representation that encodes all semantically significant features of a sentence. On both views, logical form is a level of representation where ambiguities have been resolved. It is also a representation that allows the calculation of being a logical truth or logical validity – by humans, familiar with the rules of logic.

In an influential paper, Gajewski proposed that a certain type of logical triviality has consequences for grammaticality (cf. Gajewski 2002). In particular, he proposed that there is a formally specifiable subset of trivial sentences, that he calls 'L-trivial', whose members are systematically unacceptable.[1] On the standard, Tarskian conception, logical truth and consequence are defined in terms of variation of truth across all interpretations of the logical form. L-triviality, in contrast, is calculated on modified logical forms in which all lexical material has been 'bleached', i.e., replaced non-uniformly by a fresh variable of the appropriate type. Gajewski (2002) calls these impoverished logical forms 'logical skeletons':[2]

(5) Logical skeletons are obtained from the logical form α as follows:
  a. Identify the maximal constituents of α containing no logical items.
  b. Replace each such constituent with a distinct variable of the same type.

Thus the logical form of (6) is something akin to (6-a) (where *raining* stands for the proposition that it is raining), but its logical skeleton is (6-b), where p and q stand for propositional variables:

(6) It is raining or it is not raining.
  a. $raining \vee \neg raining$
  b. p or not q

Given logical skeletons, L-triviality can be defined as follows:

(7) An LF constituent a of type t is L-trivial iff a's logical skeleton receives the denotation 1 (or 0) under all interpretations.

L-triviality and unacceptability are linked in the following way:

(8) A sentence is ungrammatical if its Logical Form contains a L-trivial constituent.

[1] Gajewski's (2002) original definition is about L-analyticity, but this might be slightly misleading, hence we use L-triviality, similarly to some other authors, e.g., Del Pinal (2019).

[2] Gajewski (2002) works with the notion of logical form that arises from generative grammar, compare with Heim and Kratzer (1998).

Under the resulting picture, grammar itself is endowed with the capacity to calculate L-triviality. This means that the grammar of natural languages has to include (or at least interact with) a system of 'natural logic', or a 'natural deductive system' (see Fox and Hackl 2007; Chierchia 2013). By assumption, this deductive system is blind to conceptual information and cannot 'see' non-logical terms, it operates only on the basis of functional terms. This in turn presupposes that terms can be sorted into two non-overlapping classes, lexical terms and functional (or logical) terms.

In the remainder of this section we first discuss two examples of applications of L-triviality. Second, we go on observing some problems that motivate us to look for an alternative explanation.

12.2.1 *Two Examples*

Exceptive constructions In his 1993 paper, von Fintel proposed that the restricted distribution of connected exceptive phrases (e.g., *but John*) can also be explained by appeal to triviality. Exceptives formed with *but* are compatible with positive and negative universal quantifiers (*every, no, none, all,* etc.), and incompatible with any other quantifier:

(9) a. *Every/No* boy but John smokes.
b. **Some/*three/*many/*most/*less than three* boys but John smoke.

The semantics that von Fintel (1993) gives for exceptive *but* assumes that the argument of *but* is the least (i.e., the unique minimal set) one has to take out of the restrictor to make the statement true:

(10) $[\![\text{but}]\!]=(C)(A)(D)(P)=1$ iff
$C \neq \emptyset$ and $D(A\text{-}C)(P)=1$ and $\forall S\ [D(A\text{-}S)(P)=1 \rightarrow C \subseteq S]$

According to the above, *Every boy but John smokes* means that C=John is the unique minimal set one has to take out of the domain A=[[boy]] of the quantifier D=[[every]] to make the statement *Every boy smokes* true. Why is it that only universal quantifiers can host *but*-exceptives? The problem is that with (almost) all other quantifiers, such least-exceptions lead to a contradiction.

Universal quantifiers are left downward monotone, while existential quantifiers such as *some, a, (at least) one, two,* are left upward monotone.[3]

[3] One problem for the theory is the quantifier *most*, as noted by von Fintel (1993) himself. See also Gajewski (2002) for further problems.

In his article von Fintel (1993) shows that modifying left upward monotone quantifiers with a *but*-exceptive always leads to a contradiction.

(11) A determiner D is left upward monotone if
for all models $M = < E, [\![.]\!] >$, and all $A \subseteq B \subseteq E$,
if $X \in [\![D]\!](A)$ then $X \in [\![D]\!](B)$.

Intuitively, left upward monotonicity captures the inference from sets to supersets on the left argument of the quantifier. For quantifiers that are upward monotone on their restrictor argument, it is always the case that if the statement D(A-C)(P) is true, then D(A)(P) is true as well, thus one could have always taken out less than C (in our case, the empty set) from A to make the statement true. But this means, that the second clause in (10) (namely $\forall S\ [D(A\text{-}S)(P)=1 \rightarrow C \subseteq S]$, that requires that every alternative subset of A that one could have taken out of A to make the statement true has to be a superset of C) is false in the case of left upward monotonic quantifiers. Therefore, modifying such quantifiers with a *but*-exceptive results in ungrammaticality. Left downward monotone quantifiers, however, do not lead to a problem.

Gajewski (2002, 2008) shows that the problem persists once the LF of an exceptive sentence is transformed into a logical skeleton:

(12)
a. Some boy but John smokes
b. Logical skeleton of (12-a): some $[P_1$ but $P_2]\ P_3$
c. Interpretation : $[\![\text{some}]\!](I(P_1)\text{-}I(P_2))\ (I(P_3))=1$
and $\forall S\ ([\![\text{some}]\!](I(P_1)\text{-}I(P_2))\ (I(P_3))=1 \rightarrow I(P_2) \subseteq S)$

Presumably, the connective *but* and the quantifier *some* are functional words and are not replaced in the logical skeleton (Gajewski uses invariance as a criterion, we come back to this shortly). Because of the left upward monotonicity of the quantifier *some*, whatever the interpretation of P_1 and P_2, if the sentence is true with $(I(P_1)\text{-}I(P_2))$ as the domain of the quantifier *some*, it will also be true with $(I(P_1))$ as its domain. For this reason, the second clause in the interpretation will always be false, at least as long as P_2 is prevented from being empty. Therefore all interpretations of (12) will map it to false and it is L-trivial, hence it will be predicted to be ungrammatical.

In contrast, observe again the case of simple contradiction:

(13) This table is red and not red
a. Logical skeleton: [this P_1 is P_2 and not P_3]

In this case, once we remove the identity of the non-logical expressions to create the logical skeleton, we cannot deduce triviality anymore. The algorithm for forming logical skeletons assigns distinct variables to the two occurrences of *red*, P_2 and P_3. Clearly the resulting logical skeleton is not L-trivial. If the interpretation of P_2 is not the same set as the interpretation of P_3, triviality does not follow.

Weak islands Abrusán (2007, 2014) discusses the problem of weak islands. Weak islands are contexts that are transparent to some but not all operator-variable dependencies. Some paradigmatic cases of weak island violations include examples of degree extraction in (14-a) and (15-a), as opposed to the acceptable questions about individuals in (14-b) and (15-b):

(14) a. *How tall isn't John?
b. Who didn't John invite?

(15) a. *How tall do you regret that you are?
b. Who does John regret that he invited to the party?

The traditional analysis derives these facts from a syntactic contrast (cf. Rizzi 1990 and much subsequent work). Szabolcsi and Zwarts (1993) noted, however, that the oddness of the a. examples above has a semantic flavor: they simply do not seem to make much sense. More recently, Kuno and Takami (1997) and Fox and Hackl (2007) showed that many weak island violations can be ameliorated by adding certain quantificational elements such as modals to the sentence, compare with example (16). The fact that modals can obviate weak islands is unexpected and hard to explain on a syntactic analysis.

(16) How much wine are you not allowed to drink?

Abrusán (2007, 2014) put forth a novel semantic theory that could explain all the basic examples of weak island violations as well as the cases of modal obviation. The central thesis of this work is that these islands arise because they are predicted to lead to a contradiction at some level. There are two ways in which a contradiction can arise. In the case of factive islands, the question always has a contradictory presupposition. Observe first that presuppositions project universally from question alternatives: (17) presupposes that you invited all these ten people:

(17) Who among these ten people do you regret that you invited?
presupposes: you invited these ten people

Universal projection is not problematic in the case of questions about individuals because the answers are independent. But in the case of degree and manner questions it leads to problems. Assume, following Schwarzschild and Wilkinson (2002), that degree predicates relate individuals and intervals of degrees.[4] Then the presupposition of (18) is that your height is contained in every interval on some scale.

(18) *How tall do you regret that you are?

a. 'For what interval I, you regret that your height is in I?'

b. $[\![Q]\!]^{C,w}=\{\lambda w.\ \text{you regret that your height} \in I \text{ in } w \mid I \in D_I\}$

(19) Presupposition of (18): $\forall I \in D_I$: your height $\in I$

'your height is contained in every interval in D_I'

This presupposition is contradictory and cannot be met in any context: assume I_1 and I_2 are two non-overlapping intervals on the scale. Since heights are points on a scale, they cannot be contained in both of these intervals.

In the case of negative and wh-islands a contradiction arises in a different manner. We illustrate here a negative degree island. Following Dayal (1996) and Fox and Hackl (2007), Abrusán (2007, 2014) assumed that questions presuppose that they have a unique maximally informative answer, i.e., a true answer that logically entails all the other true answers.

(20) *Dayal's (1996) presupposition (aka Maximal Informativity Principle (MIP))*

Any question presupposes that it has a maximally informative answer, i.e., a true answer which logically entails all the other true answers.

If the MIP cannot be met, the statement for any potential answer that it is the maximally informative answer to some question is bound to state a contradiction.

In the case of negative and wh-islands the MIP cannot be met and therefore the exhaustification of any answer *Ans* to a negative degree question *Q* expresses a contradiction.[5] Observe (21) (as above, degree predicates are assumed to range over intervals):

[4] According to this proposal, the denotation of a degree (interval) predicate such as *tall* is as follows:
(1) $[\![\text{tall}]\!] = \lambda I_{\langle d,t\rangle}$: I is an interval. $\lambda x.$x's height $\in I$

[5] Exhaustification of answers, following Fox and Hackl (2007), is defined as follows:

(21) *How tall isn't John?

a. 'For what interval I, John's height is not in I?'

b. $[\![Q]\!]^{C,w}=\{\lambda w.$ John's height is not $\in I$ in w| $I \in D_I\}$

If a degree d is not contained in an interval I, it follows that it is also not contained in any subinterval of I and nothing follows wrt. to d for any other interval in the domain. Given this entailment pattern, an exhaustive answer to (21) amounts to the following:

(22) Exh (Q)(John's height is not $\in I_1$) is true iff

a. John's height is not $\in I_1$

b. For every interval $I_2 \in D_I$, if John's height is not in I_2, then I_2 is contained in I_1.

If John has a non-zero height, then there is no answer to (21) that can be exhaustified as in (22) without leading to a contradiction.[6]

Examining the nature of the contradiction that arises in the cases of weak island violations, one might wonder whether it is an instance of L-triviality of Gajewski (2002). The logical form of the exhaustive answer is as follows:[7]

(23) Exh ({John is not I-tall | $I \in D_I$})(John is not I_1-tall)

From this, the logical skeleton of an exhaustive answer to (21) can be obtained by replacing *John* and the predicate *tall* with fresh variables a_i and P_i respectively:

(24) Exh ({ a_i is not P_i | $i \in \{1,\ldots,n\}$ })(a_1 is not P_1)

Strictly speaking, this logical skeleton is not contradictory. For it to be trivial, we have to assume that (a) the values for a_i and P_i are held constant across the question alternatives, (b) the restriction that degree predicates

(1) *Exh(Q)(Ans)* is true iff

a. *Ans* is true

b. for any alternative proposition φ in Q, if φ is true, then φ is entailed by *Ans*.

[6] If John's height is zero, then the most informative answer, the proposition that John's height is not included in $(0; +\infty)$, must be already entailed by the common ground. Abrusán and Spector (2010) argue that a maximally informative answer must also be contextually informative.

[7] NB: In the examples above the interpretation of the relevant propositions/questions was represented for better readability, not their logical form.

operate on degree scales and that they range over intervals is maintained when creating the logical skeleton.[8]

Note that these restrictions are not unique to Abrusán's account. Similar restrictions need to be assumed by Gajewski (2008), Fox and Hackl (2007), Chierchia (2013) – see Abrusán's (2014) Chapter 6 for further discussion on this. But such ad-hoc assumptions are hard to accept if we are looking for a principled account of ungrammaticality based on L-triviality.

12.2.2 *Problems with L-Triviality*

Restrictions As we have seen above, to maintain that certain ungrammatical logical truths and falsities are indeed L-trivial in Gajewski's sense, we need to make a number of ad-hoc assumptions, some relatively innocent, others more problematic. We need to assume that grammar contains various restrictions on the domains of predicates (cf. Gajewski 2008; Chierchia 2013; and Abrusán 2014). Gajewski (2008) suggests that these might be represented as semantic presuppositions, in which case failure to meet them results in undefinedness. Accordingly, the definition of L-triviality is changed as follows:[9]

(25) A sentence S is L-trivial iff S's logical skeleton receives the truth value 1 (or 0) on all interpretations in which it is defined.

We further need to assume that the process of creating logical skeletons interacts with the process of alternative generation in a special way, namely it has to use the same constants/variables in each alternative to replace non-logical words, except the focused/questioned word (Fox and Hackl 2007; Chierchia 2013; and Abrusán 2014). Even more problematically, as discussed in Abrusán (2014), lexical presuppositions would have to interact with the mechanism checking L-triviality in a rather ad-hoc fashion.

Defining logical words Gajewski's (2002) proposal hinges on distinguishing two types of vocabulary, logical and non-logical vocabulary. Finding a conceptually motivated account for this division is one of the major long-standing issues in the philosophy of logic. The most

[8] In the case of factive islands one more important assumption is needed to make these examples L-trivial, namely that the factivity of factive predicates is preserved when creating the logical skeleton.

[9] Alternatively, we need to postulate a special module of grammar, DS that enforces these restrictions and at which ungrammaticality is calculated (see Fox and Hackl 2007).

well-known account is due to Tarski (1966/86) who defined the difference between logical and non-logical words in terms of permutation invariance (see also Sher 1991; van Benthem 1989, 2002; Bonnay 2006, 2008, and various articles in this volume, among others). The intuition behind this approach is that invariant elements do not depend on the identity of the particular individuals in the domain. Gajewski (2002) also follows this tradition, more precisely van Benthem's (1989) extension of the permutation-invariance idea to typed languages.

Permutation invariance as a definition for logical items is a relatively simple idea, though not without any problems, both from the logical and the empirical perspective (see for example van Benthem 2002 and Bonnay 2006 for an overview). However, from the linguistic perspective, it seems to include both too many and too few items. It includes too many, because, as shown by Gajewski (2009), it predicts predicates such as *self-identical* and *exist* to be logical, although we have the sense that they are not. On the other hand, as mentioned by van Benthem (2002), it excludes items in natural language that intuitively should count as logical: for example the quantifiers *every* and *each* in natural language carry the restriction that they can only quantify over countable objects, hence the sentences **Every salt is on the table, *Each milk is in the fridge* are unacceptable (unless the domains of salt and milk have been somehow individuated in the context). In contrast, the quantifier *all* can combine with mass nouns as well: *All the salt is on the table, All the milk is in the fridge*. The sensitivity of some quantifiers to the countability of the predicates they combine with makes them not permutation-invariant, hence not logical on the permutation-invariance theory. Many alternative versions of the basic invariance idea exist, which characterize logicality as invariance under some other transformation (for example Feferman 1999 (relation-invariance); Bonnay 2008 (invariance under potential isomorphism)). There are also many conceptually different accounts of logicality, for example proof-theoretic or algorithmic accounts (see van Benthem 2002 and references therein), or accounts that extract logical items from consequences (Bonnay and Westerståhl 2016). Neither of these are problem-free however, or significantly better suited for the linguistic purpose that we are concerned with. An alternative possibility, one that Gajewski also considers, is to replace the logical/non-logical distinction with the functional-lexical distinction familiar from the linguistic literature (cf. Abney 1987; von Fintel 1995). This too, however, suffers from difficulties, as some words, for example prepositions or the word *there*, are not clear cases of either category.

It seems that for the moment there is no foolproof method that can distinguish logical words from non-logical ones that also makes the cut in a linguistically intuitive way. It cannot be excluded that such a property could be found in the future. But at least, the difficulties mentioned above suggest that the logical/grammatical aspects and the conceptual aspects of meaning do not map neatly onto two different classes of words. Instead, both functional and lexical words might have logical and conceptual aspects of meaning, packaged together.

Deductive system The conception of grammar and a natural deductive system that follows from Gajewski's (2002) proposal has profound implications for how we should think about the language system and its interaction with other cognitive systems in general. First, it suggests a very radical form of modularity of language: grammar is insulated not only from conceptual systems, general world knowledge, but also from most of the information encoded in lexical items. Second, if L-triviality can have implications for grammaticality, the grammar (or the language module, language organ, or whatever) needs to contain – or at least interact with – a natural deductive system (cf. Fox and Hackl 2007).

As it was emphasized in Del Pinal (2019), a deductive system that operates on logical skeletons is a rather exotic system for which most classical formulas and rules of inference are invalid. It is conceivable that the properties of the natural deductive systems, as used by grammar, could be radically different from classical systems. However, Del Pinal (2019) argues that certain key accounts depending on logical skeletons, e.g., Chierchia's (2013) account of polarity-sensitive items, cannot be maintained if the Law of Non-Contradiction is invalid at the level of representation where the deductive system determines grammaticality.

In addition, there seems to be a problem with logical skeletons insofar as we can have judgments about those. As far as we can tell, logical skeletons correspond to second order formulas, with variables both in predicate and argument position. The second order formula $\forall X \forall x(Xx \rightarrow Xx)$ is true in all second order models and hence a logically valid second order sentence, as is the formula which results from removing the universal property quantifier, while $\exists X \exists y(Xy \wedge \neg Xy)$ is true in no second order models and so logically contradictory. Yet these are no more astonishing in terms of logic than first order logical validities. It would be strange if there were such a sharp division on logicality properties.

12.3 An Alternative to L-Triviality

Our idea is that the judgments about the examples in (1) are not the result of some property of logical absurdity or validity but rather of a failure of composition. While we agree with the proponents of L-triviality that the judgments about the uninterpretability of the examples in (1) stem from a problem with their semantics, we do not believe that L-triviality poses a viable solution. First, formulating precisely the notion of L-triviality faces serious obstacles as we have seen in the last section. Second, although the idea that truth and triviality have consequences for grammar is an interesting and provocative one, we believe that natural language examples do not provide evidence for it.

So for us, the judgments about the sentences in (1) have nothing to do with truth or triviality. Instead, we propose that the reason why examples in (1) are uninterpretable is that in building up the semantic representation for such examples, we encounter an insuperable semantic problem so that we can't fit the semantic pieces together to build up the full representation. This failure of composition comes about because some predicates in uninterpretable sentences presuppose that their arguments have a type or obey certain semantic principles that they in fact do not or cannot obey. This view thus claims that the sentences in (1) are uninterpretable for semantic reasons and provides an easy to understand intuition about their uninterpretability: we can't even say what the sentences in (1) mean or could mean because we can't put the meanings of their components together into a coherent whole.

12.3.1 Types for Semantic Composition

To flesh out this idea, we need to have a theory of the semantic principles that are operative in composition. These sort of questions have been pursued in formal semantic frameworks that use some notion of semantic typing to investigate cases of apparent meaning shift in phenomena like coercion and aspect selection (Cruse 1986; Jackendoff 1992; Nunberg 1993; Pustejovsky 1995; Asher 2011) *inter alia*. The idea of semantic types is familiar from Montague Grammar, but the authors just cited above attempt to extend the Church Montague system of functional types over the base types E (type of entities) and T (the type of truth values) to a much richer system incorporating subtypes of E and T, or PROP, the type of propositions.

But what sorts of meanings are types? TCL distinguishes two types of semantic content: *external* content and *internal content*. External content of an expression corresponds to the model-theoretic meaning that determines its appropriate extension (at points of evaluation). This is the usual notion of content in formal semantic theories. In addition, however, each word in TCL has a type. Types are semantic objects and encode the *internal meaning* of the expression associated with it. So for instance, the external semantics or extension of the word *wine* is a set of wine portions at some world and time, while the type or internal meaning of *wine* is given by the features we associate with wine – e.g., it's a liquid, a beverage, has alcohol and a particular taste). Internal semantics can also make use of multi-modal information; so olfactory and gustatory features can also play a role.

TCL's characterization of internal content yields a natural link between internal content and external, model-theoretic content. The internal semantics 'tracks' the external semantics, in that in the majority of cases or in normal circumstances, the internal semantics determines appropriate truth conditions for sentences. The features encoded in the internal semantics enable speakers to correctly judge in normal circumstances whether an entity they experience falls under the extension of a term. The internal content given by the types doesn't determine the expression's extension in all cases, as philosophical, externalist arguments show (Kripke 1980; Putnam 1975). But assuming speaker competence, internal content should normally yield the correct extensions for expressions.[10,11]

Before proceeding to introduce how composition works in this system, we briefly sketch what lexical semanticists and logicians assume about the structure of types. In a richly typed system, there are often difficulties if one assumes a universal type (Luo 1994), but it is standard to assume a most specific type, $\bot$, that is a subtype of all types above it. Thus, the set of types forms a semi-lattice ordered by the subtyping relation. For each syntactic category there is a maximal type. For instance, questions and propositions are both maximal types; there is no higher type that unifies the type of questions, which are sets or families of propositions, and the type of propositions. Similarly, there is no higher type above that of entity or E, though there are many subtypes of E. The same thing, we assume, holds of DP types and first order property types (Asher 2011). These highest types have semantic properties that are important for grammaticality. In the next section we see how DS buttresses this component of our view as well.

[10] Note that internal content in TCL is not the same as the intension of an expression, if the latter is understood as a function from indices to truth values.

[11] Types, conceived as justifications, are formally modeled as defeasible proofs that can combine to create more complex types. This is possible by exploiting a deep relation between proofs and types known as the Curry-Howard correspondence (Howard 1980).

12.3.2 Composition and Its Failures

Types and internal content play an important role in our story about composition and failures of composition. Internal content is mainly responsible for guiding the construction of semantic logical forms. When composition succeeds, type presuppositions of predicate and argument are completely compatible, and a logical form for a sentence is constructed, all the internal meaning constraints given by the lexical expressions that make up that sentence have been satisfied given how those expressions compose together.

The process of integrating a lexical meaning into an interpretive context can also shift a lexical meaning or introduce new meaning components as a result of the integration, depending on the type presuppositions of predicates and their arguments. Coercions such as (26) are an example of a construction where type presuppositions get satisfied, because the predicate licenses the addition of material needed to accommodate the type presuppositions of the argument; here the shift from *the artichokes* denoting a set of entities to it denoting events of eating artichokes.

(26) Marta enjoyed the artichokes.

But there are also cases of so called *semantic anomaly*, where composition fails because of an irreparable type mismatch. Some cases of semantic anomaly are well known and staple examples of introductory linguistic classes:

(27) #Colorless green ideas sleep furiously.

(28) #Tigers are Zermelo Frankel sets.

In such examples, it's relatively easy to see what's gone wrong; predicates like *are Zermelo Frankel sets* presuppose that their argument is some sort of abstract entity, while *tigers* contributes an argument whose semantic type must be a physical object. Given that these two types have incompatible individuation principles and essential properties (Asher 2011), the argument cannot compose with the predicate and we get a semantic anomaly. In these cases, the predicate does not license a coercion, and so the sentences like (27) and (28) are predicted to be semantically anomalous. And this gives us at least in principle a way of linking the ungrammaticality of (1) with semantic anomalies discussed in the coercion literature: type presuppositions can lead to irresolvable type conflicts when the type presuppositions of a predicate and its arguments are fundamentally incompatible. In that case the construction of logical form fails, the composition of meaning cannot go through, and we have a sentence that doesn't make sense.

Semantic anomaly is a graded phenomenon, see for example Magidor (2013). Some examples are interpretable in the right contexts but difficult outside of them, like:

(29) Squirrels are human.

Some, like (27) or (28) are difficult to interpret in any context. Nevertheless, even with these examples, we can glean what could have been said. The contradictions between type presuppositions in these examples is somehow localized. And by relaxing those type presuppositions, for instance by supposing that the predicate in (28) simply is seeking an entity of general type E, the composition could actually have succeeded. Context might shift type presuppositions, but only to a limited degree.

The problems of composition in (1) are worse in that no relaxation of the type presuppositions seems possible. Not only are the semantic principles involved in the composition of the examples in (1) unshiftable in practice, we can't even comprehend how they could be shifted. They are somehow constitutive of the construction in a way that simpler type presuppositions of open class expressions, nouns, verbs and adjectives, are not. Our hypothesis is that the type presuppositions that are violated in (1) are constitutive of the type of denotation at the highest level of the type hierarchy. For instance, it is something about the semantic type of questions that leads to the uninterpretability of weak island sentences; it is something about the semantic type of constructions of the form *A but B* that leads to the uninterpretability of certain exceptive constructions; it is something about the nature of quantification that leads to the uninterpretability of *there is every girl*.

12.3.3 Meaning Shift in Co-composition

To get clearer on this intuition, we need to understand what it is to have a shiftable (internal) meaning, and if the hypothesis that shiftability depends upon the structure of the type system, what that structure is. In previous work (Asher et al. 2016), we have shown that there is shifting even when the type presuppositions of predicate and argument are completely compatible. While phenomena like coercion and aspect selection might seem like rather special linguistic phenomena, another sort of meaning shift is very common. For example, the adjective *heavy* has slightly different senses in each of the examples below:

(30) a. *heavy* box
b. *heavy* bleeding
c. *heavy* rain
d. *heavy* smoker

Pustejovsky (1995) calls such meaning shifting compositions *co-compositions*.[12] (Asher (2011) shows that the type shifting in co-composition is plausibly different from coercion.) There is no reason to suppose that type presuppositions, which are very general, are involved here. However, in addition to general type presuppositions, an expression in TCL also has a more specific, 'fine-grained' type that encapsulates the internal content specific to the term. It is this fine-grained content that TCL exploits in co-composition. In TCL the noun and adjective meanings affect each other, and the output of an adjective-noun composition is the conjunction of a modified adjectival meaning and a modified noun meaning, which are both first order properties and apply to individuals, as in (31).[13] The *adjective-noun composition* schema (31) introduces functors that potentially modify both the adjective and the noun's internal content in co-composition and then conjoins the modified contents. In the schema below, A is the adjective, N the noun, $\mathcal{O}_A$ the functor on the noun given by the adjective and $\mathcal{M}_N$ the functor on the adjective induced by the noun:

(31) $\lambda x\ (\mathcal{O}_A(N)(x) \wedge \mathcal{M}_N(A)(x))$

Even logical truths have shiftable content. The predicates in them can undergo meaning shift that render informative an otherwise logical trivial sentence (cf. Chierchia and McConnell-Ginet 2000; Kamp and Partee 1995; Abrusán 2014).

[12] Note that the issue is not simply finding the right scale or a contextually specified cutoff on the scale for heaviness (as in *heavy mouse* vs. *heavy elephant*. Intuitively, what we need to capture is that the 'flavor' of the adjective changes depending on the context. Modeling this variation as pervasive ambiguity (e.g., via disjoint union types in a type system) is clearly unattractive as it would not capture that there is a 'common core' of the meaning of *heavy* that seems to be present in all cases. A disjoint union type might be right for homonymously ambiguous expressions (such as *bank*) but not for logically polysemous ones, expressions whose senses have some logical or metaphysical connection.

[13] TCL's approach to adjective-noun co-composition is quite different from a standard Montagovian approach. In standard semantic treatments, an adjectival meaning is a functor taking a noun meaning as an argument and returning a noun phrase meaning; composition is a matter of applying the adjective meaning as a higher order property to the noun meaning.

(32) a. This table is red_1 and not red_2.
b. Every $woman_1$ is a $woman_2$.

In the above examples, we can (non-uniformly) shift certain aspects of the meaning of the sentence to get rid of the triviality and make the sentence informative. In particular, the two occurrences of the lexical predicates *red* and *woman* could get slightly different interpretations, i.e., their meaning could be shifted to express a particular aspect of the meaning of the predicates involved.[14]

12.4 What Types of Meaning Can Shift?

TCL alone does not supply detailed information about particular types, which is crucial to determining meaning shifts, nor does it tell us anything about the meaning functors in co-composition. Distributional Semantics, a computational approach to natural language semantics, can throw new light on meaning shifts in co-composition, as we have shown in Asher et al. (2016). This paper outlined a close correspondence between TCL and DS methods. Further, results of our more recent work suggest that DS can also help us distinguish which aspects of meaning can shift and which ones cannot (cf. Abrusán et al. 2018).

The view from the DS approach connects to a growing body of work that assumes that the meaning of lexical words can be shifted or modulated in one way or another: either within the semantics (cf., e.g., Martí 2006; Stanley 2007; Asher 2011; Alxatib and Sauerland 2013) or within the pragmatics (Kamp and Partee 1995; Recanati 2010; Lasersohn 2012). Since we assume that meaning shift diagnosed by DS approaches happens at the (co-)compositional level, the view from DS is more in line with semantic approaches.

[14] One might take this to be the grounds of another explanation of the data in (1) (see Abrusán 2014; Del Pinal 2019). While the examples in (32-a) are rescuable from triviality because of the possibility of meaning shift, in examples such as the ones in (1), triviality results from aspects of the meaning that cannot shift. Of course, this proposal assumes that some form of logical triviality is still the key to the ungrammaticality of the examples in (1). We don't believe that, and we think this explanation faces problems, as we could insist on making the two predicates in (32-a) be identical, as in:

(i) This table is red_1 and not red_2 and the property red_1 is identical to the property red_2.

Our criticism takes identity to be a logical operator, which already conflicts with Gajewski's proposal, but we believe that treating identity along with the truth functional and quantification expressions of first order logic on a par is defensible. If this is right then examples like (i) pose another problem for solutions that attempt to explain (1) through appealing to logical triviality.

In this section we first briefly sketch how to compute meaning shifts more precisely using methods from computational semantics. Then we proceed to speculate about the question of shiftable vs. unshiftable meanings. We argue that DS can help us diagnose this as well. Aspects of the meaning that correspond to (or interact with) semantic dimensions uncovered by distributional semantics methods are in principle shiftable. In contrast, aspects of the meaning that are invisible for DS are unshiftable. We propose that semantic anomaly occurs only with unshiftable content. The reason why different types of semantic anomaly give rise to different intuitions is because unshiftable contents come in various flavors.

Distributional semantics can pick up the aspects of lexical meaning that vary with the context: these are the aspects of the meaning that are affected by changes in the distribution. They can perhaps also pick up aspects of possible variation at least among supertypes. However, those that can't possibly vary without destroying the highest type are (by definition) not sensitive to changes in the context (even counterfactual ones) and so are not discoverable by distributional methods.

In practice this means that there are (at least) two reasons why certain aspects of meaning are not shiftable: (a) The meaning is present in all contexts, and so it is invisible for DS; it will not show up in dimensions of the latent space where certain contexts are operative; (b) The meaning is attached to a type so high in the hierarchy that the type has no neighbors that share the same syntactic/semantic dependencies and so there is *nowhere in the space* for its meaning to shift. Such a meaning is also invisible to DS methods, but for a different reason. As a result, context invariant aspects come in different flavors.

Thus we argue that understanding the nature of meaning shifts helps us explain the intuitive difference between classic examples of semantic anomaly and the examples in (1). We close this section by applying our idea to the two examples discussed in Section 12.2: exceptives and weak islands.

12.4.1 Distributional Semantics and Meaning Shifts

Let us provide a very quick introduction to DS and briefly describe a particular approach to capturing meaning shifts within this framework, Asher et al. (2016).

DS is based on the so-called 'distributional hypothesis' by Harris (1954), according to which one can infer a meaning of a word by looking at its context. One way of thinking about word meaning within DS is to assume that it is a vector in some space **V** whose dimensions are contextual features. For example, the meaning of the word *raspberry* might be given by the vector that captures its co-occurrence frequencies with all the words and/or grammatical features or dependency relations within a predefined context window in a corpus. Recording the vector meaning for each word (and possibly grammatical features) results in a word by context matrix. Such matrices are very large and very sparse. In order to bring out the 'information content' in them, dimensionality reduction techniques are applied. Dimensionality reduction reduces the abundance of overlapping contextual features to a limited number of meaningful, latent semantic dimensions. One such technique is non-negative matrix factorization (NMF; Lee and Seung 1999). As it turns out, reducing word-context matrices using NMF is particularly useful for finding topical, thematic information: the latent dimensions brought out by NMF can be interpreted as semantic features, or topics. Factorization also allows a more abstract way of representing the meaning of a word: we can now say that the meaning of a word is represented by a vector of size k whose dimensions are latent features.

Thus DS can generate vectors to capture individual word meaning and bring out latent dimensions that might correspond to semantic features. But in order to capture meaning shift as in the examples in (30), the meaning of the adjective needs to be adapted to the context of the particular noun that it co-occurs with. In the TCL approach this means that the distributional model needs to provide us with the functors $\mathcal{O}_A$ and $\mathcal{M}_N$. In Asher et al. (2016) we have chosen two different approaches that meet this requirement: one based on *matrix* factorization (Van de Cruys and Korhonen 2011) and one based on *tensor* factorization (Van de Cruys and Korhonen 2013).

For example, the approach based on tensor factorization that we applied factorizes a three-way tensor[15] that contains the multi-way co-occurrences of nouns, adjectives and other dependency relations (in a direct dependency relationship to the noun) that appear together at the same time. A number of tensor factorization algorithms exist; we opted for an algorithm called Tucker factorization in which a tensor is decomposed into a core tensor, multiplied by a matrix along each mode.

Given the results of this factorization, we proceeded to compute a representation for a particular adjective-noun composition. In order to do so, we first extracted the vectors for the noun ($\mathbf{a}^i$) and adjective ($\mathbf{b}^j$) from the

[15] A tensor is the generalization of a matrix to more than two axes or *modes*.

corresponding matrices **A** and **B**. We multiply those vectors into the core tensor, in order to get a vector **h** representing the importance of latent dimensions given the composition of noun *i* and adjective *j*. By multiplying the vector representing the latent dimension with the transpose of the matrix for the mode with dependency relations ($\mathbf{C}^T$), we are able to compute a vector **d** representing the importance of each dependency feature given the adjective-noun composition. The vector **d** is in effect the DS version of TCL's functor $\mathcal{O}_A$, which we now have to combine with the original noun meaning. This last step goes as follows in DS: we weight the original noun vector according to the importance of each dependency feature given the adjective-noun composition, by taking the point-wise multiplication of vector **d** and the original noun vector **v**.

Finally, observe an example illustrating the unshifted meaning of the adjective *heavy* vs. the shifted meaning of the same adjective in the context of the noun *traffic* as computed by our tensor method. In the examples below we list the ten closest adjectives (as computed by cosine similarity) to the unmodified and the modified adjective, respectively:

(33) $\mathbf{heavy}_A$: $heavy_A$ (1.000), $torrential_A$ (.149), $light_A$ (.140), $thick_A$ (.127), $massive_A$ (.118), $excessive_A$ (.115), $soft_A$ (.107), $large_A$ (.107), $huge_A$ (.104), big_A (.103)

(34) $\mathbf{heavy}_A$, $\mathrm{traffic}_N$: $heavy_A$ (.293), $motorized_A$ (.231), $vehicular_A$ (.229), $peak_A$ (.181), $one\text{-}way_A$ (.181), $horse\text{-}drawn_A$ (.175), $fast\text{-}moving_A$ (.164), $articulated_A$ (.158), $calming_A$ (.156), $horrendous_A$ (.146)

There is an evident shift in the composed meaning of *heavy* relative to its original meaning; there is no overlap in the lists (33) and (34) above except for *heavy*. We see this also in the quantitative measure of cosine similarity, sim_{cos} between the original vector for *heavy* $\mathbf{v}_0$ and the modified vector for *heavy* $\mathbf{v}_1$ as modified by its predicational context: With the tensor model, on average, $sim_{cos}(\mathbf{v}_{orig}, \mathbf{v}_{mod})$ was 0.2 for adjectives and 0.5 for nouns.

12.4.2 *Constraints on Shiftability from DS*

The distributional method that we described above for calculating meaning shift adapts the vector of the original predicate to its predicational context using the latent dimensions derived during dimensionality reduction. All of this depends on the fact that the original predicate's distributional meaning, the functor on it, and the result of the application of the functor to the distributional meaning, are defined in the same vector space. This has immediate consequences for types that are at a maximally general level, like questions, whose type is a family or set of propositions, propositions themselves, the general type E of entities, the general type of determiner phrases DP, quantifiers or second order properties, the general type of first order properties and so on. These elements don't have any neighbors in a vector space, in the way that say a common noun like *traffic* does, because other expressions that are not of say DP type will belong to a different syntactic category with different syntactic/semantic dependencies; they will be in a different space (though they may share the same latent dimensions as particular DPs). As a result, there is *nowhere in the space* for such a type to shift and preserve its corresponding syntactic category. DS methods *can't shift* those types, in the way that it can shift *heavy* in the context of *traffic*. So these types perforce have semantic principles that are invariant, if they have any semantic principles at all. Furthermore, because these types don't have neighbors in the vector space if indeed they inhabit a vector space at all, these semantic principles will not show up in particular latent dimensions of the vector space. DS methods, at least of the sort we have employed, will not be able to see these principles. For expressions that have type presuppositions *at this level*, it follows that they cannot be shiftable.

Of course not all type presuppositions are of such a general type; many as we have seen are subtypes of E. Nevertheless, DS puts constraints on the shiftability of these types as well. From our studies of adjective-noun composition we have found that almost all modification of noun meanings is subsective; that is, either an ADJ N is an N or it looks like an N, smells like an N, etc. We hypothesize then that non-subsective type shifts are not allowed in co-composition. Given that type presuppositions are very general, requiring a shift over type presuppositions would require in effect a non-subsective type shift – e.g., from a physical object to an abstract object type. We note that this is not necessarily the case for adjectival type shifts, only for type shifts involving nominals and the basic types of objects. If this is not allowed, then we would predict that most normal type presuppositions cannot be shifted during composition. That is, one cannot shift type presuppositions to get composition to succeed (we've already seen that coercion in TCL isn't modeled as a shifting of types of the argument or predicate, but rather a shift in the predicational environment).

On the other hand, one can imagine another kind of shift – one that rescues some sort of content from a predication in which there is a type clash between a predicate and its argument. (27) or (28) are examples of this. If we simply move to the supertype of the type presuppositions, we can see that the author of (27) or (28) was predicating a property of some object; she was just confused about or willfully misusing the meaning of the property or the object expression. The distance between the type presuppositions of the predicate and arguments may be great, but it is still defined, as *tiger* and say *ordinal* are both in the same vector space DS defines for common nouns. And thus DS and TCL together provide us with a means to distinguish (27) and (28) from the examples in (1).

12.4.3 Logical Meaning and DS

Logical meaning is present in all contexts, and so we expect it to be invisible for DS; in particular we expect that it will not show up in dimensions of the latent space where certain contexts are operative. The way the distributional method calculates meaning shifts implies that meaning shift crucially depends on the latent dimensions that we find during tensor factorization: it is the semantic features implicitly present in the latent dimension that drive the meaning shift. Thus whether or not we get logical meaning to shift depends on whether we find latent dimensions with our dimensionality reduction methods that correspond to logical meaning. In recent work (Abrusán et al. 2018) we performed a number of preliminary experiments similar to the ones described in Section 12.4.1 but this time with determiner-noun compositions. Specifically, we looked at four determiners, *a, any, some* and *every* using two different corpora: Wikipedia, and a corpus of unpublished novels collected from the web (Zhu and Fidler 2015).[16] In the resulting factorization, determiners and nouns as well as dependency relations were all linked to same latent dimension. An intuitive evaluation of the semantic coherence of each of the thirty dimensions was conducted, and we have found that many of these seem to capture interesting semantic features, albeit not logical features.

For example, in the case of *any* we get a dimension that captures its peculiar distribution. Most interestingly, perhaps, the dimensions we find with the quantifier *some* correspond to non-logical aspects of its use that have puzzled semanticists since a long while. The first of these is uncertainty about identity, also known as the *epistemic* aspect of indefinites, compare with Kratzer and Shimoyama (2002), Alonso-Ovalle and Menéndez-Benito (2015), among others. Other aspects of the determiner

[16] The former corpus contains about 1 billion words, the latter about 1.5 billion words.

some include measure and kind readings. In contrast, in the case of the determiner *a*, we mostly found topical dimensions, e.g., legal, publishing, building construction, political campaigns, people; in some dimensions *a* appeared within prepositional modifier phrases (*in a chair, with a grin*) and the rest of the dimensions were uninterpretable to us.

What we have described above is still work in progress, but it is already clear that we are not getting any dimensions via tensor factorization that correspond to logical meaning. As a consequence, we are not going to get logical meaning to shift. This is not surprising given that logical meanings are supposed to validate logical deductions universally regardless of context. Thus the fact that logical meaning shouldn't shift with content comes with the definition of logical meaning and the universally valid inferences it purports to underwrite. On the other hand the dimensions that we do get correspond to the lexical/distributional aspects of the conceptual meaning of quantifiers. In the light of this, one way to interpret our results with the determiner *a* is that this determiner does not have any extra conceptual content beyond its logical meaning.

These results suggest that we cannot distinguish logical and non-logical items based on invariance with respect to meaning shift: both lexical and functional words can have shiftable (i.e., co-composing, context-sensitive) aspects to their meaning and also stable (not-co-composing, context-invariant) aspects.

12.4.4 Examples

We are now in the position to come back to the examples discussed in Section 12.2 and examine them in the light of our proposal.

Exceptives As we have seen in Section 12.2, *but*-exceptives are unacceptable with certain quantifiers:

(35) *Some boy but John smokes

Using TCL's internal contents and notion of a type constrained composition gives a different sort of way to use unshiftable contents to predict the ungrammaticality or rather uninterpretability of the sentences in (1). Consider this alternative explanation to that of von Fintel (1993) in the case of exceptive constructions. For simplicity, we'll look at the special case of exceptive constructions where *but* links two DPs, as in DP_1 *but* DP_2, where the second is a proper name or definite description. But we can generalize the following analysis to the more general pattern. Notice that what von Fintel says about the construction amounts to the following entailment:

(36) $[\textit{everyA but B}](C)$ holds iff $\textit{every}(A \setminus B, C)$ and $\textit{every}(A \cap B, \neg C)$.

We also have:

(37) $[\textit{NoA but B}](C)$ holds iff $\textit{No}(A \setminus B, C)$ and $\textit{No}(A \cap B, \neg C)$.

As a general principle then we should have:

(38) $[Det_1 A \textit{ but } B](C)$ holds iff $Det_1(A \setminus B, C)$ and $Det_1(A \cap B, \neg C)$.

That is, the exceptive construction entails that a property C is the argument of DP_1 *but* DP_2 iff $\{A \setminus B, A \cap B\}$ forms a partition of A that corresponds exactly to the partition of A, $\{C \cap A, C^c \cap A\}$, where C^c is the complement of C. But this means that the determiner in DP_1 must distribute to form a partition over A when it combines with DP_2; i.e.,

(39) for every C, if $Det_1(A \setminus B, C)$ and $Det_1(A \cap B, \neg C)$, then every element of A is determined with respect to C.

We can say in this spirit that DP_1 *but* DP_2 holds of properties that induce a partition over the restrictor set A, what we could call Ramsey properties. This puts a type presupposition on the determiner that A combines with, as well as a type presupposition on A and B. In particular, it presupposes the type of object that is a B must be a subtype of the type of object that is an A and that whatever type of determiner A combines with must entail a partition of A in order for composition to succeed, for the Ramsey property to be built. But only determiners that are downward entailing in the restrictor like *every* and *no* can do this. Any other determiner will not set up the partition match and so *A but B* will not compose with them. According to this explanation, the problem with exceptive constructions is a type conflict that cannot be resolved ('shifted away') during composition: the type of the Ramsey property is constituted by the invariant meaning of determiners and their monotonicity properties. If the conflict cannot be resolved, then composition fails and so we predict the sentence to be uninterpretable.

We note that this type conflict is of a much more abstract nature than the one underlying (27) or (28). For those type conflicts, we can still imagine shifting functors like those in co-composition that would shift the noun *tiger* say to some radically different type of object, like *number* or *ordered pair*. The type clash here in TCL is immediate, because the type of the argument, in this case the individual variable introduced by the subject's head noun, is transmitted to the predicate in composition; and the incompatibility is simply checked. It's also relatively obvious how to fix it, even though type shifting can't shift the type outside of the space of types covered by the noun without generating some at least mild form of semantic anomaly: by allowing the co-composition functors to do their work on the head noun of the subject. Our functors make sense at an individual word level, and this is compatible with the idea that type checking for semantic well-formedness is a *local phenomenon*. With the exceptive construction, the type clash has to do with a much more abstract property of determiners, monotonicity, and further the type of the determiner is not transmitted to the conjunction; only the type of DP is. Given the nature of type presuppositions, we then *can't* check composition locally. And in fact, *some boy* could compose with *but* (as in *some boy but no girl*); but given that it has a positive DP in its second argument, the composition will fail because we have an exceptive construction and the type of the determiner can't yield the right input to *but*. Without this local checking, we can't invoke the shifting functors to try to make sense of the exceptive construction. And this explains why such sentences are so mysterious, even in comparison with semantically anomalous sentences like (27) and (28).

Weak Islands As for weak islands, let's consider the case of negative weak islands as an illustration. We can almost use Abrusán's account verbatim, appealing to type conflicts instead of logical contradictions. Recall that (21) has a true answer that entails all the true answers if and only if John's height is 0, which in turn means that (21) can be felicitous only when it is common knowledge that John's height is 0. This already conflicts with the requirement on the question type at issue, since such a question has in its answer set meaning only non 0 answers. Hence, composition fails and we cannot construct a coherent logical form for sentences like (21).

The only thing we need to check in this explanation is that the derivation of the presupposition for (21) indeed follows from internal meaning postulates. In particular we can impose constraints on questions types *à la* Dayal: a question type, which is a set of propositional types, must have a uniquely most informative element. As deduction is reflected in internal meaning via proof-theoretic rules, if we interpret this constraint on questions proof theoretically, we can derive the presupposition for (21) in terms of internal unshiftable meanings. That presupposition understood as a type restriction imposes constraints on the types of other variables and conflicts with the type requirements of questions. Once again as in the case of exceptives, the type clash can't be diagnosed locally. In consequence, we can't make use of our shifting functors to rescue the sentence or counterfactually reconstruct a good semantic composition.

The explanation for the semantic incoherence of positive weak islands is a bit different but follows the general outlines of the type conflict sketched for negative weak islands.

12.5 Conclusion

In this paper, we've taken a new look at certain types of unacceptable sentences. We've argued that these sentences are ungrammatical for semantic reasons, but our analysis differs from that of Gajewski (2002) and those following him in that we locate the semantic problem at the level of semantic composition, not at the level of logicality. We have argued that our position is preferable for several reasons, and we have illustrated this approach with a brief analysis of weak island sentences and exceptive constructions. What we find really interesting about our approach, however, is what it tells us about meaning and composition. Unacceptable sentences on our story offer evidence for a rich system of types that semantic composition makes use of. And they force us to investigate why, when meanings apparently can and do shift with context, some meanings can't. In this investigation, we have fashioned an interesting and fruitful marriage of symbolic and statistical techniques for analyzing meaning shift. The picture of meaning that emerges cuts across old distinctions like the lexical/functional content distinction and borrows both from statistical and formal notions.

Bibliography

Abney, S. 1987. *The English Noun Phrase in I its Sentential Aspect*. Ph.D. dissertation, MIT.

Abrusán, M. 2007. *Contradiction and Grammar: The Case of Weak Islands*. Ph.D. dissertation, MIT.

Abrusán, M. 2014. *Weak Island Semantics*. Oxford UK: Oxford University Press.

Abrusán, M. and B. Spector. 2010. An interval based semantics for negative degree questions. *Journal of Semantics*. doi:10.1093/jos/ffq013.

Abrusán, M., N. Asher and T. V. de Cruys. 2018. Content vs. function words: The view from distributional semantics. In: Uli Sauerland and Stephanie Solt (eds.), *Proceedings of Sinn und Bedeutung 22*, vol. 1, ZASPiL 60, pp. 1–21. ZAS, Berlin.

Alonso-Ovalle and Menéndez-Benito. 2015. Epistemic Indefinites: Exploring Modality beyond the Verbal Domain. USA: Oxford University Press.

Alxatib, S., P. Pagin and U. Sauerland. 2013. Acceptable contradictions: Pragmatics or semantics? *Journal of Philosophical Logic* 42: 619–634.

Andréka, H., J. v. B. and I. Németi. 1998. Modal languages and bounded fragments of predicate logic. *Journal of Philosophical Logic* 27: 217–274.

Andréka, H., J. van Benthem, N. Bezhanishvili and I. Németi. 2014. Changing a semantics: Opportunism or courage? In *The Life and Work of Leon Henkin*, eds. I. S. M. Manzano and E. Alonso, 307–337. Zürich: Birkhaüser.

Aristotle. 1966. *Metaphysics*. Edited by H. Apostle. Bloomington: Indiana University Press.

Asher, N. 2011. *Lexical Meaning in Context: A Web of Words*. Cambridge University Press.

Asher, N., T. van de Cruys, A. Bride, and M. Abrusán. 2016. Integrating type theory and distributional semantics: A case study on adjective-noun compositions. *Computational Linguistics* 42(4): 703–725.

Bach, E., E. Jelinek, A. Kratzer, and B. H. Partee. (Eds.) 1995. *Quantification in Natural Language*. Dordrecht: Kluwer.

Baldwin, J. T. 2018. *Model Theory and the Philosophy of Mathematical Practice: Formalization without Foundationalism*. Cambridge University Press.

Baltag, A. and S. Smets 2006. Dynamic belief revision over multi-agent plausibility models. In *Proceedings LOFT'06*, eds. W. v. d. H. G. Bonanno and M. Wooldridge, 11–24. University of Liverpool: Department of Computing.

Barrio, E., L. Rosenblatt, and D. Tajer 2016. Capturing naive validity in the cut-free approach. *Synthese*. doi: https://doi.org/10.1007/s11229-016-1199-5

Barwise, J. 1975. *Admissible Sets and Structures: An Approach to Definability Theory*. Berlin: Springer.

Barwise, J. 1985. Model-theoretic logics: Background and aims. In *Model-Theoretic Logics*, eds. J. Barwise and S. Feferman, 3–23. Berlin: Springer.

Barwise, J. and R. Cooper. 1981. Generalized quantifiers and natural language. *Linguistics and Philosophy* 4: 159–219.

Barwise, J. and S. Feferman. (Eds.) 1985. *Model-Theoretic Logics*. Berlin: Springer-Verlag.

Barwise, J. and J. Perry. 1983. *Situations and Attitudes*. Cambridge: MIT Press.

Barwise, J. and J. van Benthem. 1999. Interpolation, preservation and pebble games. *Journal of Symbolic Logic* 64: 881–903.

Beall, J. and G. Restall. 2006. *Logical Pluralism*. Oxford: Oxford University Press.

Beall, J. and G. Restall. 2009. Logical consequence. In *The Stanford Encyclopedia of Philosophy* (Fall 2009 ed.), ed. E. N. Zalta. Metaphysics Research Lab, Stanford University.

Beall, J. and B. C. van Fraassen. 2003. *Possibilities and Paradox*. Oxford: Oxford University Press.

Beall, J. and J. Murzi 2013. Two flavors of curry paradox. *The Journal of Philosophy* 110(3): 143–165.

Beck, S. 2011. Comparison constructions. In *Semantics: An International Handbook of Natural Language Meaning*, eds. K. v. H. C. Maienborn and P. Portne, 1341–1389. Berlin: De Gruyter Mouton.

Belnap, N. 1962. Tonk, plonk and plink. *Analysis 22*(6): 130–134.

Belnap, N. and G. Massey. 1990. Semantic holism. *Studia Logica* 49(1): 67–82.

Benzmüller, Christoph, C. E. B. and M. Kohlhase. 2004. Higher-order semantics and extensionality. *Journal of Symbolic Logic* 69: 1027–1088.

Beth, E. W. 1963. Constanten van het wiskundige denken. *Mededelingen van de Koninklijke Nederlandse Akademie van Wetenschappen* 26: 231–255.

Blackburn, P., M. de Rijke, and Y. Venema. 2001. *Modal Logic*. Cambridge: Cambridge University Press.

Bonnay, D. 2006. *Qu'est-ce qu'une constante logique?* Ph.D. dissertation, University Paris 1.

Bonnay, D. 2008. Logicality and invariance. *Bulletin of Symbolic Logic* 14(1): 29–68.

Bonnay, D. 2014. Logical constants, or how to use invariance in order to complete the explication of logical consequence. *Philosophy Compass* 9: 54–65.

Bonnay, D. and D. Westerståhl. 2016. Compositionality solves Carnap's problem. *Erkenntnis* 81(4): 721–739.

Boolos, G. 1971. The iterative conception of set. *Journal of Philosophy* 68(8): 215–231.

Boolos, G. 1975. On second order logic. *Journal of Philosophy* 72: 509–527.

Boolos, G. 1984. To be is to be a value of a variable (or to be some values of some variables). *Journal of Philosophy* 81(8): 430–449.

Boolos, G. 1985. Nominalist Platonism. *Philosophical Review* 94(3): 327–344.
Boolos, G. 1993. *The Logic of Provability*. Cambridge: Cambridge University Press.
Burgess, J. P. 1999. Which modal logic is the right one? *Notre Dame Journal of Formal Logic* 40: 81–93. repr. in his *Mathematics, Models and Modality* (2008), Cambridge University Press, 169–184.
Burgess, J. P. 2003. Which modal models are the right ones (for logical necessity)? *Theoria* 18: 145–158.
Burgess, J. P. 2015. *Rigor and Structure*. New York: Oxford University Press.
Button, T. and S. Walsh. 2018. *Philosophy and Model Theory*. Oxford: Oxford University Press.
Caramazza, A. and A. E. Hillis. 1989. The disruption of sentence production: Some dissociations. *Brain and Language* 36: 625–650.
Carnap, R. 1937/2001. *The Logical Syntax of Language*. London: Routledge.
Carnap, R. 1943. *Formalization of Logic*. Cambridge, MA: Harvard University Press.
Carnap, R. 1946. Modalities and quantification. *The Journal of Symbolic Logic* 2: 33–64.
Carnap, R. 1952. Meaning postulates. *Philosophical Studies* 3: 65–73.
Casanovas, E. 2007. Logical operations and invariance. *Journal of Philosophical Logic* 36: 33–60.
Charlow, S. and Y. Sharvit. 2014. Bound 'de re' pronouns and the lf of attitude reports. *Semantics and Pragmatics*.
Chierchia, G. 2013. *Logic in Grammar*. Oxford UK: Oxford University Press.
Chierchia, G. and S. McConnell-Ginet. 2000. *Meaning and Grammar: An Introduction to Semantics* (second ed.). Cambridge: MIT Press.
Chomsky, N. 1981. *Lectures on Government and Binding: The Pisa Lectures*. Dordrecht: Foris.
Church, A. 1944. Review of Carnap 1943. *Philosophical Review* 53(5): 493–498.
Ciná, G. 2017. *Categories for the Working Modal Logician*. Unpublished Ph.D. thesis.
Cohen, P. J. 1963. The independence of the continuum hypothesis. *Proceedings of the National Academy of Sciences of the United States of America* 50(6): 1143–1148.
Cook, R. 2014. There is no paradox of logical validity. *Logica Universallis* 8: 447–467.
Cook, R. T. 2002. Vagueness and mathematical precision. *Mind* 111: 225–248.
Cooper, R. 1983. Quantification and semantic theory. *Synthese*.
Correia, F. and B. Schnieder. (Eds.) 2012. *Metaphysical Grounding: Understanding the Structure of Reality*. Cambridge: Cambridge University Press.
Cresswell, M. 2013. Carnap and McKinsey: Topics in the pre-history of possible-worlds semantics. *Proceedings of the 12th Asian Logic Conference*, R. Downey, J. Brendle, R. Goldblatt and B. Kim (eds), World Scientific, pp. 53–75.
Cresswell, G. E. H. M. 1996. *A New Introduction to Modal Logic*. Routledge.
Cresswell, M. and A. von Stechow. 1982. De re belief generalized. *Linguistics and Philosophy* 5: 503–535.

Cruse, D. A. 1986. *Lexical Semantics*. Cambridge University Press.
Davidson, D. 1967. Truth and meaning. *Synthese* 17: 304–323.
Dayal, V. 1996. *Locality in WH Quantification: Questions and Relative Clauses in Hindi. Studies in Linguistics and Philosophy.* Kluwer Academic Publishers.
Del Pinal, G. 2019. The logicality of language: A new take on triviality, "ungrammaticality", and logical form. *Noûs*, 53: 785–818.
Doets, H. 1996. *Basic Model Theory*. Stanford: CSLI Publications.
Dogramaci, S. 2015. Why is a valid inference a good inference? *Philosophy and Phenomenological Research* 94: 61–96.
Dorr, C. 2016. To be F is to be G. *Philosophical Perspectives* 30: 39–134.
Dowty, D. R. 1979. *Word Meaning and Montague Grammar*. Dordrecht: Reidel.
Ebbinghaus, H.-D. and J. Flum. 2005. *Finite Model Theory*. Berlin: Springer.
Engström, F. 2014. Implicitly definable generalized quantifiers. In *Idées Fixes: A Festschrift Dedicated to Christian Bennet on the Occasion of His 60th Birthday*, ed. M. Kaså, 65–71. Göteborg: University of Gothenburg.
Etchemendy, J. 1990. *The Concept of Logical Consequence*. Cambridge: Harvard University Press.
Etchemendy, J. 2008. Reflections on consequence. In *New Essays on Tarski and Philosophy*, ed. D. Patterson, 263–299. Oxford: Oxford University Press.
Evans, G. 1985. *Collected Papers*. Oxford: Oxford University Press.
Ewald, W. 2019. The emergence of first-order logic. In *The Stanford Encyclopedia of Philosophy* (Spring 2019 ed.), ed. E. N. Zalta. Metaphysics Research Lab, Stanford University.
Feferman, S. 1999. Logic, logics, and logicism. *Notre Dame Journal of Formal Logic* 40(1): 31–54.
Feferman, S. 2010. Set-theoretical invariance criteria for logicality. *Notre Dame Journal of Formal Logic* 51(1): 3–20.
Feferman, S. 2015. Which quantifiers are logical? a combined semantical and inferential criterion. In *Quantifiers, Quantifiers, and Quantifiers: Themes in Logic, Metaphysics and Language*, ed. A. Torza, 19–31. Springer International Publishing.
Field, H. 2008. *Saving Truth from Paradox*. Oxford: Oxford University Press.
Field, H. 2009. What is the normative role of logic? *Proceedings of the Aristotelian Society, Supplementary* 83: 251–268.
Fine, K. 2012. Guide to ground. In *Metaphysical Grounding: Understanding the Structure of Reality*, eds. F. Correia and B. Schnieder, 37–80. New York: Cambridge University Press.
Florio, S. and L. Incurvati. 2019. Metalogic and the overgeneration argument. *Mind* 128(511): 761–793.
Fox, D. and M. Hackl. 2007. The universal density of measurement. *Linguistics and Philosophy* 29: 537–586.
Fox, J. G., S. W. Barthold, M. T. Davisson, C. E. Newcomer, F. W. Quimby, and A. L. Smith. (Eds.) 2007. *The Mouse in Biomedical Research* (second ed.). Amsterdam: Elsevier.
Freitag, W. 2009. *Form and Philosophy: A Topology of Possibility and Representation*. Heidelberg: Synchron.

Freitag, W. and A. Zinke. 2012. The theory of form logic. *Logic and Logical Philosophy* 21: 363–389.
Friedman, N. and Y. Grodzinsky. 1997. Tense and agreement in agrammatic production. *Brain and Language* 56: 397–425.
Frigg, R. and S. Hartmann. 2009. Models in science. In *The Stanford Encyclopedia of Philosophy* (Summer 2009 ed.), ed. E. N. Zalta. Metaphysics Research Lab, Stanford University.
G. Aucher, J. van Benthem and D. Grossi. 2018. Modal logics of sabotage revisited. *Journal of Logic and Computation* 28: 269–303.
Gajewski, J. 2002. *On Analyticity in Natural Language*. Manuscript MIT.
Gajewski, J. 2008. More on quantifiers in comparative clauses. In *Semantics and Linguistic Theory XVIII* ed. T. Friedman S. Ito, 340–357. NY: Ithaca.
Gajewski, J. 2009. *L-triviality and Grammar*. Handout, UConn Logic Group, University of Connecticut.
Gallin, D. 1975. *Intensional and Higher Order Modal Logic*. Amsterdam: North Holland.
Garson, J. W. 2013. *What Logics Mean: From Proof Theory to Model-Theoretic Semantics*. Cambridge: Cambridge University Press.
Gentzen, G. 1934. Untersuchungen über das logische schließen. *Math. Zeitschrift* 39: 405–431.
Giannakidou, A. 2011. Positive polarity items and negative polarity items: Variation, licensing, and compositionality. In *Semantics: An International Handbook of Natural Language Meaning*, eds. C. Maienborn, K. von Heusinger, and P. Portner, 1660–1721. Berlin: de Gruyter.
Glanzberg, M. 2014. Explanation and partiality in semantic theory. In *Metasemantics: New Essays on the Foundations of Meaning*, eds. A. Burgess and B. Sherman, 259–292. Oxford: Oxford University Press.
Glanzberg, M. 2015. Logical consequence and natural language. In *Foundations of Logical Consequence*, eds. C. R. Caret and O. T. Hjortland, 71–120. Oxford University Press.
Gödel, K. 1939. Consistency proof for the generalized continuum hypothesis. *Proceedings of the National Academy of Sciences of the United States of America* 25(4): 220-224.
Goldblatt, R. 1973. A new extension of s4. *Notre Dame Journal of Formal Logic* 14: 567–574.
Gómez-Torrente, M. 1996. Tarski on logical consequence. *Notre Dame Journal of Formal Logic* 37(1): 125–151.
Gómez-Torrente, M. 1998. On a fallacy attributed to Tarski. *History and Philosophy of Logic* 19(4): 227–234.
Gómez-Torrente, M. 1999. Logical truth and Tarskian logical truth. *Synthese* 117: 375–408.
Gómez-Torrente, M. 2000. *Forma y Modalidad. Una Introducción al Concepto de Consecuencia Lógica*. Buenos Aires: Eudeba.
Gómez-Torrente, M. 2002. The problem of logical constants. *Bulletin of Symbolic Logic* 8: 1–37.
Gómez-Torrente, M. 2003. Logical consequence and logical expressions. *Theoria* 18(2): 131–144.

Gómez-Torrente, M. 2009. Rereading Tarski on logical consequence. *The Review of Symbolic Logic* 2(2): 249–297.
Graedel, E., T. Wolfgang and T. Wilke. (Eds.) 2002. *Automata, Logics, and Infinite Games*. Heidelberg: Springer.
Griffiths, O. and A. Paseau. 2016. Isomorphism invariance and overgeneration. *Bulletin of Symbolic Logic* 22(4): 482–503.
Griffiths, O. and A. Paseau. 2022. *One True Logic*. Oxford University Press.
Grossi, D. and P. Turrini. 2012. Short-sight in extensive games. *Proceedings of the 11th International Joint Conference on Autonomous Agents and Multi-Agent Systems*: 805–812.
H. van Ditmarsch, W. van der Hoek and B. Kooi. 2007. *Dynamic-Epistemic Logic*. Cambridge UK: Cambridge University Press.
Hacking, I. 1979. What is logic? *Journal of Philosophy* 76(6): 285–319.
Halbach, V. 2018. Substitutional analysis of logical consequence. *Noûs*. published online. https://doi.org/10.1111/nous.12256
Hale, B. and A. Hoffman. (Eds.) 2010. *Modality: Metaphysics, Logic, and Epistemology*. Oxford: Oxford University Press.
Halldén, S. 1962. A pragmatic approach to modal theory. *Acta Philosophica Fennica* 16: 53–64.
Halle, M. and A. Marantz. 1993. Distributed morphology and the pieces of inflection. In *The View from Building 20*, eds. K. Hale and J. Keyser, 111–176. Cambridge, MA: MIT Press.
Hanson, W. 2002. The formal-structural view of logical consequence: A reply to gila sher. *Philosophical Review* 111: 243–258.
Hanson, W. H. 1997. The concept of logical consequence. *Philosophical Review* 106(3): 365–409.
Hanson, W. H. 1999. Ray on Tarski on logical consequence. *Journal of Philosophical Logic* 28(6): 605–616.
Hardy, G. 1921. Srinivasa Ramanujan. *Proceedings of the London Mathematical Society*: xl–lviii.
Harris, Z. S. 1954. Distributional Structure. *Word* 10(23): 146–162.
Heim, I. and A. Kratzer. 1998. *Semantics in Generative Grammar*. Oxford: Blackwell.
Helmholtz, H. 1878. *The Facts of Perception*. Middletown, Conn.: Wesleyan University Press.
Hellman, G. 1989. *Mathematics without Numbers: Towards a Modal-Structural Interpretation*. Oxford: Oxford University Press.
Higginbotham, J. 1998. On higher-order logic and natural language. In *Philosophical Logic*, ed. T. Smiley, 1–27. Oxford: Oxford University Press.
Hjortland, O. T. 2014. Speech acts, categoricity, and the meanings of logical connectives. *Notre Dame Journal of Formal Logic* 55(4): 445–467.
Hodes, H. 2004. On the sense and reference of a logical constant. *Philosophical Quarterly* 54(214): 134–165.
Hodes, H. T. 1984. Logicism and the ontological commitments of arithmetic. *Journal of Philosophy* 81: 123–149.

Hodges, W. 1997. *A Shorter Model Theory*. Cambridge: Cambridge University Press.

Howard, W. A. 1980. The formulas-as-types notion of construction. In *To H. B. Curry: Essays on Combinatory Logic, Lambda Calculus and Formalism*, eds. P. Seldin and J. R. Hindley, 479–490. Academic Press. First appeared in *Language* 29 (1953): 47–58.

Hughes, G. E. and M.J. Cresswell. 1996. *A New Introduction to Modal Logic*. Routledge.

Humberstone, L. 2011. *The Connectives*. Cambridge, MA: MIT Press.

Isaacson, D. 1987. Arithmetical truth and hidden higher-order concepts. In *Logic Colloquium '85: Volume 122 of Studies in Logic and the Foundations of Mathematics*, 147–169. Amsterdam: North Holland.

Isaacson, D. 1992. Some considerations on arithmetic truth and the ω-rule. In *Proof, Logic, and Formalization*, ed. M. Detlefsen, 94–138. London: Routledge.

Isaacson, D. 1994. Mathematical intuition and objectivity. In *Mathematics and Mind*, ed. A. Goerge. Oxford: Oxford University Press.

Jackendoff, R. 1992. *Semantic structures*. MIT Press.

Jackson, B. B. 2017. Structural entailment and semantic natural kinds. *Linguistics and Philosophy* 40: 207–237.

Johnson, W. E. 1921. *Logic*. Dover.

Kamp, H. and B. Partee. 1995. Prototype theory and compositionality. *Cognition* 57(2): 129–191.

Kant, I. 1781. *Critique of Pure Reason*. Cambridge University Press.

Kaplan, D. 1968. Quantifying in. *Synthese* 19: 178–214.

Keenan, E. and D. Westerståhl. 1997. Generalized quantifiers in linguistics and logic. In *Handbook of Logic and Language*, eds. J. van Benthem and A. ter Meulen, 837–893. Amsterdam: Elsevier.

Keenan, E. L. and D. Paperno. (Eds.) 2012. *Handbook of Quantifiers in Natural Language*. Berlin: Springer.

Keenan, E. L. and J. Stavi. 1986. A semantic characterization of natural language determiners. *Linguistics and Philosophy* 9: 253–326. Versions of this paper were circulated in the early 1980s.

Keisler, H. 1970. Logic with the quantifier "there exist uncountably many". *Annals of Mathematical Logic* 1(1): 1–93.

Kennedy, C. 2007. Vagueness and grammar: The semantics of relative and absolute gradable adjectives. *Linguistics and Philosophy* 30: 1–45.

Kennedy, J. 2015. On the "logic without borders" point of view. In *Logic Without Borders: Essays on Set Theory, Model Theory, Philosophical Logic and Philosophy of Mathematics*, eds. A. Villaveces, R. Kossak, J. Kontinen, and Å. Hirvonen, 1–14. Berlin: De Gruyter.

Ketland, J. 2012. Validity as primitive. *Analysis* 72(3): 421–430.

Klein, F. 1872. *A Comparative Review of Recent Researches in Geometry*. Erlangen: Verlag von Andreas Deichert.

Klein, F. 1987. *A comparative Review of Recent Researches in Geometry*. Erlangen: Verlag von Andreas Deichert.

Koellner, P. 2010. Strong logics of first and second order. *Bulletin of Symbolic Logic* 16(1): 1–36.

Kratzer, A. 1981. The notional category of modality. In *Words, Worlds, and Contexts*, eds. H.-J. Eikmeyer and H. Rieser, 38–74. Berlin: de Gruyter.

Kratzer, A. and J. Shimoyama. 2002. Indeterminate pronouns: The view from Japanese. *The Proceedings of the Third Tokyo Conference on Psycholinguistics*: 1–25.

Kreisel, G. 1967. Informal rigour and completeness proofs. *Studies in Logic and the Foundations of Mathematics* 47: 138–186.

Krifka, M. 1999. Manner in dative alternation. *Proceedings of the West Coast Conference on Formal Linguistics* 18: 260–271.

Kripke, S. 1980. *Naming and Necessity*. Cambridge: Harvard University Press.

Kuno, S. and K. Takami. 1997. Remarks on negative islands. *Linguistic Inquiry* 28(4): 553–576.

Lahiri, U. 1998. Focus and negative polarity in Hindi. *Natural Language Semantics* 6: 57–123.

Lasersohn, P. 2012. Contextualism and compositionality. *Linguistics and Philosophy* 35: 171–189.

Läuchli, A. 1970. An abstract notion of realizability for which the predicate calculus is complete. In *Intuitionism and Proof Theory*, eds. A. K. J. Myhill and A. Vesley, 227–34. Amsterdam: North-Holland.

Lee, D. D. and H. S. Seung. 1999. Learning the parts of objects by non-negative matrix factorization. *Nature* 401(6755): 788–791.

Lepore, E. 1983. What model-theoretic semantics cannot do. *Synthese* 54: 167–187.

Lindström, P. 1966. First order predicate logic with generalized quantifiers. *Theoria* 32(3): 186–195.

Lindström, P. 1969. On extensions of elementary logic. *Theoria* 35: 1–11.

Linnebo, Ø. 2003. Plural quantification exposed. *Nous* 37: 71–92.

Liu, F. 2009. Diversity of agents and their interaction. *Journal of Logic, Language and Information* 18: 23–53.

Löding, C. and P. Rohde. 2003. Model checking and satisfiability for sabotage modal logic. *Proceedings FSTTCS 2003*, eds. P. K. Pandya and J. Radhakrishnan, 302–13. Heidelberg: Springer.

Luo, Z. 1994. *Computation and Reasoning: A Type Theory for Computer Science*. Oxford, UK: Oxford University Press.

Lycan, W. G. 1989. Logical constants and the glory of truth-conditional semantics. *Notre Dame Journal of Formal Logic* 30: 390–400.

MacFarlane, J. 2005. Making sense of relative truth. *Proceedings of the Aristotelian Society* 105: 321–339.

MacFarlane, J. 2017. Logical constants. In *The Stanford Encyclopedia of Philosophy* (Winter 2017 ed.), ed. E. N. Zalta. Metaphysics Research Lab, Stanford University.

MacFarlane, J. G. 2000. *What Does It Mean to Say That Logic Is Formal?* Ph.D. thesis, University of Pittsburgh.

Magidor, O. 2009. The last dogma of type confusions. *Proceedings of the Aristotelian Society* 109: 1–29.
Magidor, O. 2013. *Category Mistakes*. Oxford University Press.
Makowsky, J. 1995. The impact of model theory on theoretical computer science. In *Studies in Logic and the Foundations of Mathematics*, Volume 134, 239–262. Elsevier.
Mancosu, P. 2006. Tarski on models and logical consequence. In *The Architecture of Modern Mathematics*, eds. J. Ferreiros and J. J. Gray, 209–237. Oxford: Oxford University Press.
Mancosu, P. 2010. Fixed- versus variable-domain interpretations of Tarski's account of logical consequence. *Philosophy Compass* 5: 745–759.
Mancosu, P. 2014. *The Adventure of Reason: Interplay between Philosophy of Mathematics and Mathematical Logic, 1900–1940*. Oxford: Oxford University Press.
Manders, K. L. 1987. Logic and conceptual relationships in mathematics. In *Studies in Logic and the Foundations of Mathematics*, Volume 122, 193–211. Elsevier.
Martí, L. 2006. Unarticulated constituents revisited. *Linguistics and Philosophy* 29.
Mautner, F. I. 1946. An extension of Klein's Erlanger program: Logic as invariant-theory. *American Journal of Mathematics* 68: 345–384.
McCarthy, T. 1981. The idea of a logical constant. *Journal of Philosophy* 78(9): 499–523.
McCarthy, T. 1987. Modality, invariance, and logical truth. *Journal of Philosophical Logic* 16: 423–443.
McGee, V. 1991. *Truth, Vagueness, and Paradox: An Essay on the Logic of Truth*. Hackett Press.
McGee, V. 1992. Two problems with Tarski's theory of consequence. *Proceedings of the Aristotelian Society* 92: 273–292.
McGee, V. 1996. Logical operations. *Journal of Philosophical Logic* 25(6): 567–580.
McGee, V. 1997. How we learn mathematical language. *Philosophical Review* 106(1): 35–68.
McKay, T. J. 2006. *Plural Predication*. Oxford: Oxford University Press.
Montague, R. 1970. Universal grammar. *Theoria* 36: 373–398. Reprinted in *Formal Philosophy*, ed. R. Thomason (1974). New Haven: Yale University Press, 222–246.
Montague, R. 1973. The proper treatment of quantification in ordinary English. In *Approaches to Natural Language*, eds. J. Hintikka, J. Moravcsik, and P. Suppes, 221–242. Dordrecht: Reidel. Reprinted in *Formal Philosophy*, ed. R. Thomason (1974). New Haven: Yale University Press, 247–270.
Mostowski, A. 1957. On a generalization of quantifiers. *Fundamenta Mathematicae* 44: 12–36.
Murzi, J. and L. Rossi. 2017. Naïve validity. *Synthese*. https://doi.org/10.1007/s11229-017-1541-6
Murzi J. and L. Shapiro. 2015. Validity and truth-preservation. In *Unifying the Philosophy of Truth. Logic, Epistemology, and the Unity of Science, vol 36*, eds.

T. Achourioti, H. Galinon, J. Martínez Fernández, and K. Fujimoto, 431–459. Dordrecht: Springer. https://doi.org/10.1007/978-94-017-9673-622
Nicolai, C. and L. Rossi 2017. Principles for object-linguistic consequence: From logical to irreflexive. *Journal of Philosophical Logic* 47(3): 549–577.
Nour, K. and C. Raffalli. 2003. Simple proof of the completeness theorem for second-order classical and intuitionistic logic by reduction to first-order mono-sorted logic. *Theoretical Computer Science* 308(1): 227–237.
Novaes, C. D. 2014. The undergeneration of permutation invariance as a criterion for logicality. *Erkenntnis* 79(1): 81–97.
Nunberg, G. 1993. Indexicality and deixis. *Linguistics and Philosophy* 16(1): 1–43.
Oliver, A. and T. Smiley. 2013. *Plural Logic*. Oxford: Oxford University Press.
Pacuit, E. 2017. *Neighborhood Semantics for Modal Logic*. Beijing: Springer.
Parsons, C. 2013. Some consequences of the entanglement of logic and mathematics. In *Reference, Rationality, and Phenomenology. Themes from Føllesdal*, ed. M. Frauchiger, 153–178. De Gruyter.
Partee, B. 1986. Noun phrase interpretation and type-shifting principles. In *Studies in Discourse Representation Theory and the Theory of Generalized Quantifiers*, eds. D. d. J. J. Groenendijk and M. Stokhof. Dordrecht: Foris.
Paseau, A. 2014. The overgeneration argument(s): A succinct refutation. *Analysis* 74(1): 40–47.
Paseau, A. and O. Griffiths. forthcoming. *One True Logic*. Oxford University Press.
Peacocke, C. 2004. Understanding logical constants: A realist's account. In *Studies in the Philosophy of Logic and Knowledge*, eds. T. J. Smiley and T. Baldwin, 163–209. Oxford: Oxford University Press.
Pearson, H. 2010. How to do comparison in a language without degrees: A semantics for the comparative in Fijan. *Proceedings of Sinn und Bedeutung*, Volume 14, eds. S. Schmitt and V. Zobel, 356–372.
Peters, S. and D. Westerståhl. 2006. *Quantifiers in Language and Logic*. Oxford: Clarendon.
Pinker, S. 1989. *Learnability and Cognition*. Cambridge: MIT Press.
Plotkin, G. 1980. Lambda definability in the full type hierarchy. In *To H.B. Curry. Essays on Combinatory Logic, Lambda Calculus and Formalism*, eds. J. Seldin and J. Hindley, 363–373. Cambridge Mass.: Academic Press.
Potter, M. 2004. *Set Theory and Its Philosophy. A Critical Introduction*. Oxford: Oxford University Press.
Priest, G. 1995. Etchemendy and logical consequence. *Canadian Journal of Philosophy* 25(2): 283–292.
Prior, A. N. 1960. The runabout inference-ticket. *Analysis* 21(2): 38.
Pustejovsky, J. 1995. *The Generative Lexicon*. MIT Press.
Putnam, H. 1975. The meaning of "meaning". In *Language, Mind and Knowledge. Minnesota Studies in the Philosophy of Science*, Volume 7, ed. K. Gunderson, 131–193. Minneapolis: University of Minnesota Press.
Quine, W. V. 1956. Quantifiers and propositional attitudes. *The Journal of Philosophy* 53: 177–187.

Quine, W. V. 1970. *Philosophy of Logic*. Cambridge, MA: Harvard University Press.

Rantala, V. 1982a. Impossible worlds semantics and logical omniscience. *Acta Philosophica Fennica* 35: 106–15.

Rantala, V. 1982b. Quantified modal logic: Non-normal worlds and propositional attitudes. *Studia Logica* 41: 41–65.

Rappaport Hovav, M. and B. Levin. 2008. The English dative alternation: The case for verb sensitivity. *Journal of Linguistics* 44: 129–167.

Ray, G. 1996. Logical consequence: A defense of Tarski. *Journal of Philosophical Logic* 25(6): 617–677.

Rayo, A. 2002. Word and objects. *Noûs* 36(3): 436–464.

Rayo, A. and G. Uzquiano. 1999. Toward a theory of second-order consequence. *Notre Dame Journal of Formal Logic* 40(3): 315–325.

Rayo, A. and T. Williamson. 2003. A completeness theorem for unrestricted first-order languages. In *Liars and Heaps*, ed. J. Beall, 331–356. Oxford: Oxford University Press.

Rayo, A. and S. Yablo. 2001. Nominalism through de-nominalization. *Noûs* 35(1): 74–92.

Read, S. 1994. Formal and material consequence. *Journal of Philosophical Logic* 23: 247–265.

Recanati, F. 2010. *Truth-Conditional Pragmatics*. Oxford: Clarendon Press.

Restall, G. 2000. *An Introduction to Substructural Logics*. London: Routledge.

Rips, L. J. 1994. *The Psychology of Proof*. Cambridge: MIT Press.

Rizzi, L. 1990. *Relativized Minimality*. MIT Press.

Rosen, G. 2010. Metaphysical dependence: Grounding and reduction. In *Modality: Metaphysics, Logic, and Epistemology*, eds. B. Hale and A. Hoffman, 109–136. Oxford: Oxford University Press.

Routley, R. and R. K. Meyer. 1973. Semantics of entailment. In *Truth, Syntax, and Modality*, ed. H. Leblanc, 194–243. Amsterdam: North-Holland.

Rumfitt, I. 2000. Yes and no. *Mind* 109(436): 781–823.

Sacks, G. E. 1972. *Saturated Model Theory*. Reading, Massachusetts: W. A. Benjamin.

Sagi, G. 2014. Formality in logic: From logical terms to semantic constraints. *Logique et Analyse* 57: 259–276.

Sagi, G. 2015. The modal and epistemic arguments against the invariance criterion for logical terms. *Journal of Philosophy* 112: 159–167.

Sagi, G. 2017. Extensionality and logicality. *Synthese*: 1–25.

Sagi, G. 2018. Logicality and meaning. *Review of Symbolic Logic* 11: 133–159.

Schwarzschild, R. and K. Wilkinson. 2002. Quantifiers in comparatives: A semantics of degree based on intervals. *Natural Language Semantics* 10(1): 1–41.

Scott, D. 1970. Advice on modal logic. In *Philosophical Problems in Logic*, ed. K. Lambert, 143–173. Dordrecht: Reidel.

Scroggs, S. 1951. Extensions of the Lewis system S5. *The Journal of Symbolic Logic* 16: 112–120.

Shapiro, L. 2010. Deflating logical consequence. *Philosophical Quarterly* 61: 320–342.
Shapiro, L. 2013. Validity curry strengthened. *Thought* 2(2): 100–107.
Shapiro, L. 2015. Naive structure, contraction, and paradox. *Topoi* 34: 75–87.
Shapiro, S. 1987. Principles of reflection and second-order logic. *Journal of Philosophical Logic* 16(3): 309–333.
Shapiro, S. 1991. *Foundations without Foundationalism: A Case for Second-Order Logic*. Oxford: Oxford University Press.
Shapiro, S. 1998. Logical consequence: Models and modality. In *The Philosophy of Mathematics Today*, ed. M. Schirn, 131–156. Oxford: Clarendon Press.
Shapiro, S. 2014. *Varieties of Logic*. Oxford University Press.
Sher, G. 1991. *The Bounds of Logic*. Cambridge: MIT Press.
Sher, G. 1996. Did Tarski commit 'Tarski's fallacy'? *Journal of Symbolic Logic* 61: 653–686.
Sher, G. 2003. A characterization of logical constants is possible. *Theoria: Revista de Teoria, Historia y Fundamentos de la Ciencia* 18: 189–197.
Sher, G. 2008. Tarski's thesis. In *New Essays on Tarski and Philosophy*, ed. D. Patterson, 300–339. Oxford: Oxford University Press.
Sher, G. 2013. The foundational problem of logic. *Bulletin of Symbolic Logic* 19: 145–198.
Sher, G. 2016. *Epistemic Friction: An Essay on Knowledge, Truth, and Knowledge*. Oxford: Oxford University Press.
Smiley, T. 1982. The schematic fallacy. *Proceedings of the Aristotelian Society* 83: 1–18.
Speitel, S. G. and D. Westerståhl. July 2019. *Carnap's Problem for Languages* $\mathcal{L}(Q_1, \ldots, Q_n)$. Unpublished manuscript.
Stalnaker, R. C. 1984. *Inquiry*. Cambridge: MIT Press.
Stanley, J. 2007. *Language in Context: Selected Essays*. Oxford: Clarendon Press.
Statman, R. 1982. Completeness, invariance, and λ–definability. *Journal of Symbolic Logic* 47: 17–26.
Stevenson, J. 1960. Roundabout the runabout inference-ticket. *Analysis* 21(6): 124–128.
Strawson, P. 1957. Propositions, concepts, and logical truths. *Philosophical Quarterly* 7: 15–25.
Sudo, Y. 2012. *On the Semantics of Phi-Features on Pronouns*. Ph.D. dissertation, MIT.
Suppes, P. 2002. *Representation and Invariance of Scientific Structures*. Stanford: CSLI Publications.
Szabolcsi, A. 2012. *Quantification*. Cambridge: Cambridge University Press.
Szabolcsi, A. and F. Zwarts. 1993. Weak islands and an algebraic semantics for scope taking. *Natural Language Semantics* 1: 235–284.
Tarski, A. 1935. Der Wahrheitsbegriff in den formalisierten Sprachen. *Studia Philosophica* 1: 261–405. References are to the translation as "The concept of truth in formalized languages" by J. H. Woodger in Tarski (1983a). Revised version of the Polish original published in 1933.

Tarski, A. 1936. O pojciu wynikania logicznego. *Przegląd Filozoficzny* 39: 58–68. References are to the translation as "On the concept of logical consequence" by J. H. Woodger in Tarski (1983b).

Tarski, A. 1936. Über den Begriff der logischen Folgerung. *Actes du Congrès International de Philosophie Scientifique* 7: 1–11. Translated by J. Woodger as On the Concept of Logical Consequence and repr. in Tarski's *Logic, Semantics, Metamathematics*. Clarendon Press, 2nd ed. (1983), J. Corcoran (ed.), 409–420.

Tarski, A. 1960. Truth and proof. *Scientific American* 6: 63–77.

Tarski, A. 1983a. *Logic, Semantics, Metamathematics* (second ed.). Edited by J. Corcoran with translations by J. H. Woodger. Indianapolis: Hackett.

Tarski, A. 1983b. On the concept of logical consequence. In *Logic, Semantics, Metamathematics* (second ed.), ed. J. Corcoran, 409–421. Indianapolis: Hackett Publishing Company.

Tarski, A. 1986. What are logical notions? Text of a 1966 lecture edited by J. Corcoran. *History and Philosophy of Logic* 7: 143–154.

Tarski, A. 1987. A philosophical letter of Alfred Tarski. *Journal of Philosophy* 84: 28–32. A 1944 letter of Tarski to Morton White, published with a preface of the latter.

Tarski, A. and S. Givant. 1987. *A Formalization of Set Theory without Variables*. Providence (R.I.): American Mathematical Society.

Tarski, A. and R. L. Vaught. 1957. Arithmetical extensions of relational systems. *Compositio mathematica* 13: 81–102.

Tharp, L. H. 1975. Which logic is the right logic? *Synthese* 31(1): 1–21.

Trueman, R. 2018. Substitution in a sense. *Philosophical Studies* (175): 3069–3098.

ten Cate, B., J. van Benthem and J. Vaananen. 2007. July. Lindstrom theorems for fragments of first-order logic. In 22nd Annual IEEE Symposium on Logic in Computer Science (LICS 2007) 280–292. IEEE. Vancouver

Urquhart, A. 1972. Semantics for relevant logics. *Journal of Symbolic Logic* 37: 159–169.

Uzquiano, G. 2003. Plural quantification and classes. *Philosophica Mathematica* 11: 67–81.

Väänänen, J. 2011. *Models and Games*. Cambridge UK: Cambridge University Press.

Väänänen, J. and T. Wang. 2015. Internal categoricity in arithmetic and set theory. *Notre Dame Journal of Formal Logic* 56(1): 121–134.

van Benthem, J. 1982. The logical structure of science. *Synthese* 51: 431–472.

van Benthem, J. 1984. Questions about quantifiers. *Journal of Symbolic Logic* 49(2): 443–466.

van Benthem, J. 1986a. *Essays in Logical Semantics*. Dordrecht: Reidel.

van Benthem, J. 1986b. *Modal Correspondence Theory*. Unpublished Ph.D. thesis.

van Benthem, J. 1991. *Language in Action*. Amsterdam: North-Holland.

van Benthem, J. 1996. *Exploring Logical Dynamics*. Stanford: CSLI Publications.

van Benthem, J. 1998. *Dynamic Bits and Pieces*. Institute for Logic, Language and Computation, University of Amsterdam.

van Benthem, J. 2002. Invariance and definability: Two faces of logical constants. Essays in honor of Sol Feferman. In *Reflections on the Foundations of Mathematics. ASL Lecture Notes in Logic 15.*, eds. W. Sieg, R. Sommer and C. Talcott, 426–446. Nattick, Massachusetts: A.K. Peters, Ltd.
van Benthem, J. 2010. *Modal Logic for Open Minds.* Stanford: CSLI Publications.
van Benthem, J. 2011. *Logical Dynamics of Information and Interaction.* Cambridge: Cambridge University Press.
van Benthem, J., B. ten Cate, and J. Väänänen. 2009. Lindström theorems for fragments of first-order logic. *Proceedings LICS* 5: 1–27.
van Benthem, J. 2014. *Logic in Games.* Cambridge Mass.: The MIT Press.
van Benthem, J. and G. Bezhanishvili. 2008. Modal logics of space. In *Handbook of Spatial Logics*, eds. M. Aiello, I. Pratt-Hartmann, and J. van Benthem M. Aiello and J. van Benthem, 217–298. Heidelberg: Springer.
van Benthem, J. and G. Bezhanishvili. 2016. Tracking information. In *Michael Dunn on Information-Based Logics*, ed. K. Bimbó, 363–389. Dordrecht: Springer.
van Benthem, J. and D. Bonnay. 2008. Modal logic and invariance. *Journal of Applied Non-Classical Logics* 18: 153–173.
van Benthem, J. and D. Ikegami. 2008. Modal fixed-point logic and changing models. In *Pillars of Computer Science: Essays Dedicated to Boris (Boaz) Trakhtenbrot on the Occasion of His 85th Birthday*, eds. N. D. A. Avron and A. Rabinovich, 146–65. Berlin: Springer.
van Benthem, J. and E. Pacuit. 2011. Dynamic logics of evidence-based beliefs. *Studia Logica* 99: 61–92.
van Benthem, J., J. van Eijck, and B. Kooi. 2006. Logics of communication and change. *Information and Computation* 204: 1620–1662.
van Benthem, J., N. Bezhanishvili, S. Enqvist and J. Yu. 2016. Instantial neighborhood logic. *Review of Symbolic Logic* 10: 116–144.
van Benthem, J., N. Bezhanishvili, and S. Enqvist. 2019. A new game equivalence, its logic and algebra. *Journal of Philosophical Logic* 48(4): 649–684.
Van de Cruys, T., T. P. and A. Korhonen. 2011. Latent vector weighting for word meaning in context. *Proceedings of the 2011 Conference on Empirical Methods in Natural Language Processing*, 1012–1022. Edinburgh.
Van de Cruys, T., T. Poibeau, and A. Korhonen. 2013. A tensor-based factorization model of semantic compositionality. In *Conference of the North American Chapter of the Association of Computational Linguistics (HTL-NAACL)*, 1142–1151. Dordrecht: Kluwer.
Venema, Y. 2012. *Lectures on the Modal Mu-Calculus, on-line lecture notes, Institute for Logic, Language and Computation.* University of Amsterdam.
von Fintel, K. 1993. Exceptive constructions. *Natural Language Semantics* 1: 123–148.
von Fintel, K. 1995. The formal semantics of grammaticalization. *Proceedings of NELS 25* 2: 175–189.
van Benthem, J. 1989. Logical constants across varying types. *Notre Dame Journal of Formal Logic* 30(3): 315–342.

van Dalen, D. 2008. *Logic and Structure*. Berlin/Heidelberg: Springer.
Warmbrōd, K. 1999. Logical constants. *Mind* 108: 503–538.
Weyl, H. 1963. *Philosophy of Mathematics and Natural Science*. New York: Atheneum.
Whittle, B. 2004. Dialetheism, logical consequence, and hierarchy. *Analysis* 64(4): 318–326.
Williamson, T. 2003. Everything. *Philosophical Perspectives* 17: 415–465.
Williamson, T. 2013. *Modal Logic as Metaphysics*. Oxford University Press.
Wittgenstein, L. 1922. *Tractatus Logico-Philosophicus*. Kegan and Paul.
Wittgenstein, L. 1993. Some remarks on logical form. In *Philosophical Occasions 1912–1951*, eds. J. Klagge and A. Nordmann, 2–35. Hackett Publishing Company.
Woods, J. 2014. Logical indefinites. *Logique et Analyse* 227: 277–307.
Woods, J. 2016. Characterizing invariance. *Ergo* 3: 778–807.
Yi, B.-U. 2005. The logic and meaning of plurals. Part I. *Journal of Philosophical Logic* 34: 459–506.
Yi, B.-U. 2006. The logic and meaning of plurals. Part II. *Journal of Philosophical Logic* 35: 239–288.
Zermelo, E. 1930. Über Grenzzahlen und Mengenbereiche: Neue Untersuchungen über die Grundlagen der Mengenlehre. *Fundamenta Mathematicae* 16: 29–47. Translated as "On boundary numbers and domains of sets: New investigations in the foundations of set theory". In *From Kant to Hilbert: A Source Book in the Foundations of Mathematics*, Volume 2, ed. William Ewald, 1219–1233. Oxford University Press, 1996.
Zhu, Y., R. Kiros, R. Zemel, R. Salakhutdinov, R. Urtasun, A. Torralba, and S. Fidler. 2015. Aligning books and movies: Towards story-like visual explanations by watching movies and reading books. In *Proceedings of the IEEE international conference on computer vision*, 19–27. Vancouver.
Zhu, Y., R. K. R. S. Z. R. S. R. U. A. T. and S. Fidler. 2015. *Aligning Books and Movies: Towards Story-Like Visual Explanations by Watching Movies and Reading Books*. CoRR abs/1506.06724.
Zimmermann, T. E. 1999. Meaning postulates and the model-theoretic approach to natural language semantics. *Linguistics and Philosophy* 22: 529–561.
Zinke, A. 2018. *The Metaphysics of Logical Consequence*. Frankfurt a. M.: Klostermann.
Zucker, J. 1978. The adequacy problem for classical logic. *Journal of Philosophical Logic* 7(1): 517–535.
Zucker, J. and R. Tragesser. 1978. The adequacy problem for inferential logic. *Journal of Philosophical Logic* 7(1): 501–516.

Index

a priori, 40, 52
accessibility, 150
adjective, 192, 241, 266
admissible interpretations, 187
agent diversity, 138
analyticity, 70, 194, 196
animal models, 225
argument, 264
  completeness, 170
arithmetic, 97, 98, 105, 106, 111, 112
assumption
  consistency, 155
  constancy, 164
  metaphysical, 198
  uniqueness, 144
attitudes
  propositional, 41

Beall, Jc, 98
belief, 130
  conditional, 130
  de re, 241
bisimulation, 127
bivalence, 25
bound variables, 237
Burgess, John, 162

Carnap's problem, 61
Carnap, Rudolf, 162
Carnap-categoricity, 70
categoricity, 69
choice
  axiom, 7, 142
  function, 147
  operator, 58
  property, 147
circularity, 18, 67
co-composition, 265
combined criterion, 63, 70
completeness, 145
compositionality, 4, 250
computation, 123, 251
computer science, 1
conceptual analysis, 211
conditional
  material, 22
consequence, 131
consistency, 101
contextualism, 231
continuum hypothesis, 7, 59, 143, 148, 157, 213
contradiction, 258
counterexample, 39, 90
criterion, 69
  for term identity, 195
  logical constants, 22
  semantic, 63
  semantical-inferential, 65

definability, 117
definite description, 242
demarcation of logic, 58
denotation, 21, 63
downward entailing, 231

elimination
  rules, 98
entailment, 133
  lexical, 217
entanglement of logic and mathematics, 143, 148
Etchemendy, J., 202
evidence, 130
extension, 187
  hereditary, 86

Feferman, S., 63, 74
first-order logic, 100, 126
formality, 25, 38, 58, 69, 213
functional categories, 243

G-triviality, 227, 229, 233–235, 237, 240, 243

Gómez-Torrente, M., 30
Gajewski, J., 227, 249
Generalized Tarski Thesis, 214
grammatical categories, 189
grammaticality, 252

Henkin
  semantics, 7, 65
  structures, 76
higher-order semantics, 146, 148
hyperintensionality, 80
hypothesis, 230, 265

implicit definability, 61, 64
induction, 123
inference, 120, 121
  rules, 63, 74, 261
inferential constraints, 60
intensionality, 86, 240
interpolation, 132
interpretational definition, 186, 205
intuitionist, 179
invariance, 37, 42, 53, 56, 117, 118, 120, 129, 132, 139
  isomorphism, 160
  permutation, 107, 109, 221, 260
  under inclusion-isomorphisms, 124
  under isomorphism, 22, 25, 160
  under models, 20
  under permutations, 3

L-triviality, 240, 249, 250, 253, 254, 258, 259, 261
large cardinals, 143, 149
Lindström, 23
linguistics, 1
logic
  dynamic, 134
  epistemic, 135
  modal, 127, 160
  of logical truth, 108, 174
  veridicality, 17
Logic of Logical Truth, 166
Logic of Truth by Logical Form, 163
logical
  consequence, 3, 16, 20, 54–56, 178, 213, 217
  constant, 60, 164, 178
  constants, 21, 35, 44, 47, 137, 217, 240
  form, 21, 49, 164, 231, 252
  games, 136
  meaning, 272
  necessity, 42, 52, 165
  neutrality, 143, 156
  notions, 3
  truth, 42, 144, 160, 186, 194, 198, 204, 266
  validity, 142
  words, 57, 259
logicality, 16, 35, 58, 60, 70, 81, 85, 104, 110, 124, 247, 260
  foundational perspective, 29
  linguistic perspective, 13, 29
  pragmatist approach, 32
Logically universal, 164
Łos-Tarski, 133

mathematics, 1
meaning
  postulates, 221
meaning shift, 267
metaphysical necessity, 189
modal force, 25
Modal Propositional Logic, 160
modality, 108
model, 19, 20, 28, 43, 143, 158, 199, 210, 223
model theory, 1, 35, 210
model-theoretic language, 216, 224
modus ponens, 165
Mostowski, A., 23
Murzi, J., 98
muskens frames, 86

natural language, 125, 194, 216
necessitation, 173
necessity, 213
negative polarity items, 230
normative, 41

operator, 42, 93, 184, 267
overgeneration argument, 143
overgeneration problem, 58

PA, 98, 105, 111
paradox, 97–99, 104
plural
  quantification, 144
pluralism, 163, 211, 214
pragmatic characterization, 53
predicate, 102, 112
preservation theorem, 133

quantifier, 64, 72, 223, 225
  generalized, 65
quasi-apriori, 40
Quine, W.V., 194

rationality, 140
reference, 217

S4, 171
  axioms, 169

S5, 174
  axioms, 169
second-order logic, 144, 147, 158
semantic
  anomaly, 268
semantics, 45, 54, 62, 143, 192, 195, 212, 217
  distributional, 268
set theory, 142
Sher, Gila, 36, 38
soundness, 145
Strawson, Peter, 194
structural truth, 191

Tarski, A., 2, 20, 22–24, 28, 35, 45, 46, 54, 55, 80, 97, 118, 188, 190, 198, 213
Tarski-Knaster, 123
Tarski-Sher thesis, 4, 22
tautology, 165
tracking, 135
triviality, 235, 252, 259, 261
truth, 99
truth by logical form, 166
type theory, 81

ungrammaticality, 227, 264
uniform reinterpretation, 193
uniqueness, 61

Val, 98, 100, 104
validity, 97, 98, 144, 214
variable, 21, 128, 235

wide vs. narrow form, 166
words
  content, 232, 243
  function, 243, 245
world
  actual, 21, 44, 198
  possible, 44, 214

zoom level, 126, 128, 135

Lightning Source UK Ltd.
Milton Keynes UK
UKHW020924020921
389885UK00005B/43